Primate Odyssey

PRIMATE ODYSSEY

by Geoffrey H. Bourne

G. P. PUTNAM'S SONS, New York

To Benito, Bobby, Yakut
and
all my other Primate friends

Contents

Introduction   9

I.   *The Primates Enter*   13

II.   *The Lemurs Emerge*   24

III.   *The Lorisoids*   49

IV.   *The Link That Was Not Missing—The Spectral Tarsier*   59

V.   *The New World Monkeys*   66

VI.   *The Old World Monkeys—The Macaques*   107

VII.   *The Old World Monkeys—Mangabeys, Baboons, Drills, and Mandrills*   165

VIII.   *The Old World Monkeys—Guenons, Snub-Nosed Monkeys, and Leaf-Eaters*   200

IX.   *The Apes Take the Stage*   249

X.   *The Lesser Apes—The Gibbon and the Siamang*   257

XI.   *The Great Apes*   278

XII.   *Man and His Future*   427

Bibliography   463

Index   471

Introduction

THE PRIMATES, in their upward surge to *Homo sapiens,* produced many evolutionary by-products. Some of these were failures and were suppressed and superseded by superior life-forms; some were successes in their own right even though they never developed the control of their own environment or the power to destroy it that man has been successful in doing. Some forms which would not have made it in the world at large managed to achieve considerable evolutionary success sequestered in geographically isolated areas.

This is not a scientific or technical book—it is meant for the general reader, and I feel that college students would also find it of interest and value. In order to make the relationships between the various Primates more understandable, I have attempted to paint a broad picture of the stream which started many million years ago and which has flowed unremittingly to the present, carrying the progenitors of man before it on the crest and spilling off the many other Primate forms on its banks as it passed along.

These Primate forms—the lemurs, the monkeys, and the apes—are all described in the following pages, and the reader is informed about where and how they live, their mental and physical capabilities, and the relationships they make with man. In some parts of the book I have used Primates as the basis of discussions on communication and speech, self-recognition, social behavior, and the development of the brain and mental processes. The last chapter tells of the origin of man and what the future may hold for him.

The close ties between, and the evolutionary origin of, man and the other Primates are undeniable. The legislature of the state of Tennessee, not satisfied that Tennessee was the laughingstock of the civilized world nearly half a century ago at the time of the Scopes' "monkey trial," is trying to do it again by demanding that school

9

books give equal time to the divine and the evolutionary origin of man. In what used to be thought of as a progressive state, the California legislature has emulated that of Tennessee.

The theory of the evolutionary origin of man was promulgated, debated, fought, and vindicated a hundred years ago, its recognition changed the philosophy of the world, and the recognition of the close similarity between monkeys, apes, and man has had a profound and highly beneficial influence on medical research.

The increased knowledge of Primates in recent years and the growing awareness of their physiological as well as anatomical closeness to man have led to their increased use in biomedical research. The results are impressive; already, among many advances, the use of monkeys has provided us with vaccines for poliomyelitis and measles and ensures the safety of other vaccines and medical products as well. Much more spectacular breakthroughs will occur in the future.

The writings produced by the fertile minds of many other scientists and writers have been the catalysts which helped me write this book, and I have acknowledged my debt to 180 of those most constantly referred to by including their names and writings in my bibliography. For what they have contributed to me and to this book this is not much of a reward, but nevertheless I hope they, wherever they may be, will accept this modest gesture as an expression of my gratitude and homage.

The typing of this manuscript was a formidable task, and I must thank my two secretaries, Mary Mauldin and Stephanie Ours, for their fast, accurate, and very hard work.

Once again in the preparation of this book I have to thank the willing help of Mrs. Helen Wells (formerly Helen Cousar), the Yerkes Center artist, and Mr. Frank Kiernan, the center photographer. I have also to thank many persons who helped me with information and photographs. The latter are individually acknowledged in the book where their photographs are displayed. It was a pleasure to be able to deal again with G. P. Putnam Son's editor-in-chief, Mr. William Targ, in the preparation of this book.

Finally, I have to acknowledge and thank my wife for her patience and sympathy while I secluded myself night after night for two long years while this book was conceived, incubated, and delivered.

Atlanta, Georgia   GEOFFREY H. BOURNE
May, 1973

Primate Odyssey

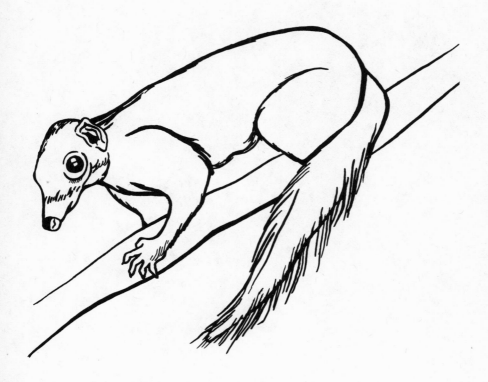

I. The Primates Enter

THERE WAS a light wind rustling through the trees; a full moon stood stark in the sky, its face obscured from time to time by scudding clouds. In the large animal wing at the Yerkes Primate Center, a few apes stirred restlessly. In one den a female chimpanzee was agitated; she stood up; she lay down; she touched her genital aperture with her fingers; she brought them forward to her nose and smelled them; she touched the aperture again and then licked her fingers. Somewhere in her subconscious self she knew something important was about to happen; she had no idea what it was. Then she felt a strange movement within her body. Automatically, she bent down, and the movements became stronger and stronger. Again she touched her genital area; suddenly there was a flow of water that

A chimpanzee baby is born—the mother lies resting—at the Yerkes Center.

The mother chimpanzee mouths away blood and mucus.

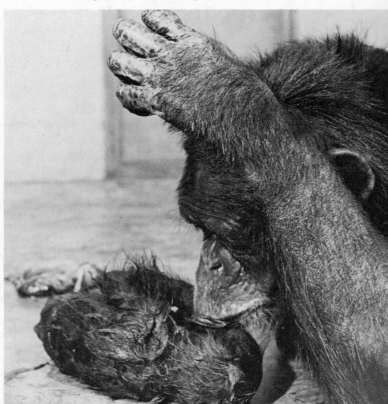

14

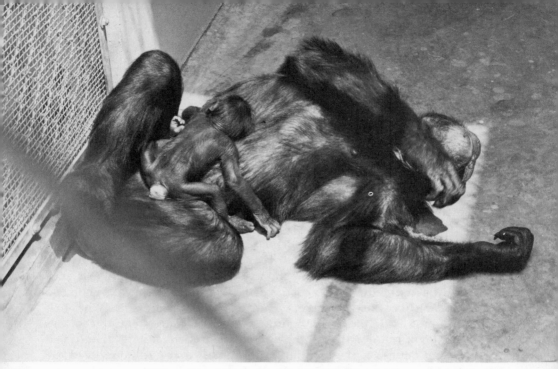

The mother chimpanzee relaxes with baby.

gushed from her vagina. The movements within her body became stronger and stronger, and a little hair-covered head, followed by an ugly little face appeared; then a shoulder, a second shoulder, and with a sudden rush, a baby chimpanzee was born.

It all happened in about half an hour. The mother, sometime afterward, passed the placenta. After the baby was born, she applied her mouth to its mouth and sucked away the mucus that may have obstructed the breathing passages. She licked the baby and pressed it to her breast. She lay down with it on one side of the cage, feeling a sudden unaccountable tiredness and sleepiness. The placenta was still attached by the umbilical cord to the baby, and it was not until some hours later that she gnawed it through. And so the cycle of production of a tiny chimpanzee baby had come a complete turn, and a new chimpanzee was living in the world.

Not far away, in Atlanta's major hospital, a woman lay tossing in turmoil. Sweat started from her brow and lips like dew. A nurse was with her and, later, a young doctor. She moaned and called out. She was in the throes of labor. For short periods she was conscious, but then she would lapse back into semiconsciousness. Always the grinding

15

pain of the contracting uterus was within her. And so, hour by hour, the pain continued; exhaustion began to come upon her. At last, the appearance of the water, then the head covered in hair, the pink face, and, after a considerable period of time, the young baby. Another human being had come into the world. In this night, two members of the highest level of animal evolution had appeared on the earth. What was the history that lay behind their origin; what was their relationship; what had happened over the millions of years which had resulted in the production of such different but such similar and closely related animal forms?

It all started many millions of years ago. The estimated time that has passed since the highest Primates developed from their primitive forebears is about 75,000,000 years, which takes us back to the Cretaceous period, and of this time, man himself has been in existence for little more than 1,000,000 years. Even earlier than 75,000,000 years, ratlike and molelike creatures, which represented the first developing mammals, were scuttling about the earth, eating the eggs of the great reptiles around them, a procedure which in itself hastened the final demise of these remarkable animals that had dominated the earth for so many millions of years.

About 60,000,000 to 70,000,000 years ago, in the period of geological time known as the Paleocene, the climate was warm and humid; there were enormous tropical forests which extended over a great area of the world. The land from France and Germany to South Africa was moist, humid, and covered in jungle, and so was much of America. Running in this jungle, poking their noses out of nests and holes in the trees, and waving their tails were some strange little creatures. Their heads, which gazed alertly around, were squirrellike in appearance. The whole animal was, in fact, built like a squirrel, but the muzzle was more pointed. An observer would have seen the animal leave its nest with a sudden jump and display a silky coat, colored gray or chestnut.

A look at the teeth of these animals would have shown that they were certainly not squirrels. A squirrel has large chisellike incisor teeth; our little animals had peglike incisors, much more primitive than those of the squirrel. They had naked, moist noses with no hair and continuous with the lip. They had two nostrils, shaped like little commas, and closely flattened against the head were two leaflike ears. On their hands and feet there were five digits, each bearing a claw. And yet in some of them the claws appeared to be flattened, almost naillike in appearance. The hind limbs appeared larger than the forelimbs. Every now and again, one of them would sit up holding

A tree shrew.

Courtesy of the San Diego Zoo

something in its hands, look at it, turn it about, and maybe try it in its mouth. An insect would fly by, and with a slap so fast that it would be almost invisible, one of the little creatures would down it, pick it up, hold it in both hands, and proceed to eat it.

Our observer would have seen the animals jump from the ground up onto a tree with quick, sharp movements. There is no doubt that its fingers would have been very mobile, curling around the smaller branches and giving the animal an extremely firm grip. From the way the tail was held when the animal was climbing, it would be obvious that it acted as a balancer. As the animal moved around, examining objects and sniffing them with its nose, it would be obvious that smell was still important in its life but that it was also very alert to movements and to strange animals and objects that came across its visual path, so that vision played an important role in its life. These small creatures were the forerunners of the Primates. These were the tree shrews, and many of their direct descendants, looking very like their original progenitors, thrive in tropical forests of Southeast Asia today.

The tree shrews used to be classified among the insectivores, to which present-day ground shrews belong, but they are not now; however, they must have developed originally from some ground animals not too different from the ground shrews, for their modern representatives are certainly unlike insectivores. Studies made over the years have demonstrated that the tree shrews see much better than the ground shrews; their sense of smell has been reduced and the area of the brain controlling smell is much bigger in the ground shrew than in the tree shrew. The ground shrew lives very much in a world of smells

17

and only vaguely in one of vision. The tree shrews resemble the lemurs (the next step up the Primate ladder) in using the incisor teeth on the lower jaw as a little comb to groom their fur. Even the parasites of the tree shrew are like those of the lemurs; the two groups of animals even have the same type of mite living on their skin underneath the fur.

Although it seems that there is still quite a gap between the tree shrews of today and the lemurs, which form the subject of the next chapter, this has been bridged by the discovery of the remains of a little animal that belonged to the Oligocene period, 35,000,000 to 45,000,000 years ago, called *Anagale* which has a structure partly like the lemurs and partly like the tree shrews. One lemurlike characteristic is that the terminal joints on the toes are rounded and flattened and have flattened nails instead of sharply curved claws.

It seems that the present-day tree shrews and lemurs have remained more or less unaltered for millions of years. This is partly because they have lived in the trees, which is a three-dimensional existence, and as the various other animal groups developed, each found its own niche at some level among the branches of the trees. A very small and primitive type of animal such as the tree shrew, very agile and lightweight, would be able to carry out its activities on the higher branches. It could lead a secluded life, unaffected by the larger animals, particularly the larger Primates, which would be forced because of their weight to occupy the larger branches lower down.

Life in the trees led to the specialized grasping type of hand. The grasping hand permits animals not only to climb and run easily on the branches, but also to develop a method of locomotion called brachiation—using the arms to swing from tree to tree. As the Primates got bigger, the larger species, such as the chimpanzee and gorilla, came to live primarily on the ground simply because they became too big and heavy to climb on anything but the lower branches of the trees.

Mammals saw the light of day during the Mesozoic era, which was nearly 200,000,000 years ago, but by the end of the Cretaceous period (the last subsection of the Mesozoic era), they had started to grow into the many different types which exist today. The Cretaceous fossils found in Mongolia include some very primitive mammal skulls which suggest that their owners looked very much like the animals we call insectivores today. By about 70,000,000 years ago (the end of the Mesozoic era), there were very many types of primitive mammals scattered about the earth. Many of the fossils of these animals had skulls and teeth that labeled them as early Primates, and they all had the grasping hand of the Primate, so it is probable that they lived in the trees. They were small; in fact, they were very little bigger than rats and mice.

Some of the primitive lemurlike animals which developed from these little creatures were really on the sidelines of evolution; they could not make it in the world in which they lived and so became extinct. Others, however, managed to hold their own and became the present-day lemurs. And still others evolved further to become the ancestors of monkeys, apes, and man. There are some scientists who think that some of these early pro-Primates evolved into animals very like the present-day tarsiers and that this was the line that led to the final development of the higher Primate and man. The period during which these generalized tarsiers and lemurs developed is known as the Eocene epoch; it lasted 25,000,000 years and ended 45,000,000 years ago. These changes, resulting in the evolution of new forms of animals, were accompanied by significant changes in the brain.

The brain of the tree shrew is bigger than that of other small mammals, and this bigger brain is due mainly to the increase in the part connected with vision. If we watch the little tree shrew make his way in the trees, we can see what an important part of his life vision has become; even his tail is used for visual signaling. He waves his tail and flicks it in the vertical position; sometimes he droops it over his head, then lets it fall back slowly to the horizontal position, only to flick it back again in the vertical position. On another branch a little way from him we see another tree shrew signaling this intruder that he had better "take off" or expect an attack, even though he is one of the same species. In fact, tree shrews are so resentful of intruders into their territory that it is very difficult to keep more than one male in an enclosure as big as 50 by 50 feet.

When tree shrews confront each other, they may fight like little devils, and the defeated animal is often eventually killed after there have been several fights. Even female tree shrews are aggressive with each other, but although the aggression is similar to that shown by the males, it is less frequent, and fights between male and female have never been seen. Once the males form a dominance hierarchy (in other words, once a male establishes himself as a leader and the others learn their places), there is a considerable sociability between the animals.

Unlike monkeys, tree shrews rarely groom each other; they nearly all groom themselves. But there is an exception; as the female enters heat, the male starts to groom her. At this time, the male uses special techniques for cleaning his own face; he first licks his palms and then rubs them on the sides of his snout and over little glandular areas on the neck and chest which produce a secretion to mark objects and also to mark his territory. Some of the Australian marsupials and also the Australian jerboa mice have little glands on the chest that they use for the same purpose. Like some of the wildcats, the tree shrew also uses urine to delineate his territory.

Tree shrews have characteristic behavior, and some of those recorded are as follows: If they see an insect, they pounce on it and immediately kill it with their teeth. Give them fruit to eat and they sit up and hold it in their hands and eat it just as a squirrel does. They have several types of calls; for example, when they are pleased about something, they produce a kind of tremolo whistling. When they are alarmed, they make shrill little cries or little grunts, and sometimes when they are handled, they utter a kind of shrill chirp, presumably in protest. When they go to sleep, tree shrews curl their bodies up and then wrap their tails around them and hold the top of the tail above the head. The top is waved softly but sometimes more vigorously when, for example, they hear a mosquito, and they do this even when asleep. In the wild, tree shrews construct nests, often in hollow branches, made of leaves, small twigs of grass, or other material that they find around.

Although tree shrews love to climb, they also spend a lot of time on the ground and normally seem to be very relaxed and placid. However, this placidity can change suddenly in an explosion of violent activity. It has been described as "going off like a rocket." They are inquisitive little animals and poke their noses into everything. If you offer your closed hand to a tame tree shrew, he will scrape at it, stick his nose at it and try to pry it open, and almost get hysterical in his attempts to find out what's inside. The zoologist Ivan Sanderson did this to his pet tree shrew, and when the animal finally found nothing inside his hand after it had gone through all the effort of opening it, it sat up on his wrist and screamed at him in what he described as "a torrent of insectivorous invective." This curiosity is probably the reason why the animals move about a good deal and why they try all kinds of things as food. This curiosity is also the basis of Primate behavior and the cause of much of its success, especially its evolution to "higher" forms, such as man. Tree shrews seem to eat everything that can be eaten and in captivity will even eat foods that are overspiced or very hot. They will also eat leaves, snails, frogs, lizards, eggs, and worms.

In the present day, there are a number of species of tree shrews scattered about the world. They are found in the Philippines,

Sumatra, Borneo, the islands of the South China Sea, India, Cambodia and other areas of Indochina, and Burma and Malaya, so they are widely spread in the Oriental part of the world.

In 1779, Captain Cook set off on his third voyage around the world. In 1780, toward the end of January, the ship dropped anchor off the coast of Cochin China and a party went ashore; included in the party was a surgeon, Dr. William Ellis. They shot a few animals, including monkeys and "squirrels." Dr. Ellis does not seem to have left any records about the monkeys he found, but he seems to have been intrigued by the squirrel. He made a drawing of one, and this, together with his notes, found its way to the British Museum. His drawing was not that of a squirrel but was obviously of a tree shrew. It had an elongated body, rather rectangular in proportions, and looked somewhat like a dachshund; it had a fluffy tail, which certainly did look like that of a squirrel, sticking up vertically and then turning forward over the back of the animal. There was an angry-looking head and some nasty teeth, but they were not squirrels' teeth. Dr. Ellis, in fact, had drawn a tree shrew so very well that there is no doubt it is not a squirrel, even though he labels it that. This seems to be the first time that a tree shrew was recorded. Dr. Ellis never published the results of his observations on the animals he found in Cochin China, so nobody was aware of the discovery of this little unsquirrellike squirrel for many years.

Some years later a Frenchman, Leschenault de la Tour, was visiting Java and was very intrigued by certain little animals that he found there that also looked like squirrels. He sent a stuffed specimen of one to the Paris Natural History Museum. But even this lay unnoticed for a number of years, and it was some time before the distinguished biologist Etiénne Geoffroy St.-Hilaire, working in the museum one day, was suddenly attracted by the little stuffed animal and was led to examine it in great detail. However, Sir Stamford Raffles was ahead of St.-Hilaire, because he had written an account of the zoological collection he had made on the island of Sumatra and its surroundings, in which he had described the tree shrew in great detail. He had not only seen the animals and described them, but had kept a tame tree shrew at liberty in his house. The animal always presented itself at the breakfast and dinner table and was given fruit and milk, which it ate with great relish.

One of the characteristics that "outlines" the development of the Primates is the progressive increase of intelligence. The development of what we call intelligence is related to the size and complexity of the cerebral cortex. The lower part of the brain (the brain stem) is

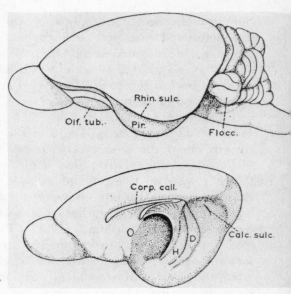

Brain of a tree shrew. Notice the smooth surface of the cerebral hemispheres.

From W. E. Le Gros Clark, *The Antecedents of Man* (Edinburgh, Edinburgh University Press, 1959)

responsible for the primitive emotions of animals—hunger, fear, anger, sex, and aggression—whereas it is in the cerebral hemispheres that the higher functions of the brain are concentrated. Here is where logical and abstract thought and consciousness are located. Under normal circumstances, the primitive urges of the lower brain are kept under some kind of control by the cerebral hemispheres, and they are a development which are a highlight of mammalian and especially of Primate evolution. We know that even in human beings this control is not complete, and under certain circumstances (for example, under the stress of extreme anger or hunger) a human may do things that his higher brain tells him are illogical or wrong. At every step upward, the Primates show an increase in brain size and a more complex folding of the cerebral cortex of the brain, which in the great apes and man becomes very complex indeed.

At this level of evolution, practically the only difference between ape and human brain is size. The ape brain has a volume of 300 or 400 cubic centimeters, and the human brain has a volume of 1,000 to 1,500 cubic centimeters. It has been said by a famous anatomist that the ape brain has no structure that the human brain does not have and the human brain has no structure that the ape brain does not have. The porpoise brain is larger than the human brain and is more complexly folded. Does this mean that the porpoise is more intelligent than man? Probably not. The increase in size and complexity of the porpoise brain is due to the overgrowth of the part of the brain that is concerned with hearing. There is no evidence that the part of the brain concerned with abstract thought and reasoning is any larger than that of the human brain. The porpoise brain is designed for the reception and integration of complex sounds having a range up to 100,000 cycles per second (middle C is 256 cycles per second), which is far beyond what the human ear can receive.

All animals with well-developed olfactory lobes in the brain live in a world which is predominantly a world of smell. They identify animals and food by smell. This is a fairly effective form of life so long as the animal remains on the ground. Once it takes to the trees, however, the sense of smell is not so useful. To avoid falling and breaking its neck and to be able to move with reasonable speed, the sense of vision and the sense of touch become of paramount importance. A sense of hearing will also help in avoiding predators. So once the animal is in the trees, the senses of vision, touch, and hearing should be taking over from the sense of smell. This is what we find happening in the tree shrew, and this is the route that leads on to higher Primates and man.

There are many things about the brain of the tree shrew which suggest that it is related more to Primates' brains than to those of other mammals, and not the least of these are weight and the greater development of the visual region. The cerebral cortex of the tree shrew shows signs of being divided vertically into layers; this and the large number of cells present in the cortex are also Primate characteristics.

So in the tree shrew, we have an animal in which the development of the cerebral cortex has started to outdistance that of other mammals. It has still not reached the levels of development even of the dog, but nevertheless, it has certain characteristics which relate more to the Primates than to any other type of animal. It seems that the present-day tree shrew, because of the developments in its brain and because of various other anatomical characters, is closely related to the originators of the Primate odyssey and the other Primates of today.

II. The Lemurs Emerge

ABOUT 60,000,000 years ago an ancestor split off from the main stream of Primate evolution and produced the group of animals we now know as prosimians. The Germans have named the prosimians *Halbaffen*, which means "half-apes" or "half monkeys"; they are monkeylike creatures which have not quite made it into the monkey kingdom. The prosimians developed rapidly and for many millions of years were the principal Primates in the forests. Not only were they present in Africa and Asia, but they also roamed Europe and even North and South America. There were many types of lemurs, a form of prosimian, at that time, and we know of at least 60 different fossil types. They disappeared completely from North and South America and from Europe at the end of the Eocene epoch

40,000,000 years ago and were eventually replaced by the monkeys. Most of them even disappeared from Europe and Asia, but one or two lemurlike species are still left in Africa.

Many millions of years ago when the island of Madagascar, located off the east coast of Africa, was part of the African mainland, prosimians ranged all over it as well as over the rest of Africa. When Madagascar was cut off about 50,000,000 or 60,000,000 years ago by the disappearance of the land bridge connecting it with the African continent, the prosimians there were left isolated on the island; they survive there and also in a few small islands in the Indian Ocean. How the lemurs got to these smaller islands, we do not know.

At the time the lemurs were isolated, the great predators—the lions, tigers, leopards, etc.—had not evolved. They evolved only after Madagascar was cut off. The lemurs that lived in the rest of the world not only were exposed to many types of predatory carnivore but were subjected also to fierce competition from the developing monkeys. The monkeys first appeared on the earth during the Oligocene epoch, something like 45,000,000 years ago, and disputed with and fought the lemurs, finally eliminating most of them and pushing the rest back to Madagascar. Then, with its separation from the mainland, Madagascar became a lemur paradise. Isolated on an island with a warm climate and plenty of lush vegetation, the lemurs flourished and multiplied and radiated out to develop a variety of different forms. Although lemurs became extinct in Europe and disappeared from the whole of America, there were still some traces of them in East Africa in the Miocene epoch, which was somewhere around 20,000,000 or 30,000,000 years ago. Some of the forms that developed in Madagascar were giant animals that subsequently became extinct.

There are something like 25 species of lemurs now living in Madagascar. Some of them have changed so little in the 50,000,000 or 60,000,000 years that they have been isolated there that they are like living fossils. The Madagascar lemurs vary in size; some of them are as big as a large dog, and some are extremely small and mouselike; some are jumpers, and most of them walk on all four limbs. One quality which the prosimians inherited from their tree shrewlike ancestors was the ability to grasp. All these animals and the monkeys and apes after them climb by means of grasping trees or branches and pulling themselves along. This is very different from other tree-living mammals who move in the trees with the aid of sharp claws which they use to dig into the barks of the trees.

The ability to climb by grasping was an important advance for the Primates to make, and because of it, they all eventually became masters of the trees. Grasping also provided a hand that one day could be used for manipulating and fashioning tools.

One big change is found between the tree shrews and the prosimians. In the tree shrew, as with other mammals including the dog, rat, squirrel, shrew, etc., the eyes are on either side of a long snout; but if you look a prosimian in the face, you will notice his eyes have moved to the front of the face. Here for the first time in evolution there is a possibility for true stereoscopic vision to develop, which was not possible when the eyes were on opposite sides of the snout and there was only a limited overlap of the fields of vision of the two eyes. With the eyes now shifted around to the front of the skull, there can be a complete overlap of the visual fields. This becomes extremely important in adapting to life in the trees, because with stereoscopic vision, much more accurate jumping and catching of branches are possible.

It is of interest that the word "lemur" comes from a Latin name *lemures* which means "spirits of the dead." There is no doubt that when the lemurs were first seen in Madagascar, some of the strange noises they made in the night might easily have seemed ghostlike. Whatever the reason for using this name for these animals, it appeared around the middle of the seventeenth century. Apart from the few species of prosimians known as the lorisoids and the tarsiers which exist in Africa and Asia, Madagascar lemurs are the only representatives left today.

There are many anatomical similarities between the tree shrews and the lemurs, and one which is important in describing the true nature of a Primate is a bone in the shoulder girdle known to the doctor as the clavicle and to the layman as the collarbone. This bone articulates (joins) with the sternum (the breastbone) in the midline of the body and articulates with the scapula (the flat shoulder blade) on the side. It functions as a strut which permits muscular attachments and provides leverage, permitting free movements of the upper limbs away from and to the body. It also permits movements forward and back. It is an essential anatomical requirement for the full range of climbing activities, and in some animals—for example, horses—the type of locomotion used does not require such a strut, and so it is absent. It is, however, present in the tupaias (the tree shrews) and the lemurs, where it is important to climbing and grasping activities. Clavicles are present in a number of other mammals, but they are not nearly as well developed as they are in the tree shrews, the prosimians, and the monkeys, apes, and man.

The multiplication and radiation of lemurs into a variety of habitats after their isolation are reminiscent of what happened to the marsupials of Australia which were isolated from the rest of the world earlier than this—before, in fact, any true mammals had developed at all. The lemurs changed anatomically to fit the various habitats they occupied. The marsupials developed likewise into marsupial moles, marsupial wolves, opossums (which were a sort of marsupial monkey), marsupial flying foxes, marsupial ungulates (kangaroos), marsupial bears (koalas), and so on. Like the lemurs, the marsupials, when they were cut off from the rest of the world in Australia, were left without predators or competitors and so were able to adapt to their various environments without interference and to develop in peace into a number of bizarre forms.

The lemurs did much the same thing. Some of the forms which developed were very minute, like the mouse lemurs; some became very colorful and could carry out acrobaticlike activities (examples of these are the ruffed lemur and the sifaka). There were also some lemurs that became very large and were, in fact, as large as chimpanzees, although they are now extinct. About 2,000 years ago, man first arrived on the island of Madagascar, and it is almost certain that these very large lemurs were living at that time. They probably survived up to quite recent times, for in 1658 a French explorer by the name of Étienne de Flacourt came to Madagascar and recorded that on the island was a large animal, about the size of a two-year-old calf, that had a round head and a face like a human being, but that had both fore and hind feet like a monkey, while his ears were like those of a man. It is possible that this animal was, in fact, the extinct giant lemur called *Megalodapis*, which is known now only from its fossil remains.

Today the island of Madagascar, rich with tropical vegetation, lies 250 miles off the southeast coast of Africa in the Indian Ocean and stretches 1,200 miles from $12°-26°$ latitude, so that its southern tip lies below the Tropic of Capricorn. Madagascar is the "Island of the Moon" with its 5,000-foot-high central highlands and with winds blowing over the warm Indian Ocean depositing their moisture mostly on the east coast.

The Arabs, who called it Al Qumar, and Greeks knew of the Island of the Moon, and Marco Polo, in 1295, knew of it and talked about a giant bird which he called the roc which lived there. This may, in fact, be the extinct giant bird *Aepyornis*.

Although they are not monkeys, lemurs are certainly Primates, even if only because of their hands, which are very much Primate hands with the thumbs "opposed" to the fingers; in other words, the thumbs can touch the tips of the fingers. Another important evolutionary feature of the lemurs which emphasizes their Primate nature is that

27

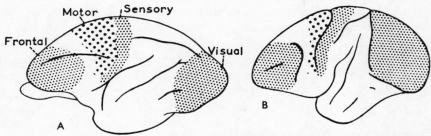

From W. E. Le Gros Clark, *The Antecedents of Man*
(Edinburgh, Edinburgh University Press, 1959)

Comparison between the brain of a lemur (A) and a macaque monkey (B). Note the larger proportion of the cerebral cortex devoted to vision in the monkey.

the hands and feet have nails instead of claws. However, the second digit of the foot still keeps its primitive claw which the animal uses for scratching and grooming.

In lemurs and their African relations the galagos, although the thumb and fingers are opposable, they are not moved separately but all at one time. For example, if they grasp a branch or an object, the four fingers and thumb close over it at once; if they pick up something, all the fingers work together. The monkeys have improved this control and can manipulate the fingers separately. They can pick up small objects, pieces of wood, insects, etc. with the forefinger and the thumb; the apes can do the same thing. It is a facility that requires a very close coordination between the eyes and the control of the finger muscles. The apes are capable of surprisingly fine control of the fingers and the thumbs, and anyone who has witnessed an ape trying delicately to remove a splinter from his finger or a freckle from his skin by bringing the nails of the forefingers of opposite hands into opposition will appreciate this fact. In view of these facts, it is not surprising that the ape and monkey brains show an increase in size of the areas that are devoted to controlling the movement of the hands, feet, and especially the fingers and toes. There is also an expansion of the area concerned with vision.

Although the lemur brain is an advance on that of lower ground-living animals, the occipital (back) part of its brain, which is concerned with the reception and integration of vision, is still small compared with other Primates, and so is the associative area of its brain. This is the region where impulses from different sense organs come together and where they are integrated and coordinated. Smell is still very important to lemurs, not only in their everyday activities

28

but also when male specimens of *Lemur catta*, the ring-tailed lemur, challenge each other with "stink fights."

The first accounts of Madagascar mammals were given by the famous biologists Alphonse Milne-Edwards and A. Grandidier, and following that, the famous classifier of living things, Linnaeus, decided in 1798 that at least one of these animals, the ring-tailed lemur, was a Primate. The famous English biologist and evolutionist Thomas Henry Huxley said of the Madagascar lemurs, "Perhaps no order of mammals presents us with so extraordinary a series of gradations as this. Leading us insensibly from the crown and summit of animal creation down to creatures from which it is but a step, it seems, to the lowest, smallest and least intelligent of placental animals."

Most of the lemurs in Madagascar live in the moist, evergreen tropical forests located on the east coast, but the ring-tailed lemur (*Lemur catta*) spends a lot of its time on the ground, and the dry thickets and woodland savanna toward the west appeal more to him. He is a pleasant animal and responds very well to captivity, where he has been bred successfully.

In all lemurs, the front incisors of the lower jaw stick forward and the animals use them as a comb. Though this activity is mainly cosmetic, it is not entirely so; it has a behavioral and a social function as well. In addition to this, especially with lemurs, which have a very dense fur, there is a possibility of a deeper, more physiological reason for it. People believe, when they watch monkeys and apes parting the hair on each other's bodies, that they are removing and eating fleas; these are not, in fact, fleas because monkeys do not carry fleas though they do sometimes carry lice. When monkeys groom each other, they often appear to pick off something from the skin and eat it. They are actually picking off little bits of scurf which they may eat.

Some years ago I suggested in the scientific journal *Nature* that possibly because monkeys and apes are covered with hair—and this would apply especially to lemurs—there is limited or no penetration to the skin of ultraviolet rays from the sun. In the case of humans, the contact of these rays with the skin causes a transformation of substances in the skin oils into vitamin D. This vitamin is necessary to the health of bones and teeth and proper mobilization of the mineral calcium in the body. In the case of monkeys and apes, I suggested that the oils in the skin leak out onto the surface and saturate small pieces of skin, or scurf, which become irradiated by the ultraviolet rays of the sun and form vitamin D. When the apes and monkeys groom each other, they eat some of the scurf and get a minimal amount of vitamin D. Also the oil, which comes from the skin and from the glands at the base of the hair follicles, spreads along the surface of the hair, becomes irradiated and forms some vitamin D, and the animal licking or mouthing the coat of another will get a mouthful of irradiated oil. In the case of the lemur, passing the fur through the dental comb represents the perfect way of getting this irradiated oil into the mouth. The relationship of the preening of birds to vitamin D intake is discussed in my book *The Ape People.*

When the Yerkes Center liberated four chimpanzees recently on a Georgia island, the animals rapidly became covered with ticks. As they began to integrate themselves into a social group and to groom each other, the ticks began to disappear until some six months afterward, very few ticks could be seen on the animals. So perhaps mutual and self-grooming also had their origin, at least in part, in keeping the animals free of ectoparasites.

The Different Sorts of Lemurs

There are about twenty-five different lemurs in Madagascar, and the baby of them all, as far as size is concerned, is the mouse lemur. It weighs only about 1 ⅓ ounces and is the smallest living Primate. Despite the mouse lemur's size, its brain has a number of characteristics which indicate that it is of an advanced nature, and the animal shows signs of being quite intelligent. It differs from many of the other lemurs in having a projecting snout like a dog and, also like the dog, has a moist muzzle. Its face has a set of sensory, or tactile, whiskers that have a sense of touch and are called vibrassae. These are used, by animals that have them, as a guide when they enter a narrow space. If the whiskers don't touch the sides of the space, the animal knows that its head and body can get through without difficulty. These whiskers are obvious in the cat. The mouse lemur not only has whiskers but also has ears like a cat.

Mouse lemurs roam in groups in small areas only about 150 feet in diameter in the forest region of Madagascar, and each group has a female core "sprinkled" with a few males which are probably the breeding group. Surrounding this nucleus is a group of males which probably do not play any part in breeding. The females nest in groups, using either hollow trees or little nests composed of leaves that they make for themselves. The males live nearby, either in solitary bachelorhood or in pairs.

The mouse lemur is fertile only at certain seasons of the year, and

30

during the infertile periods the vaginal orifice is completely closed over and covered with skin—a nonhuman Primate version of the chastity belt. The same condition occurs in galagos. About a week before opening, the vulva becomes red and swollen, and within three days of the opening, mating takes place. About ten days after it opens, the vagina starts to close, and within two weeks it is completely closed agai... The vagina then opens sixty days later to give birth to the young. Unlike most Primates, the mouse lemur has two to three babies in her litter, and she nurses them for about forty-five days. When they are born, after feeding them, the mother licks the region of the skin between the genital area and the anus to stimulate her babies to urinate and defecate. Not only does the female have a chastity belt for most of the time, but the male as well is apparently programmed to get the sexual urge only during that period when the vagina is open. When the male is ready for action, his testicles swell to eight times their size during the "resting period" and active production of spermatozoa occurs.

Smell is very important to the mouse lemur and, as with other lemurs, is apparently an important means of social communication. Lemurs use urine to establish territorial limits. Male lemurs have very strong scent glands situated on the inner surface of the forearm, just above the wrist and also at the junction of the arm and shoulder. They anoint their tails with this scent and then wave the tails actively about as if to disseminate the scent.

When lemurs are held in captivity, they mark the bars or the wall of a cage by rubbing the perineal region (the area between the anus and the genitals) against the bars. There are scent glands in this region, too, and this is another way of establishing their territory.

In general, lemurs live in troupes, and the only ones that are truly nocturnal are the lepilemurs and the spotted lemurs. The lepilemurs also appear to be solitary. There are many other types of lemurs, including dwarf lemurs, hairy-earred lemurs, snubbed-nose and gray lemurs.

One of the biggest lemurs is *Lemur variegatus*, the ruffed lemur, which may grow as long as four feet and is covered with long, black-and-white, silky fur arranged in a number of different patterns. These animals live in the treetops, mostly in the strips of forest in the east and west near the Madagascar coast, and they are more easily seen at night since they are partly nocturnal. In the mornings, they enjoy lying in the sun.

Another species is the black lemur (*Lemur macaco*); then there is the brown lemur (*Lemur fulvus*), the mongoose lemur (*Lemur mongoz*), and the red-bellied lemur (*Lemur rubiventer*). All except the first of these

Above, Mouse lemurs.

**Variegated lemurs. Bottom left, black-and-white variety;
bottom right, black-and-red variety.**

Mongoose lemurs.

Ring-tailed lemurs (*Lemur catta*).

animals are colored different shades of brown, and the arrangement of the colors in the lighter and darker areas varies so much that it is difficult to find two animals that are alike. The mongoose lemur is very sweet-natured and gentle and makes a good pet. It is found on the nearby Comoro islands and also on the northern tip of the main island of Madagascar. It is very active and lives in the trees; it is a daytime animal and a vegetarian.

One of the most extraordinary lemurs is the aye-aye whose scientific name is *Daubentonia*. It is an animal which was not, it seems, able to make up its mind whether it wanted to be a lemur or a rodent. It seems to stand close to that early time of evolutionary divergence between animals when the potentiality to develop into either a Primate or a rodent existed. It has enormous front incisor teeth, then a space, then its grinding teeth or molars. This is the same arrangement that the squirrel and other rodents have. As with rodents, the incisors are continuously growing; they grow throughout the life of the animal to replace the tooth substance that is worn away at the biting end. The aye-aye's hands and feet are typically Primate; but only the big toe has a nail, and the other toes and fingers have claws. The animals have very big tails which they wrap around themselves when they go to sleep. They were discovered in 1790 by a French explorer, who described them as squirrels, and their squirrellike nature was later confirmed by some very eminent anatomists. In 1890 a scientist suggested that they may actually be lemurs, and eventually it was established that they were, in fact, lemurs, not rodents.

Although looking a bit like a human hand, the hand of this lemur shows an outstanding difference; the middle two fingers (the third and fourth fingers) are twice the length of the other fingers on the hand, and the third finger is long and thin like a stick. The aye-aye uses this long finger to claw out wax from its ears and to pick its teeth, but its more important function is to get grubs. The aye-aye taps its way along the branches of trees until it comes to the entrance of a long passage which a grub has chewed in the wood, and it then sticks its long, thin finger down the passage and hooks the grub on its claw. If it is unable to put its finger into the passage, it will break the passage open with its teeth. The rodentlike front teeth are also used to gnaw open sugarcane, and the long, thin middle finger is used for hooking out the larvae it finds there.

The aye-aye is fearless; it strikes at intruders and bites viciously if it is caught, all the time making a very harsh noise. Fossil aye-ayes, which are twice the size of the present-day animals, are known in Madagascar. There are not many of these animals left now in the

From E. G. Boulenger, *Apes and Monkeys*
(New York, Robert McBride
and Company, 1930)

The aye-aye lemur.

wild; in 1957 ten small groups of aye-ayes were located in areas on the
northeast coast of Madagascar, and since then, one of these groups has
died out. The aye-aye is officially an endangered animal and is likely
to be lost with the advance of civilization into these areas.

Another lemur worthy of special note is the sifaka or monkey lemur.
It is shaped more like a monkey than the other lemurs. It has a black,
naked, rather flat face without the long snout that some of the other
lemurs have. It is relatively slow and deliberate in its movements and
lives mainly in the trees, but every now and again it succumbs to the
temptation to come down to the ground with its friends to play or feed.
Only recently has it been possible to keep the sifaka in captivity; part
of this is apparently because its diet is rather specialized. It is a great
jumper, and some tribes in Madagascar think it is sacred because it
enjoys sunbathing, which they interpret as sun worshiping.

Another member of this strange community of animals is the woolly
lemur, which has a similar shape and size to the sifaka, but longer
hind limbs than the sifaka. It also has very large feet. The woolly
lemurs are also known as avahis. They are solitary creatures, living
primarily in the trees, and they are completely nocturnal.

Another interesting type of lemur is the indri, which lives mainly in
the volcanic mountains and in parts of the coast of Madagascar. These
animals are about two feet in length; they have an incongruous little
two-inch tail, while their arms and legs have about the same
proportions as humans. The name "indri" derives from the word
indrizy, which is said to mean "look at that" or "there he is." The

35

The avahi (the woolly lemur).

From Ernest P. Walker, *The Monkey Book*
(New York, Macmillan Company, 1954)

Malagasy people call the animal an *amboanala* which means the "forest dog." The indri have a bit of a snout, naked black faces and black ears, and in contemporary fashion fringes of fur that cover the ears. The long, thick fur is black and white, and the top of the head is usually white, relieved by a pair of black eyes. They have black shoulders, which extend almost to the elbows, and their coloring gives the impression that they wear black socks and gloves.

These animals roam about the forest, mostly in family groups, and can make a very loud noise, particularly when they call out all together. They can also wail like a trumpeter or like an injured dog (some trumpeters sound like that, too). Their oddly human shape and their noisy behavior evoke an eerie atmosphere which causes the locals to regard them with some awe. There are rumors that they are able to catch a spear when it is thrown at them and then throw it back. Since the same story has been told about the baboon in West Africa and about the howler monkeys in South America, it obviously has to be taken with a grain of salt.

At this point, I am reminded of a rhyme once published by Professor Ernest A. Hooton, of Harvard University, and perhaps my readers would enjoy it:

> The lemur is a lowly brute,
>   its primate status, some dispute.
> He has a damp and longish snout,
>   with lower front teeth hanging out.
> He parts his hair with his comb—jaw,
>   and scratches with a single claw.

The indri.

From Ernest P. Walker, *The Monkey Book*
(New York, Macmillan Company, 1954)

That still adorns a hinded digit,
    wherever itching makes him fidget.
He is arboreal and omniverous,
    from more about him, Lord deliver us.

Allison Jolly, however, has made a determined effort to provide
"more about him," at least of two species: the sifaka, already
mentioned, which belongs to the same family as the large indri, and
which is known as Propithecus and secondly, the ring-tailed lemur
(*Lemur catta*).

The sifaka is fairly well distributed in most of the forested region of
Madagascar. It is happy to spend most of its time in the trees and
regales itself on leaves, fruits, flowers, bulbs, and bark. I was surprised
to find out that in captivity it will eat eucalyptus leaves, which are
normally very much an acquired taste. It will also eat guavas, rice,
and bananas, a taste with which I am more sympathetic, but it does
not seem to accept animal protein. Its body is about three feet long, so

Portrait of sifaka.

Courtesy of Dr. John Buettner-Janusch

that standing on its hind legs, it is about four feet tall, quite a large animal. On the ground, it walks slowly on its hind legs or hops along. When it gets into the trees, it clings to the trunks in a vertical position. The leaping of a sifaka from one tree to another is something worth observing. A sifaka, when he jumps, can leap with great suddenness,— forward, backward, or sideways as far as 30 feet. For leaps to close objects, it simply projects itself from one to another, but on a long leap the body is horizontal to the ground, with the arms stretched in front. Just before landing, the animal lays its body back so that the hind limbs actually make contact with the trunk of the tree on which it plans to land before the forelimbs do.

Monkeys are clever, but not so the sifakas. Give them anything they are unfamiliar with, and they do not know what to do with it; they show no signs of understanding it or trying to find out about it. Even in their natural surroundings, they do not manipulate objects as a monkey does. The only objects they carry are food, but sometimes they will carry branches from trees that have food on them. They never carry food very far and soon sit down to eat it. They seem to have a good sense of vision and have a disquieting ability of identifying a distant observer, even when there is no movement.

They recognize individuals in their own groups or even individuals belonging to another troupe, by the variations in the markings on their faces. They seem to have stereoscopic vision, an important evolutionary development, and there is evidence that they can see color, at least in the orange and red end of the spectrum. Their sense of smell, however, is still acute and is very important in their lives, because they leave their individual scents on branches as markers, and members of other troupes will sniff at the branches in a new area when they enter it to see if other sifakas have been there.

Sifakas like to sun themselves. At first light (6–6:30 A.M.) they start this procedure, and a whole troupe may arrange itself on a series of branches, facing the sun, usually with some body contact with each other; they also indulge at that time in self-grooming. There are legends among the Malagasy tribes that these lemurs are sun worshipers. Some of the Malagasy even believe that they are incarnations of ancestors who continue their worship of the sun in the guise of a sifaka. Allison Jolly says that their sunbathing looks to her less like religious fervor than "our indolent cult of sunbathing at the beach."

In grooming, they use their lower incisors as dental combs just as other lemurs do. When they are finished self-combing, they usually have a good scratch with their "toilet claw"—this is the one finger which still has a claw; the others all have a nail, the hallmark of the Primates. The claw is on the second toe of each foot, and it sticks up at right angles to the toe; with it they scratch themselves like a dog—on the shoulder and even behind the ear. More intriguing, however, is the fact that they also use it to clean their ears, putting the claw carefully into the earhole and turning it around, then pulling it out and licking it. This seems a pretty sophisticated bit of toiletry, but I wonder if its origin wasn't due really to the necessity of removing ticks from the ear.

When a group is ready to move on, the members perform a mass urination and defecation, which is started by one animal and continued by the others. Once this ceremony is completed, the troupe will move together into another area. There is no particular place where these excretory functions are carried out; any branch where they happen to be will do. If they are subjected to any kind of fear, a barrage of urination and defecation is likely to occur; even the sight of an observer can sometimes have this effect—not a very flattering greeting.

The reaction of a sifaka group to humans is interesting. If the human simply stands still, the group will start to move slowly forward, with some low vocalizations. They may move as close as 10 yards, and they all face the observer. They lick their noses and weave their heads from side to side while continuing to stare. Later on they become more excited, and the noise they make sounds a little like the word "sifaka," and this is the origin of their local name. They may sit and make *sifaka* noises at the observer for twenty-five minutes or even as long as two hours. Eventually they retreat by hopping backward, keeping their faces toward the observer—obviously they do not trust humans very much—and they continue to make the *sifaka* noise until they are out of sight.

Lemurs are very much "contact" animals. They love to have body contact with each other, and this is really a form of social communication. All lemurs tend to get together in groups with everyone touching someone else. Allison Jolly has described a group of sifakas, who were cuddling up together, as looking like a spherical ball with a whole lot of heads sticking out on the outside. However, when they are on the branches, they cannot form these groups, so as an alternative they line up, each animal pressing his belly against his neighbor in front. They often do this in the early morning when they have just awakened and before they move around to sun themselves. Sifakas play-wrestle, too, for body contact when they are not resting. Their play is not aggressive, however, and the sifaka troupe seems to move collectively, not following one leader.

The desire for contact with other bodies is fairly common in the animal kingdom, but in humans it tends to be restricted to those who have an emotional attachment to each other; otherwise it tends to be resented. However, in recent years Esalen in California, Anthos in New York, and similar groups across the country have formed where strangers make manual and body contact with each other. It is possible that this may be of some social therapeutic value, but it sounds like fun.

Grooming is also a form of contact; it is an important part of lemur life, just as it is with all Primates. Anyone who looks at baboons and chimpanzees will see elaborate grooming, and various forms of this exist in the human race as well. In humans grooming incorporates many activities such as brushing and combing of the hair, professional hairdressing, massage, petting, and a lot of presexual activity. The ministrations of the barber or the hairdresser are, in fact, highly ritualized grooming. The patting and stroking of dogs and cats are also a variety of grooming.

When sifakas copulate, the male leaps onto the back of the female and often holds her with his feet around her waist while he makes entry. The courtship and mating behavior is quite complicated. Males have been known to groom their own genitalia until they get an erection. Males and females will also groom each other's genitalia.

When first born, the lemur baby clings to the fur across the lower part of the mother's abdomen. The baby makes no journeys in this warm, furry environment except to crawl up to one of its mother's nipples to suck the milk. The mother just leaves the baby there and makes no effort to help it or direct its activity with her hands, but for its convenience she will often take up what has been called the living playpen position. This involves sitting with her knees apart and her tail curled between her legs. She rests her elbows on her knees and partly blocks out the gap between the thigh and the body with her arms. Within these confines, the baby can scramble around in safety and is protected from falling if the mother is sitting up in the trees. Allison Jolly has some interesting descriptions of infant play. One of them reports:

> An infant at Benala began to play the game of "leave mother and dash back"—familiar to me in *Lemur mongoz* (the mongoose lemur). The infant pulled itself up the small sloping branch where its mother sat and then dropped back onto her. The movements were hilarious; the baby pushed hard with its hind legs on the branch. In the adult, this move propels the animal several feet upward, and it catches hold with its hands and feet. The infant's uncoordinated legs only bounced its rump an inch or so away from the branch. Then it pulled itself upward, hand-over-hand, like climbing on its mother's fur. After two or three bounce-and-hauls, the baby panicked and dropped back, only to start the game again.

At about three months, the infant switches to riding on the mother's back. Her belly fur is too sparse to hold its growing weight, and anyhow, the baby is now big enough to get in the way when its mother tries any of those fancy leaps. It does not jockey ride like a young baboon but remains with ventral surface pressed against the mother, feet grasping her waist on either side and hands on her shoulder fur; in other words, it rides rather like a scared jockey.

Another interesting type of lemur in Madagascar is the *Lemur catta*. It is a beautiful black-and-white animal with thick, soft fur covering a light, slender body; in fact, it weighs only three-quarters of the weight of a cat of the same size. There are dark rings along the length of the tail which give the animal its popular name, the ring-tailed lemur. On the ground, the animal walks on all fours; but its arms are much

The sifaka—mother and four-and-a-half-month-old baby girl.

Courtesy of Dr. John Buettner-Janusch

Sifaka with one-week-old baby hanging on across the pelvis.

Courtesy of Dr. John Buettner-Janusch

The sifaka running.

Courtesy of the San Diego Zoo

shorter than its legs, and the large hindquarters give it the power to spring considerable distances through the branches of the trees. It has a relatively small head with a long and rather pointed muzzle. When the animal moves, it points its muzzle downward and is thus able to stare ahead with its two amber eyes.

Lemur catta is a social animal and in the wild lives in troupes. Individuals who are more venturesome and energetic than the others can range farther than the troupe as a whole. The middle of the day, however, is rest time. *Lemur catta* not only is at home in the trees and uses them to move from one place to another but also travels on the ground. It is up with the sun or even before dawn; then, like the sifakas, it takes the sun when it appears and with that ceremony complete, takes breakfast.

If you have a *Lemur catta* in captivity and you want to find out how clever it is, the results are disappointing; the poor creature does very badly at psychological tests which are used to estimate the intelligence of other Primates. For example, it cannot see that if it pulls in a string with food attached to the other end, it can get the food. But in the wild, it has no problem in getting the idea that if it pulls in a branch with food on its end, it can have a meal. In captivity, *Lemur catta* treats inanimate objects as though they were pieces of food. Strange objects are tasted and smelled; if they taste and smell good, they get chewed up—if not, they get thrown away. Allison Jolly has never seen *Lemur catta* in the wild play with, investigate, or pick up an inanimate object unless it looks like a piece of food. Anything live, however, gets an immediate reaction from them. They investigate insects and small animals like mouse lemurs and have also been known to investigate human beings. Presumably, they have a sense of humor, because one of their sports is to tease the sifakas.

They have a variety of calls, varying from a scream to a purr, with howls and barks intermixed. Scent is very important to them, and anatomical studies show that a large part of the brain is concerned with smell, and scents and odors play an important part in their lives.

The waving of the tail appears to be one method of broadcasting their odor and attracting another animal's attention. They also leave odor when marking leaves, branches, and territories where other lemurs will be likely to encounter it or in territory where they want to establish their own particular preserve.

The eyesight of these animals does not appear to be as good as that of the sifaka; but they can see very well at night, and they probably have both stereoscopic and color vision.

They move in troupes, and the troupe may be spread out over as much as 50 yards. When a troupe is moving, the juveniles, dominant males, and females take the lead, while the subordinate males bring up the rear. The signal for a move is a series of clicks, but often the animals will make some noises which sound like moaning or wailing before they actually move off. The ring-tailed lemurs do a good deal of mutual grooming, and body contact is very important to them; they also indulge in a good deal of play in which they often chase each other in circles. They jump on each other and wrestle; they may sit facing each other, resting on their hands and feet and biting each other. All these reactions help establish social bonds between them.

I have mentioned that the sense of smell plays an important part in the life of lemurs, especially in their signaling system. *Lemur cattas* produce an odor that humans find hard to detect but which is obvious to them. Allison Jolly has noted that when two lemurs engage in what is obviously a stink fight or a territorial battle, it looks like a silent movie. She could not hear the method of communication, but she was able to follow it through the various gestures. The nature of the scent probably differs between the various subspecies of lemur, and this is one way they recognize each other. The scent also appears to change with the sexual state of the animal and possibly also with the season.

There are many behavioral gestures which the lemurs make that are very typically Primate in nature. They will stare to make a threat, they will stroll with the head up confidently, they will bend over and crouch in a submissive attitude, and they will stand up on their hind legs and raise their hands to fight.

To some extent, Primates demonstrate their intelligence by the relations they make with animate objects and the use of the inanimate objects as tools. So far it seems that only the apes or the cebus monkey, the little organ-grinder monkey, have the ability to invent new tools. This must be because they can perceive new relationships and then can invent a tool to help them out with a particular problem. An example of this is the use of sticks and stems of grass by chimpanzees to fish for termites. Primates also demonstrate their intelligence by the social organizations they develop and the complexity of the social signaling the group produces. In intelligence tests which demonstrate symbolic learning, which the rhesus monkey can do very well, the lemur actually scores badly. This is so whether the simplest of insight tests are used or whether tests like object discrimination and delayed response learning provide the mental hurdle. One captive lemur could not solve a simple puzzle over a ten-day period.

The social relationships of the lemur appear to be less sophisticated than those of higher Primates, despite the fact that they resemble the

Racial discrimination among the lemurs. A brown lemur joins the two ring-tailed lemurs that were sunning in picture above. One ringtail has already left, and the other is departing in picture below.

From J. A. Allen, "Primates of the Congo," *Bulletin of the American Museum of Natural History*. Vol. 47, No. 283 (1925)

Demidoff's lemur.

Ring-tailed lemurs (*Lemur catta*).

higher Primates in having permanent troupes of mixed sexes and ages. Their social grooming resembles that of higher Primates, and the reaction of individual members of the troupes to each other is a function of their relative dominance.

The ring-tailed lemur is a very beautiful animal with its black-and-white fur and its fluid motion. Its fairylike lightness and its soft fur make picking it up quite an event. Some of the poses it adopts, often with its long tail wrapped around its neck, charm artists and sculptors. The ring-tailed lemur is able to jump with very little effort, sailing up to a height of about 10 feet, and it can then grip and cling to a wall with very little more than half an inch of picture molding for support.

Edward H. Duro, who is in charge of the care of the prosimian colony at the Oregon Primate Research Center in Portland, Oregon, has given a fine description of lemurs in the publication *Primate News*:

Would lemurs make good pets? Yes, except for the house breaking. Most people who have an animal like that will set aside one room for it. One man had a monkey that he toilet trained. Is this possible with a lemur? I think it depends on the animal. Like people, they have a wide range of individualities. I don't think I'd want to have one in my house. Pug, the Mongoz lemur, good as he is, when he has to go. . . . They have no way of letting you know. The ring-tails are cleaner than most lemurs, but they are highstrung and get nervous diarrhoea. For the most part, you can't force yourself on them but ring-tails really want to be the ones who do the making up.

Some of the lemurs will talk in their own way to you. How? François sort of barks. The only sound Moonbeam makes is that very soft, subdued cat cry. Lemurs also call to each other. Have you seen pictures of coyotes baying? Lemurs do that too; stand up on their hind feet and bay. The males especially. The females don't have the volumes to their howling that males do. Invariably, they do it before feeding time and after people leave, and they are left alone. When the Fulvus lemurs are scared, they do it too. This is not the reason ring-tails do it. A tap on the wall makes the Fulvus go ape, but not the ring-tails. They talk back and forth to each other and they'll talk to you in a greeting sense.

III. The Lorisoids

ALTHOUGH THE TRUE lemurs disappeared from the rest of the world while they thrived in Madagascar, some lemurlike animals survived. It is possible that these animals, called lorisoids, are actually more primitive than the lemuroids. They are found in Africa, India and Ceylon, and Indochina.

Little or nothing is known about the fossil ancestors of the lorisoids, but the present-day members have been divided into two main groups: the bush babies, which are called by the scientific name of *Galago,* and the lorises, which include four types—the pottos, the angwantibos, the slender lorises, and the slow lorises.

The dwarf bush babies are so tiny that one is surprised to find them so small and thrilled to find them so sweet and attractive. They are also known as Demidoff's bush baby and rejoice in the most impressive scientific name of *Galagoides demidovii*. They fit comfortably on the palm of the hand and though they may be sitting gazing innocently around with their bug eyes, they are capable without warning of a fantastic 10- or 12-foot leap in any direction, even vertically up.

These diminutive pre-Primates occupy forests along the equator in Africa, and they are so tiny and so well camouflaged that they are very difficult to see, even by native hunters. These tiny animals live on insects, tree snails, tree frogs, and green nuts; they have a partiality for green almonds.

The large galagos are quite different from the dwarf galagos. They are galloping and hopping animals. They are very much night animals; in captivity they may be observed to sleep all day. They wake up at sundown and expect to be fed at that time. They use their hands to sort the pieces of food with great care. After their meal, they carry out their toilet. In the wild they then go on the prowl; in captivity they start a restless pacing, turning and climbing, doing it over and over again like a machine. After an hour or two of this, it is time for another snack, followed by some more toilet. This was the routine followed by the galago that my wife and I owned.

We found it a very sweet and gentle creature. When we attempted to pet it, it would stretch one arm out straight so that we could tickle it under its armpit. It would remain like this almost as long as we were prepared to go on tickling. Sometimes, when we wanted to stroke it and it was not in the mood, it would grab one of our hands and push it away with its own hand in the gentlest fashion. Although it normally made little noise, sometimes in the middle of the night it would let out an almost incredibly loud wail which must have really caused the neighbors to wonder what on earth we had inside our fence.

I have found a very similar type of behavior in the South American kinkajou. It sticks out its arm in exactly the same way as the galago so it can be tickled under the armpit, and it will also push your hand away in the same gentle fashion if it does not want to be petted; if you persist, it will mouth your finger or hand with its teeth but never actually bite. This very stereotyped bit of behavior appeals to me as being extraordinary; these two animals are not placed by the zoologists close to each other at all and have originated thousands of miles apart, millions of years ago, on different continents.

The galagos have an anatomical peculiarity, a cartilaginous protuberance underneath the ordinary fleshy tongue which can almost be described as a second tongue, and this is used in conjunction with the front teeth in the process of grooming. For example, the animal

Courtesy of Mr. Harry Wohlsein and Dr. William Montagna, Oregon Primate Center

Galago—note the nails on the fingers and toes.

uses its front teeth as a comb to curry the fur and then uses the cartilaginous underpart of the tongue to clean the teeth of the fur and other accumulated material. Galagos also have a tendency to hang upside down from a branch, and they fold and unfold their legs, in this way raising and lowering themselves. Another surprising habit is the use of their urine to moisten their hands and feet; presumably this helps them get a good grip on trees and branches, although I suspect that it also helps mark the branches they travel along and lets other galagos that use the same branches know that they are in that territory. Some of us spit to get a grip, but more sophisticated professionals, like circus acrobats, use rosin. My father, one of the pioneers who opened up the goldfields in Western Australia, talked of people urinating on their blistered hands when they were digging for gold to make it possible to keep going; he said it hardened the skin of the hands. Whether they got a better grip on the pickax or shovel as a result, he never said.

Another interesting anatomical feature of the galago is that its tarsus, a bone in the foot which extends from the heel and ankle to the instep, is greatly elongated so that it becomes one-third the length of the shinbone. This greatly elongates the foot so that it is larger than the animal's forearm. This is an adaptation to a hopping gait, for the animals hop along on their hind legs like kangaroos. They have considerable jumping ability, and they have been described as jumping from one palm tree to another with great agility and of adhering to anything they land on, like a lump of wet clay. At the London Zoo many years ago, one was let loose in the superintendent's apartment, and it had no difficulty in leaping several feet in the air, clearing the furniture with ease. Natives in Africa are said to set out pots of palm wine for it to drink after which it can easily be captured.

As far as reproduction is concerned, the bush babies belong to the lower classes among Primates. Instead of the sophisticated behavior of having one infant at a time or at the most twins, they often have a litter of three babies.

There are about eleven different races of great galago; they are found along the east coast of Africa and in Angola and the southern part of the Republic of Zaire, formerly the Congo.

All lorisoids have an unusual anatomical feature—the second, or index, toe bears a long, curved claw, which sticks up and is used to scratch the skin. All the other fingers and toes have the flat nails characteristic of the Primate. The animal that gave the name to this group of animals is the slender loris, which is found climbing the trees in the forested part of southern India and all over Ceylon. The Dutch word *loris,* from which it is derived, means "clown." The slender loris is quite small, about a foot in length, and usually weighs about 400 grams—approximately 12 ounces. The lorises move in a very deliberate fashion, but they can move fast when they need to. They prowl the trees at night and sleep during the day. They eat the usual shoots and leaves of plants but will also eat the flowers, and nuts are a popular article of diet, particularly if they are not ripe. Like most Primates, the slender loris relishes a good fat insect, nice bird's egg, or a small animal that can't fight back or run away fast enough.

In Hugh Craig's *The Animal Kingdom,* there is a graphic account of the slender loris creeping up on its prey:

> Alas for the doomed bird that has attracted the fiery eyes of the Loris! No Indian on the war-path moves with stealthier step or more

53

Slow loris.

Courtesy of Mr. Harry Wohlsein and Dr. William Montagna, Oregon Primate Center

deadly purpose than the Loris on its progress towards its sleeping prey. With movements as imperceptible and silent as the shadow on the dial, paw after paw is lifted from its hold, advanced a step and placed again on the bough, until the destroyer stands by the side of the unconscious victim. Then, the hand is raised with equal silence, until the fingers overlay the bird and nearly touch it. Suddenly the slow caution is exchanged for lightning speed, and with a movement so rapid that the eye can hardly follow it, the bird is torn from its perch, and almost before its eyes are opened from slumber, they are closed forever in death.

There are many other stories about the slender loris, a number of them of doubtful authenticity.

The hands and feet of the loris are rather peculiar; both the thumbs and the big toes are very widely opposed—that means they can extend out at a wide angle from the rest of the hand and the fingers—but their index fingers are reduced to a mere stump.

Slender loris.

From Ernest B. Walker, *The Monkey Book*
(New York, Macmillan and Co., 1954)

The other well-known type of loris is the slow loris, which inhabits Malaya, the Indonesian islands, and the Philippines.

The slow lorises are found in a bewildering variety of sizes with different colors and color patterns. They are covered with very long woolly fur and curl up in a ball to sleep in the daytime, but with the onset of night they open their big, round eyes and gad about the treetops, looking for shoots, fruit, birds and their eggs and insects. In a pinch, they will fill up on leaves. These animals are said never to leave the trees if they can avoid it, although when they are on the ground, they can move with a sort of wavering trot.

There are many legends about the slow loris: Its fur is said to speed the healing of wounds; if a ship carries one as a mascot, it is never becalmed; dead or alive, the loris has power over the lives of human beings; the animal itself is unhappy since it is supposed to be continually seeing ghosts and this is the reason why it always buries its head in its hands.

Another lorisoid is the potto. It is about the size of a small cat, and although it lives in Africa, it is fairly well known in Europe and to a lesser extent in the United States. It also rejoices in the confusing name of honey bear, a name which in the United States is used for the kinkajou. Other names for the potto are bush bear or tree bear, but perhaps it is most appropriately called by its West African pidgin name, "softly softly." There seems to be a good deal of confusion of names between the potto and kinkajou, because the Latin name for the kinkajou is *Potos*. The kinkajou, however, is a South American animal related to the raccoon, and as mentioned, is often called colloquially the honey bear. The Latin name for the potto (softly softly) is actually *Periodicticus*.

55

Potto and baby.

Courtesy of the San Diego Zoo

The abode of the "softly softly" is in Africa, especially in Nigeria and the Congo. It moves much more slowly than any of the galagos and never seems to jump. In fact, the potto never lets go with more than one foot or hand at a time, so it always has three hands or feet holding on. Like the rest of the lorisoids, pottos are nocturnal. They do not lie down to sleep but sit up with their heads lowered onto their abdomens with each of their four limbs hanging onto something, so they are well anchored in position. At night their movements are slow but sure, and they move continuously.

The most famous potto is called Bosman's potto after the Dutchman who first described it as long ago as 1704. Some old descriptions are entertaining, if not exactly accurate:

> A creature, by the Negroes called Potto, but known to us by the name of "Sluggard," doubtless from its lazy, sluggish nature, a whole day being enough for it to advance ten steps forward.
>
> Some writers affirm that when this animal has climbed upon a Tree, he doth not leave it until he hath eaten up not only the Fruit, but the leaves entirely; and then descends fat and in good case in order to get up into another Tree; but before his slow pace can compass this he becomes as poor and lean as 'tis possible to imagine; and if the Tree be high, or the way anything distant, and he meets nothing on his journey, he invariably dies of Hunger betwixt one tree and the other.

Another legend about the pottos is that they strangle animals. They are said to creep up on their victim with infinite slowness and care,

56

Portrait of a potto.

approaching from the back and suddenly encircling the animal's throat with their hands and squeezing it until it is dead. To what extent they kill larger animals in the wild we do not know, but they are well equipped with teeth and could certainly give a good account of themselves in any encounter.

Pottos have only a small tail but have a very unusual anatomical development which serves for defense. The last two vertebrae of the neck and the first thoracic vertebra have horny spines attached to them, which penetrate through the skin. When the animal is attacked, it stands up and at the last instant drops its head so that it brings these spines into a forward position and the attacking animal is likely to get a painful reception.

The potto has only one relative in Africa. It is known as the Calabar potto and is a small creature with a face like that of a tiny fox, with no tail, and a bright orange-rust woolly coat. Its ears are naked, and its hands and feet have tiny fingers and toes which are very small and poorly developed. The lower joints of the hands and feet are joined by a web. The Calabar potto is also known as the angwantibo.

The angwantibo was discovered back in 1869, and its original description was published by T. H. Huxley. It is much more active than the potto, and much quicker at getting about, although its individual body movements appear slow. It is a fairly gentle little creature, exceedingly clean with a very elaborate toilet. After every meal, the whole coat is meticulously combed with the lower front teeth, and the ears, face, and chest are smoothed off with saliva applied by the hands. An unusual activity perhaps, but I have seen humans spit on their hands and then use them to smooth down their hair, and I have heard of Indian maidens of some tribes using urine as a hairdressing.

Angwantibos are hypersensitive to sound and move instantly into a defensive retreat. They are carnivorous and insectivorous, wolfing down all kinds of soft-bodied insects and worms but being a little more fussy with hard-shelled insects and snails. They thrive on raw meat, especially bird meat. Their hands are even more modified than those of the potto, the index finger having vanished altogether.

IV. The Link That Was Not Missing—The Spectral Tarsier

BACK IN THE 1930's, I had the privilege of working with one of the world's outstanding anatomists and anthropologists, Professor Frederick Wood Jones. He had startled the scientific world by claiming that man and apes were descended from a little animal, scarcely more than the size of a rat, with big eyes, which was found mainly in Borneo and the Philippines. It was called the spectral tarsier. His theory was described by some scientists who did not approve it as the Tarsian Heresy. However, subsequent studies by distinguished anatomists such as Sir Wilfrid Le Gros Clark of Oxford University demonstrated that man might indeed have descended from a *Tarsius*-like ancestor. Later the finding of a variety of related fossil remains by Dr. Louis Leakey and others served to make the hypothesis enunciated by Professor Wood Jones highly probable.

"The smallest long-tailed monkey in Luzon" was the description given to the *Tarsius* when it was first discovered in the seventeenth century by the Jesuit priest Camelli, who was working in the Philippines. An English naturalist, James Petiver, described the animal in 1702 and illustrated his article with a humorous drawing of it. Petiver used the native name *Magu* for *Tarsius,* which was and still is used in parts of the Philippines. A few years later, the great Buffon saw a museum specimen—which he thought might be a rodent rather like a jerboa. *Tarsius* did not reach Primate status until 1777, when it was officially classified among the lemurs, where it remained for quite a time. However, it is not a lemur, and it occupies a niche of its own insofar as the evolution of man is concerned.

Some 50,000,000 or more years ago, in Eocene times, there were some twenty different *Tarsius,* and most of them were found in North America. There is only one species of *Tarsius* today, and it is similar to the fossil animals which were grouped together under the name of Anaptomorphidae. *Tarsius* has thus survived millions of years on this earth with very little change and has been described as a living fossil. It is said to be the "oldest" mammal now living on the earth.

Although inclined to be excitable, tarsiers are fundamentally gentle creatures and make good pets; they seem to identify closely with their owners and have a disconcerting way of looking them straight in the eye. They are certainly not lacking in courage, but they do not attack animals which are too big for them. They make good pets and, if teased, will stand up on their legs, clench their fists, and spar like a boxer. They also use this technique to convey to their master that they wish to eat.

They are very conservative animals and resist any changes. Once they become accustomed to their surroundings, it may take a day for them to get over the shock if a piece of furniture is moved in a room that they are familiar with.

In the wild they do not organize themselves in social groups but tend to form breeding pairs.

I had the opportunity of seeing a large number of tarsiers in captivity in the Philippines. These were, of course, not pet animals but were kept in a cage with the object of getting them to breed. I noticed

Tarsius leaping.

Courtesy of Barbara Harrisson and the Sarawak Museum

Courtesy of Barbara Harrisson and the Sarawak Museum

Portrait of *Tarsius* (Charles).

Tarsier—note pads on its
outstretched fingers.

that if a finger were held out to the animals, they would usually dash
for it and attack it, biting it if they could. On occasion, when I held a
small stick into the cage, it was quickly bitten. This group of tarsius
were being fed grasshoppers and lizards, and an attendant spent all his
time collecting quantities of these creatures to keep the colony
supplied. However, they had to be alive for the tarsiers to accept them,
and only when a grasshopper moved would the animal pounce, pick it
up in its two hands, put it in its mouth, and chew it up.

Professor Wood Jones has kept tarsiers, and he has described one
that would spring blindly at the face of an intruder in the house and
try to bite whatever part of the face presented itself. However, Wood
Jones points out that they can be tamed very easily; they enjoy being
caressed, and they will also, when they get very friendly, lick one's
face. They are not very vocal but do have a very shrill call, though
none of the caged animals I saw made any noise at all.

One of the very interesting anatomical similarities between the
tarsier and the human is that both have similar round noses; the hair

62

Tarsius climbing tree.

grows right up to the edges of the nostrils, and some hair can even be found inside the nose. They differ from the lemurs, where the opening of the nostril is in the shape of a comma and there is a snout or muzzle, the end of which is completely free of hair, so that from this point of view alone, the tarsier would be classified with the apes and man, rather than with the lower Primates such as the lemurs. Tarsiers pick up food with their hands instead of with their mouths, there is no doubt that this has a survival value because the face and eyes are less likely to be injured. An animal that picks food up in its hands can explore it with its eyes and fingers and drop it if it appears to be unpleasant or dangerous. This ability to identify prey by sight suggests some degree of stereoscopic vision in the tarsier. In man and apes, there is a small, highly sensitive part of the retina of the eye known as the macula on which an image that needs to be seen with maximum clarity is focused. In the tarsier, a central, fairly well-developed area is present in the retina which is a precursor of the macula but does not give the animal the extreme clarity of vision of the higher Primates.

63

Tarsiers also lack the coordinated muscular control of the eyes necessary for stereoscopic vision, but they overcome this by holding the eyes steady and moving the head and so bringing the eyes in the direction that they want them to focus. This has resulted in a remarkable degree of mobility of the skull and neck, and the tarsier has the ability to rotate the head through 180 degrees. I have seen a tarsier sitting with his back toward me, clinging with his arms and legs to a tree trunk and swinging his head completely around so that it looked as though it had been stuck on the animal backward. Dr. Ashley Montagu has given an interesting account of the eyes of the tarsier. He says the eyes are actually quite mobile when the animal is not gazing intently and that they can roll sideways and vertically, showing the white eyeball. He says that a weirder and more ghostly countenance could not be found outside a witch's family album.

The tarsier has very long legs, and it progresses by jumping along the ground. It can leap over six feet in one hop and can leap up to four feet high. It is a startling experience to see these animals leaping away in the wild. These animals have developed muscular pads on their toes and fingers which permit them to exert a kind of suction on anything they cling to with a smooth surface.

Tarsiers have a single baby, and it takes about two months before the baby can leap. At birth, the baby weighs just under one ounce, and it clings to the fur of the mother's belly for most of those two months and then quickly learns the leaping process.

When *Tarsius* is looking for prey, it cannot see any closer than four inches, but beyond this limit, anything moving will catch its attention. It keeps its eyes fixed on its prey, insect or lizard, as it moves, turning its head continuously but slowly from side to side as if it were using it to scan the visual field. When the prey is properly placed—in other words, when *Tarsius* knows that one sharp grab will secure it—it acts like lightning; one scarcely visible movement and the lizard or insect is in its long fingers, and it starts his meal by biting off its head—unless the prey is a bit too large. It rarely eats the head but consumes the body from neck to tail. It pulls off the wings and legs from grasshoppers and discards the innards from lizards. If they find many animals for food, tarsiers tend to kill more than they need and come back later to eat the rest. However, if in captivity you offer them a dead lizard or insect, they will ignore it. In the wild tarsiers will eat eggs and young birds still in the nest, and probably the young of small mammals; they will certainly gobble up newborn mice in captivity.

They are capable of producing a strong odor which may be used for territorial marking or for some type of communication. They are not

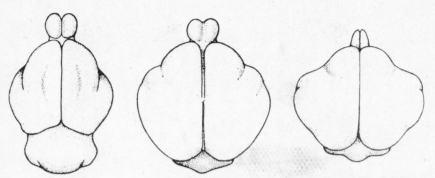

From W. E. Le Gros Clark, *The Antecedents of Man* (Edinburgh, Edinburgh University Press, 1959)

Brain of *Tarsius* on right compared with brain of fossil *Tarsius*-like animal center—note great similarity between the two. Brain on left is from fossil lemur (Adapis).

very vocal but have one or more birdlike vocalizations. When a female is in heat, the male will give a series of birdlike chirps. It is a soft call and is used when he is courting the female. It has been described as an "effortless bubbling."

It is in the brain of the tarsiers that we begin to see the relationship between the higher Primates and ourselves. The part of the brain connected with smell is much reduced, compared with lemurs, and a great part of the cortex of the cerebral hemispheres is concerned with vision. This causes a considerable development of the back part of the brain which projects over the cerebellum (which controls the animal's balance) and gives it an appearance more like that of a monkey or ape than any other animal. In fact, it is hard to tell its brain, on superficial examination, from that of a small monkey. The tarsier still has the primitive brain—which even apes and man still have—but in addition to the increase of the part associated with vision, that which plays an important part in higher mental processes is better developed in this animal than in any other examined so far.

V. The New World Monkeys

THE ANCESTORS OF the New World or South American monkeys were isolated in America when it was eventually cut off from Europe by the loss of the northern land bridge. Then the various ice ages forced them down into Central and South America, where they have persisted to this day. The South American monkeys include animals such as the little cebus (or organ-grinder monkey), the squirrel monkey, the howler monkey, the owl monkey, the spider monkey, and a group of monkeys called marmosets, a term which is often loosely, but inaccurately, used to include tamarins and titi monkeys.

The marmoset group has a rather uncertain position in the evolutionary line of monkeys. Very little is known of their origins, and

Pygmy marmosets.

Courtesy of Mr. Frank DuMond and the Monkey Jungle, Miami, Florida

White-eared marmoset.

White-lipped marmosets.

Courtesy of Mr. Frank DuMond and the Monkey Jungle, Miami, Florida

very few of their fossils are known. It seems that they spun off very early from the mainstream of Primate evolution and have remained more or less in a primitive condition ever since. Very little is known about their relationship to other monkeys, but one of the things that distinguishes them from monkeys is that most of the marmoset group have digits with claws. In fact, there is only one digit (the big toe) that carries a nail. Like all other South American monkeys, they have an extra premolar which is not found in the African monkeys, the Indian monkeys, or the apes.

The marmosets belong to a group of monkeys known as the hapalines, which also includes the tamarins and a group called the pinches. Then there are the titi tamarins, which include Goeldi's marmoset, and the titi monkeys themselves.

Among the marmosets, one of the most interesting is the pygmy marmoset, a little creature the size of a medium-sized rat, able to move with incredible speed. The marmosets have been used for centuries as pets by the South American Indians, and when the Spanish and Portuguese invaded the Western world, they were intrigued by these attractive little animals. As a result, they took them back to Europe, where they soon became very popular among the aristocracy, not only in Spain and Portugal, but also in France. Women often found themselves recipients of these little animals as gifts, and they became popular throughout Europe. The French women carried them around in the sleeves of their jackets and blouses, in pockets, or wherever the animals could snuggle.

While we have never kept marmosets as pets, we have had some personal experience with the pygmy marmoset. On one occasion, we bought about a dozen of these tiny creatures to see if they would be useful as laboratory Primates. Since they are very small, their food requirements would be little and their accommodation would be very lightweight and inexpensive. They came in a packing case with wire on the front, and we opened it very carefully, getting ready to transfer them by hand into the cages, and in a flash, they were all out. They moved so fast that they were scarcely visible. In no time, they were up on the ceiling in the maze of pipes in the basement where we opened up the crate, and there seemed to be no hope of catching them. However, with two weeks of patient and tedious trapping, we managed to get them all back into their cages. We concluded that animals able to move faster than we could see them were not going to be very useful as laboratory Primates.

More recently I have acquired four pygmy marmosets which grace my office in a big cage made specially for them and placed near the window so that they can look out. Inside the cage we have placed plastic foliage which they have learned not to eat but which they use for scrambling on. They jump about and chirrup all day long taking a great interest in everything that goes on and dashing into the swinging gourd in which they sleep at night if anyone strange comes into the room. Twice a day they get a special treat of mealworms which they snatch and chomp up happily. Strained baby food and banana are their standbys, however. They are bright and cheerful little creatures, and I was glad to get an opportunity to know them better.

These animals have a silky fur with a greenish tone, but the presence of yellow in the hair gives a mottled appearance to the coat. Ivan Sanderson says that "they are rather furious little beasts that have a habit of swaying from side to side when you look at them and making threatening expressions by lowering their eyebrows and raising the hair on their heads. The males, tiny as they are, will erect the hair all over their bodies and scream at anything, however big it may be." They are very difficult to see in the wild because they move around to the opposite side of anything they are clinging to so that it is always between the observer and them. The animals we had kept up a continuous twittering. Various observers have also noted that they could make a variety of noises, including some which are supersonic and cannot be heard by humans much like "silent dog whistles."

Most Primates bear only one offspring at a time, although multiple births do occur. The pygmy marmoset and the other marmosets have not quite made the Primate status in this respect; they do not have litters, so they compromise and nearly always have twins. When the twins are born, there is a curious division of labor. The father takes them up as soon as they are born and carries one on the upper part of each leg. The mother is allowed to have them for nursing, but he is the one who weans them and teaches them how to behave and sees that they keep their place in the family group.

The common marmoset, which bears the name *Hapale jacchus,* is restricted to the tropical forests of Central and South America, especially Panama and Brazil. The common marmoset is about the size of a large rat, with a long body and a long, hairy tail; it runs up and down trees, looking very much like a squirrel. The hair is soft and silky and is quite thick and speckled. *Hapale* means "tender," and the second name, *jacchus,* means a "leaper," a good description for these animals, which not only leap well but do so with quick, staccato movements. They live mostly on insects, which they eat by grasping the insect in one hand and biting its head off.

Grooming is an important part of their lives, and they comb their fur with their hind feet. In captivity they are very fond of owners with beards, which provides them something to hang onto. They are very vocal, with quite a range of notes. Many of their calls have been recorded; but their significance has not been worked out, although the sounds for contentment, hunger, and fear have been identified.

Despite their attractive appearance, marmosets do not demonstrate much intelligence. They appear to be gentle creatures but can bite very severely and are particularly nervous with strangers. Their life-span is not accurately known, but most estimates are somewhere around ten or twelve years.

The common marmosets often fight among themselves, sometimes engaging in real violence. In most cases fighting starts from an infringement of the social code by one of the animals in a junior position. Sanderson has actually seen a senior animal grasp an animal that has offended in some way and break its arm or leg over a branch. The punished animal then has to hide in a corner, eating food scraps and presumably plotting vengeance, which it often exacts when the broken limb has healed.

The next group of hapalines is the tamarins. There are three or four different sorts—the bald-headed tamarin, the black-faced tamarin, and the white-faced or mustached tamarins. In addition, there is a red-handed tamarin, with orange hands and feet. The black-faced tamarins usually seem to have some kind of association with capuchin monkeys, and they travel together in neighboring troupes. When any danger appears, the tamarins give the alarm and then tear off at high speed through the trees, while the monkeys follow as fast as they can. These black tamarins do not appear to be tamable and can bite viciously.

Another group of tamarins is known as the golden lion or maned tamarins, or *Leontocebus*. Their faces are hairless; they have beautiful golden-yellow manes and coats and long, slender limbs. At their best, they are said to be more vividly colored than any other mammal. The hairs are not only golden but appear to be iridescent and metallic as well.

John Perry, of the National Zoo in Washington, D.C., has drawn attention to the fact that these animals are in serious danger of extinction. In 1968, however, according to Perry, the lion marmoset did not appear on any of the official lists of endangered species, and there were, apparently, only a few hundred of the animals left in the wild. They live in the coastal forests, close to Rio de Janeiro, and were being collected there by trappers for foreign zoos and even for the pet trade. They are very attractive exhibits for any zoo. Perry makes the following comments about the sad fate of those that went to zoos: "'Less than one hundred were scattered through the world's zoos; however, because they are delicate and difficult to keep, many died in transit, others soon after arrival. Several zoo groups were wiped out by disease, and some were killed in fights with cage mates. In captivity, vivid red-gold coats often faded to creamy-yellow. Infant mortality was high, commonly from rickets."

The zoo in Rio de Janeiro was concerned, however, about the drastic decrease in numbers of golden marmosets. Dr. A. F. Coimbra-Filho gave a talk about the lion marmoset and the case for its conservation at an international meeting in 1968. He pointed out that the original forest in which the animals existed had been broken up by the early settlers. Rio de Janeiro and the surrounding towns had occupied much of what was marmoset territory, and in addition, a good deal of it had been cleared for farming and for grazing of cattle and other purposes. The only pieces of forest land left on which the golden marmosets were still ranging were privately owned. At that time approximately 600 golden marmosets survived. Dr. Coimbra-Filho suggested that the lower part of the João River Basin be preserved for maintaining these animals. As a result of Coimbra's speech, the name of the golden marmoset was added to the Red Book which lists the endangered animal species of the world. The International Union for the Conservation of Nature then suggested that zoos should cease buying this marmoset, and in this way the trappers and traders would find themselves deprived of a market. Following this action, many zoos pledged not to buy the animals, and a number of animal dealers refused to supply them.

Like other marmosets, the golden marmoset usually gives birth to twins. They cling to the mother's fur almost from the time they are born, but the mother only holds the infant for about five days, and then the father takes over the parental care. After five days the female gets restless and approaches the male, repeatedly rubbing against him. It takes only twenty-four hours for him to get the message and take over the twins. Thereafter, for most of the day, the twins cling to their father. Within a few weeks, the young start moving about on their own and become independent quite soon. However, even when the next young ones are produced, the infants still keep up the association with the family.

Perry makes the following observation about them:

> In nature, they can run more freely; other adults accept their presence. Thus, they encounter other maturing juveniles, eventually

Golden lion tamarins.

select mates and a new pair sets up housekeeping in a territory of its own. Zoo cages provide little opportunity for acting out this sequence of events. Young have been born, often, but separation from parents is more likely to be abrupt than gradual. And, once removed from the prenatal cage and placed in another, juveniles are not in free association with prospective mates.

This is part of the reason why the lion marmosets have not bred well in zoos. Another possibility is the fact that originally the zoos had thought that these animals lived extensively on fruit, whereas in actual fact they need a good deal of protein and probably also require vitamin D_3, like other South American Primates. Animals form vitamin D in their skin or in the natural oils of their fur, under the influence of sunlight. If there is insufficient exposure to sunlight, vitamin D must be included in the diet. There are a number of types of vitamin D. Most animals can get along with vitamin D_2, but South American monkeys need vitamin D_3 and cannot survive on vitamin D_2.

Frank DuMond, of the Monkey Jungle, south of Miami, has

Courtesy of Mr. Frank DuMond and the Monkey Jungle, Miami, Florida

Cottontop pinche.

described how he put pairs of golden marmosets in cages close to each other, separated only by the wire mesh. The pairs which he had selected did not mate, but the males began to court the females through the wire; in other words, the female on the other side of the wire was more attractive than the female he had. So Frank DuMond helped with a little wife swapping and exchanged the females.

Attempts are now being made by the Brazilian government to purchase 8,000 acres of land under private ownership that it plans to use as a preserve for the lion marmoset. The timber on this land, however, is being steadily cut by the owners, so swift action is necessary. John Perry has the following comments to make about the future of the golden marmoset:

> Even if all goes well, the species will never again be entirely safe. An 8,000 acre preserve will not support many marmosets—Coimbra hesitates to predict how many—when the single population is subject to disaster. Poaching may not have been entirely stopped; there are occasional rumors of golden marmosets being smuggled across

Brazil's border, and certain dealers say privately that they can supply them to anyone willing to buy contraband animals. Zoo breeding may succeed, but the available time is definitely short—the wild caught stock are fast vanishing. Intensive research should have begun five years ago. Now, if one hypothesis proves wrong, there may not be time to test another. And even if zoos are on the right track, the centers can accommodate only a few dozen animals. Something will have been gained, however, even if the golden marmoset is lost. Brazil, a huge and diverse nation, has many acute conservation needs. Because of the golden marmoset, more Brazilians, students and ordinary citizens as well as leaders, now realize that their nation's wildlife is in danger and that it is worth saving.

The third type of hapalines includes the pinches; these are the white-handed marmoset, the cottontop marmoset, and Geoffroy's marmoset. They all bear the scientific generic name of *Oedipomidas*. The pinche marmosets have very long tails, are big leapers, and are almost entirely carnivorous, having a particular fondness for birds. Many of these animals come into the United States every year, and people buy them and take them home, put them in special cages, give them a great deal of care and affection, and then feel concerned when they die for no obvious reason. The reason is that they have to be fed properly to survive. In the absence of birds, which form a significant part of their natural diet, they have been kept alive and in good health by being fed mice, insects, chopped liver with raw eggs, dried shrimp, and cheeses.

Other hapalines are the titi tamarins, which include Goeldi's marmoset, known scientifically as *Callimico goeldii* ("Goeldi's beautiful monkey"), and titi monkeys.

There exist in Central and South America a number of rather primitive Primates, of which only one really looks like a monkey. These include the owl monkeys, which are also called the douroucoulis, the sakiwinkis, the uakaris, and the bearded sakis. Finally, there is the squirrel monkey, with its yellowish-green fur and pretty little "mask" of dark hair over its face; it is *the* one out of the whole group that looks most like a monkey. All these monkeylike creatures are relatively primitive but are definitely more advanced toward monkeyhood than the marmosets.

The douroucoulis have a thick, fine, soft fur and are about twice the size of a squirrel. They have a very unusual way of sitting; they fold their hands under the chest and rest them on their feet, with the back arched; this gives them a sort of spherical appearance. The neutral kind of fur they have serves very well to camouflage them in the wild.

75

The owl or night monkey.

The underside, the belly, however, may be brightly colored. They have very large golden-red eyes, which are very obvious since they protrude from the head, and because of this, they are probably very effective in collecting light, a facility which would aid their seeing in the dark; in fact, they are also called night monkeys. They have very long, thin fingers with swollen pads, like those of the monkeys, apes, and man, and this is the first time that these appear in our evolutionary progression. The terminal pads carry ridges, which, as in the humans, are curved into whorls, so they can make fingerprints just as humans can.

Douroucoulis make a variety of noises. They twitter and chirp; they can also make loud booming noises and even a deep noise that resembles a gong. There doesn't seem to be any consistency in the way they use their vocal repertoire, and each animal seems to use whatever sound it likes on any occasion. It is hard to believe that there can be much meaningful communication between individuals on an auditory level. I have wondered, however, whether they have the ability to

Courtesy of Mr. Frank DuMond and the Monkey Jungle, Miami, Florida
Sakiwinki monkey.

produce supersonic sounds and if they use these more consistently for communication with each other.

One captive douroucouli gave a characteristic call when a black cat came on the scene; it was something like "tckkk-tckkk'er-[Boom]." This last noise resembled a drawn-out bang on a gong. Nothing else called forth this sound, not even a cat of another color; it was restricted only to the appearance of a black cat.

Douroucoulis are very catholic in their diet, gobbling up fruits of various sorts, nuts, snails, frogs and insects, honey, and also birds' eggs if they can get them. They like to build their nests in holes in trees and have a limited territorial range around their nests, a behavior which is quite different from true monkeys, which may range over considerable distances. The douroucoulis resemble marmosets in producing twins and share another characteristic with them—they are both especially sensitive to red light.

Sakiwinkis are about the same size as the douroucoulis, and their body shape is similar; but their fur is coarser and longer. Instead of

Bearded saki.

Courtesy of the San Diego Zoo

**Portrait of a
uakari monkey.**

Courtesy of Mr. Frank DuMond and the Monkey Jungle, Miami, Florida

Courtesy of Mr. Frank DuMond and the Monkey Jungle, Miami, Florida

**Visitors to the Monkey
Jungle watch an uakari.**

79

having one digit sticking out from the hand as a thumb, the sakiwinkis and the sakis have the first and second fingers sticking out in this way and able to work together against the other three like two thumbs.

The bearded sakis live mainly in the dense, tropical forests around the Amazon. They live in large families, and the families may come together where there is a concentration of special food in a particular area. Sanderson has a very interesting description of them running in the treetops:

> They are remarkably agile animals and can leap chasms among the treetops that would defy any Galago, Lemur or even Pinché. They also have a most comical way of running along horizontal branches on their hind legs with their arms held wide above their heads and all the fingers extended. It is impossible to keep a straight face when you see a line of these glum-faced, shaggy haired creatures performing thus, because one cannot help but expect to see a diminutive "G" man coming along behind them, holding a rifle at the ready.

The uakaris, which have the scientific name of Cacajao, are grotesque animals, with faces which certainly look like monkeys, but the whole of the head and the face is completely naked and bright pink in color and blushes to a bright red when they get excited; the full development of this red color depends on the animal's being exposed to sunlight. Long, uneven hair covers the rest of the body, giving them a straggly appearance, especially with their "cut-off" tail. If you can put up with the uakaris' appearance, they are said to make good pets, although they can yell in a raucous fashion. Normally they are not very active animals, but they are good jumpers and can show a very good turn of speed when required. If disturbed, they can produce very serious bites by the use of their long canines. There are two other species of uakari—the white-headed, which also has a red face, and the black uakari, which has a black head, arms, hands, and feet and a mantle of black hair on the shoulders.

The last but one of this group of half monkeys are the bearded sakis, most commonly seen with a reddish-brown color and a black beard. They have developed a very human way of drinking water by cupping it in their hands and bringing it up to their lips. They live mainly in the treetops. Their coats are very dense, and even the heavy rain of the tropics cannot penetrate to wet the skin of the animal. Relatively little is known about these creatures.

80

Squirrel Monkeys

The last member of this so-called group of half monkeys is an animal that actually looks very much like a monkey. In the pidgin English of the Guianas, they are called the monkey-monkey-monkey, presumably because they look so much like monkeys. In the English-speaking world, they are called squirrel monkeys. There is no particular reason for calling them squirrel monkeys, because they do not bear the faintest resemblance, in color, markings, shape, or behavior, to a squirrel. The body form is surprisingly similar to that of douroucoulis and sakiwinkis, but the skull is unique in that it has a backwardly directed bulge. The face is very small, and although the animal is, in general, orange-yellow, it has a pretty little black mask that extends over its face and even on to the lower jaw. It has a relatively large brain for an animal of its size, but it is a very simple brain in the sense that it has very few convolutions and is not at all like the brain of the other true Primates.

Squirrel monkeys exist in enormous numbers in South America. Many of them come into the United States for medical research and also, unfortunately, for the pet trade. A few of these animals are fortunate enough to get into a household which understands monkeys —that is, understands how to provide a proper diet for them and how to look after them when they become ill—but not many.

They extend from Costa Rica, in Central America, down to the La Plata River, which serves as a boundary between Uruguay and Argentina, in the south. They are also found at very high altitudes. They will eat anything, including insects and frogs, tree crabs, and beetles. They move about in great bands amounting to 100 or even 500 in number.

Squirrel monkey hordes have been known to descend to the ground and invade the camps of travelers. They do this with great urgency and great speed, chattering all the time. They open everything and have been known to grab bread from a hot oven. When they finally depart, it is as if the locusts had been through the camp and indeed they have—Primate locusts.

Many squirrel monkeys are used at the Yerkes Center for a variety of studies that are important to the welfare of human beings. We find them charming little animals. Their tails, unlike those of other South American monkeys, seem only weakly prehensile. They use them mainly as a stabilizing organ when they are sitting down, and when

Courtesy of Mr. Frank DuMond and the Monkey Jungle, Miami, Florida

Squirrel monkey group.

A squirrel monkey relaxes.

Courtesy of Mr. Frank DuMond and the Monkey Jungle, Miami, Florida

they are jumping in the trees, they use them as a kind of balancer. They are gentle and affectionate and make a twittering noise. The male uses an unusual method of threatening; he throws his legs apart and exposes his erect penis. He uses this technique when he approaches an inferior or when he wishes submission from a female. The threat of an erect penis also used by some humans—some New Guinea tribes tie a long tube to their penises to simulate erection when they go to war.

The squirrel monkey has been described as having "the physiology of a child, an innocent expression, a playful smile, facile transitions from joy to sorrow, eyes suffused with tears when afraid." The latter comment is a piece of pure fantasy, man among the Primates being the only one that can produce tears emotionally.

Squirrel monkeys are very lightweight and rarely tip the scale over 12 ounces. In captivity they are full of fun—their sort of fun, like jumping unexpectedly on your shoulder, scrambling up curtains, and stealing food. One very famous, if not the most famous, squirrel monkey is Miss Baker. She was born in the jungle in Peru some time in 1958 and was captured when she was quite young by some hunters who had set a net which caught both her and her mother. They tore at the net trying to find a way out, then realized that something very unusual was happening to them and began to give their alarm calls. Eventually they were picked from the net, put in boxes, and shipped off to a pet shop in Miami. Miss Baker was among twenty-six squirrel monkeys which were sent from the shop to the Naval Aerospace Medical Center in Pensacola, where research was being carried out on the physiological problems of space travel. America was moving into the space age and was carrying out a series of investigations using animals, prior to launching man into space.

When Miss Baker arrived at Pensacola, they began conditioning her by taking her out of her cage for various periods of time each day and stroking and handling her. Each day she was kept out a little longer. She was given some particular tidbit of food to make her feel happy outside the cage, then was allowed to run about the room, to climb up on the windowsills, to meet people and also to come back to her handler when he called her. She was introduced to doctors and naval personnel, including Dr. Dietrich Beischer, who was in charge of the project. There were a number of possible flight candidates, and Miss Baker was at the head of the list. She had then to be trained to lie still in a confined space without moving. Gradually she began to accept restraining straps on her arms and legs; every time she was released from this, she was given a special treat so that she began to accept this treatment without feeling too unhappy about it. She had to be used to darkness, so she had to learn to be restrained in the dark, and this was done several times a week, until she accepted it. She was taught to accept the implantation of electrodes which would record her heartbeat, her blood pressure, and the electrical waves in her brain. She also had to learn to go without food or water for twenty-four hours. It took a number of months for her to be conditioned to all these circumstances.

The capsule in which she was to take her flight was to be composed of two sections. One was made of fiber glass and was cylindrical in form; she would lie inside this. It was insulated from heat and cold, and in the center of the cylinder was a place lined with foam rubber which was approximately her size and shape. That section fitted very well into a boxlike section of the capsule, not unlike a portable typewriter in appearance.

It was finally decided that the Jupiter missile would be launched on May 28, 1959, from Cape Canaveral in Florida. The Navy was told that its squirrel monkey should be ready to go into the nose cone of the missile. On May 27, Miss Baker was held firmly while little electrodes, which would record her physiological functions in flight, were inserted under her skin. Electrodes which would record the electrical waves from her brain were also inserted under her scalp. These electrodes were held in place by a little helmet and strap. In this condition, she was put into the fiber-glass cylinder, and her arms and legs were strapped down with her tail wound around her shoulders out of the way.

At midnight, between the twenty-seventh and twenty-eighth, Miss Baker was handed over to the Army, which was firing the Jupiter rocket. The rhesus monkey who was going with her weighed seven pounds—a much bigger and heavier animal than Miss Baker—and he was also attached and strapped to a foam couch and instrumented. Although we have been referring to Miss Baker as "Miss Baker," she had, up until this moment, been called TLC, the TLC standing for "tender loving care." Just prior to launch, the two animals had been named, the rhesus monkey being called Able and she, Miss Baker. The two monkeys went up 360 miles, traveling at the rate of 100 miles per minute. When she got up into space, she shot off down range at a speed of 200 miles per minute. The nose cone separated from the body of the missile and continued on by itself along a planned course. At the end of its powered run, it curved over and began a free fall. During this period, as during all free falls, the animals were, in effect, weightless. They were in this condition for nine minutes, came down

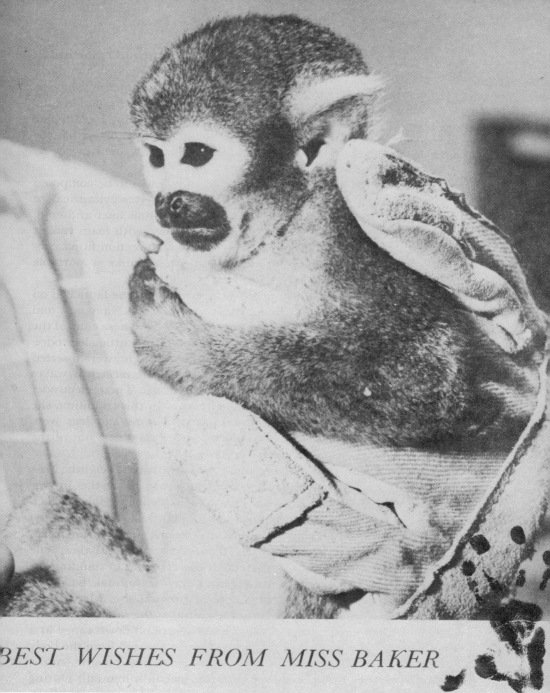

BEST WISHES FROM MISS BAKER

Photo and hand print courtesy of Alabama Space and Rocket Center, Hunstville, Alabama

Miss Baker—first monkey in space.

through the atmosphere, and then parachutes opened to slow them down. When Miss Baker struck the ocean, a number of flares were automatically released. It was a very accurate shot, down range from Cape Canaveral 1,700 miles, landing about 250 miles southeast of San Juan, Puerto Rico. The whole thing had taken only fifteen minutes. A Navy ship, the *Kiowa*, was waiting for her.

When the capsule was retrieved and brought back to the Navy ship, the capsule containing Miss Baker was withdrawn and was taken off by the Navy doctors to the hospital on the ship. When it was opened, they saw her eyes looking at them. She immediately squeezed them shut in response to the light, for she had been in the dark for quite a while. She gave them a little audible greeting, and it looked as though she was quite well. They lifted her out of the capsule and removed her jacket and the instrumentation. She was as calm as she could be. Miss Baker's photograph appeared in newspapers all over the world. She came back to the Naval Institute of Aerospace Medicine at Pensacola, and it was decided to build her a very special house.

Miss Baker lived in a little bungalow for many years at the Naval Institute of Aerospace Medicine. The walls were suitably covered with plastic and she had a stainless steel cage where she could sleep. She had a one-way glass window through which people could see her, but Miss Baker could not see them. This saved her from the strain and excitement of many visitors. She was given a male companion called Spice. She lived there, as I said, for many years and more recently was transferred to the Alabama Space and Rocket Center in Huntsville, Alabama, where a very beautiful accommodation was built for her and her new male friend, George. She is on view there to the members of the general public who visit the center. Recently, she seemed not to be doing very well, and the authorities from the Space and Rocket Center brought her over to Atlanta to visit our veterinarians at the Yerkes Primate Center. We gave her a detailed examination, drew some blood, had her tested, and recommended some changes in diet. She went back to Alabama and seems to have been very happy and healthy since then. We hope that she has many more years to go as a famous monkey personality. Anyone who would like to read the detailed story of Miss Baker should read the book *Space Monkey*, written by Olive Burt and published by the John Day Company (New York, 1960).

The remainder of the monkeys that live in South America are all animals that have developed an amazing ability of grasping things with their tails; in other words, they have *prehensile* tails. This is a capacity which no monkeys in Africa and Asia have developed. Some of these animals have this ability so well developed that they are able to support their whole bodies by the tail. Some of them can carry things in the tails. A few inches at the tip of the tail are naked and endowed with "touch" nerve endings and even have whorls like the skin of the fingers and toes (they have tail as well as fingerprints). These four groups of animals include the capuchins (the little cebus monkey), which used to be most commonly seen as an organ-grinder monkey; the spider monkey, which may be black or gold in appearance; and the woolly monkey. The remaining type is the howler monkey, which has the rather attractive generic name of "*Alouatta.*"

Capuchin Monkeys

There are a number of different forms of the capuchin, varying in skin and hair color. The name "capuchin" was derived from the crest of dark hair on its head, which gave it a fanciful resemblance to the capes of the Capuchin monks. They seem to travel around in groups, but never in the numbers which can be seen in the squirrel monkey, and unlike many monkey groups, there do not appear to be clear-cut dominant animals in their groups. They have strongly prehensile tails and are sometimes called ringtail monkeys. They make very delightful, affectionate, and gentle pets. They move relatively slowly in the trees but can move fast on occasions for short periods. I have described a number of the interesting experiences I had with my pet capuchin, Benito, in my book *The Ape People*. The capuchin monkey has a highly developed mechanical sense; in fact, he has been called the monkey mechanic. Some scientists rate their intelligence very highly and refer to them as not being too different in intelligence from the chimpanzee. Mechanically, this is probably true, and it may be of significance that the capuchin is one of the few monkeys that will scribble or paint like a great ape. I recently had a chance to see some capuchin drawings from Germany, and they are quite impressive.

The capuchin brain is large by comparison with the weight of the body and is well convoluted.

The mechanical ability of the capuchin is shown by his use of all kinds of tools to take in food. Dr. Heinrich Klüver, in 1933, carried out a number of tests on capuchins and gave the animals a pat on the back for their achievements. One capuchin in particular spontaneously used the following implements to take in food: an L-shaped stick, socks, pieces of wire (straight, hooked, and in rings), a belt, a

Woolly monkey at play with humans.

From Leonard Williams, *Man and Monkey* (London, Andre Deutsch, 1967)

rope, a floor brush, a steel strip, and pieces of cardboard. It also fabricated its own tool by tearing off a piece of paper and rolling it up so that it could be used for raking. These tools were used not only when they were in plain view near the food, but also when they were inconspicuously located around the room or even when they were wrapped up in paper. The animal would also use one tool to get another with which it could rake up food. For example, it would use a wire hook to rake in a small stick, which it would use to take in a longer stick, which it would then use to take in food. It would also use its prehensile tail to pull a box under a piece of food that was hanging out of reach so it could climb on it to get the food. It also "messed around" with the tools a good deal in the process of raking in the food, but this was due not to lack of motor ability but rather to an interest in the tools themselves and in manipulating them.

The animal also showed great powers of concentration, working for more than an hour on a problem without losing interest. Dr. Klüver points out that both macaques and capuchins will tear apart any object that can be dismantled, but that while the macaque is purely destructive, the capuchin takes it apart with much more care and

Jojo today.

'My playbiting days
with Jojo are over.'

From Leonard Williams, *Man and Monkey* (London, Andre Deutsch, 1967)

examines it; in fact, it demonstrates an interest in the object and its component parts. When moving pictures were shown to Dr. Klüver's capuchins, they reacted actively to what was being shown, as evidenced by the movements of the eyes, vocalization, and approaching to or retreating from the picture. The animal which showed greatest ability as a tool user vocalized at the movie, and when a python or a lion appeared on the screen, it called out, defecated, and ran to a part of the room from which the picture could not be seen.

My capuchin, Benito, demonstrated his mechanical ability by the way he used to get out of the cage where he was kept. This cage had a lock with a horizontal lever that could not be lifted until a vertical lever had been pushed out of the way. It seemed a pretty effective kind of lock; but it was a problem that Benito solved in a matter of seconds, and he was out of the cage and running about the room. One of Benito's activities, which always brings me a smile of remembrance, is the occasion when he got out of his cage in the basement of the house

(1) Afraid

(2) Disapproving

(3) Intrigued

(4) Laughing

(5) Perturbed

(6) Enraged

(7) Apprehensive

(8) Happy

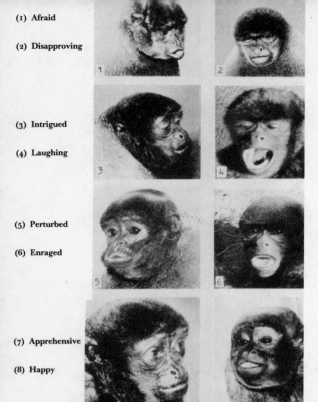

Woolly monkey expressions.
From Leonard Williams, *Man and Monkey* (London, Andre Deutsch, 1967)

and started running about on the heating ducts. The pipes were just hot enough to make it uncomfortable for his feet, but not hot enough to burn him. He was running up and down on these pipes and cursing away in some kind of monkey language, picking up his feet and shaking them, but it did not seem to occur to him that he should get off them, or if it did, he probably decided that the discomfort of the warm pipes was better than being caught by me and put back in his cage.

There is a capuchin monkey that appears on the streets of "Underground Atlanta" in the evenings. It is dressed in a little suit and stands upright with a little chain around its neck, while its owner holds it. It has a little cap and apron, and it accepts coins from passersby, which it puts in the pocket of its little apron, and then it touches its hat in thanks. Dressed up in this way, capuchins used to be much more common years ago, when their owners carried little hand organs. While the owners played the hand organ, the monkey would go around soliciting cash from the people who gathered to watch.

Woolly monkeys at play.

From Leonard Williams, *Man and Monkey* (London, Andre Deutsch, 1967)

R. L. Garner, who wrote in 1900 about his experiences with apes and monkeys, describes his attempts to understand the language of communication between monkeys. He seems to have spent a good deal of time listening to the sounds capuchin monkeys make. He describes his experiences with one he met in Charleston, South Carolina, whose name was Jokes. When he first met this animal, he mimicked the capuchin alarm call; immediately the capuchin sprang to a perch at the top of its cage and became almost wild with fear. It could not be coaxed back, retired to the back of its cage, and could not be persuaded to make friends again. When another capuchin, called Jack, was drinking from a saucer of milk Garner was holding, the monkey tried to pull it away from him. It was not allowed to have the milk and got very angry, whereupon Garner jerked it by its chain and slapped it. The animal then laid the side of its head on the floor, put its tongue out, and made a sound which was interpreted as a sign of surrender.

On another occasion, visiting a small capuchin called Banquo that lived in Cincinnati, Garner made a noise which he had identified as a "want to drink" noise. The animal's attention was attracted immediately. It came to the front of the cage and made the same noise. Garner repeated the noise, and the animal responded a second time and then turned to a small pan that was in the cage. It took up the pan and put it near the door through which the keeper normally passed food, then turned and made the drink noise again. Working with another capuchin, Garner found that the animal made a noise similar to but different from "drink" which meant specifically "milk." There also seemed to be some slight variation in the calls made for different types of food. The calls for apple, carrot, bread, and banana were all slightly different from one another.

Studies on the reactions of these capuchins to color showed that bright green was the favorite, and later studies have shown that they are relatively insensitive to red. One interesting event on which Garner comments is that on one occasion, he gave a capuchin monkey a drink from a cup decorated with bright pictures of flowers and green leaves. Every now and again the monkey would stop drinking the milk and try to pick the flowers off the cup and was rather annoyed when it was not able to remove them.

Very few studies have been made of the reaction of monkeys and apes to music, but Garner tried playing music to his capuchins and found that some of them were attracted by musical sounds but seemed rather to enjoy a single note repeated a number of times as opposed to a melody; notes of a certain pitch seemed particularly attractive. Some animals were indifferent to any kind of musical noise, and some actually showed an aversion to music.

Garner describes a very interesting encounter between the young Helen Keller and a small capuchin monkey. Miss Keller was, of course, blind, deaf, and dumb, but one of her great desires had been to see a live monkey or, as Garner puts it, to see one with her fingers. There was a capuchin monkey with the group that was kept at Central Park in New York City, which was called Nellie. Garner recounts this story as follows:

When anyone except myself had put hands upon Nellie, she had growled and scolded and showed temper. I took her from the cage. When the little blind girl first put her hands on Nellie, the shy little monkey did not like it. I stroked the child's hair and cheeks with my own hands and then with Nellie's. She looked at me and uttered one of those soft, flute-like sounds. Then she began to pull at the cheeks and ears of the child. In a few minutes, they were old friends and playmates, and for that time, they separated with reluctance. The little Simian acted as if conscious of the sad affliction of the child, but seemed at perfect ease with her. She would decline the tenderest approach of others. She looked at the child's eyes, then at me, as if to indicate that she was aware that the child was blind. The little girl appeared not to be aware that monkeys could bite. It was a beautiful and touching scene, and one in which the lamp of instinct shed its feeble light on all around. . . .

The capuchins are reputed to have very good memories. I certainly can verify this from my relations with Benito. He is now at the Yerkes Field Station, which is 25 miles away from the main Yerkes Center, and was, until recently, housed in a compound with a number of other monkeys. There are occasions when it may be six weeks or two months before I get a chance to get out to the field station, but Benito always remembers me. He comes running to the wire when he hears my voice. If he can see me from a distance, he will call to me. He wants me to come up to the wire, and he leans against it so that I can groom him, stroke him, rub his head and his ears, put my finger in his mouth, and so on. Sometimes he will do this for a few minutes and move away from the wire, not so much with the idea of getting away from my attentions, but with the hope that this will lead me to come into the compound. When my youngest son, Merfyn, came on a visit to Atlanta, he paid a courtesy call on Benito, who, although he had not seen Merfyn for four or five years, recognized him immediately and was very excited.

Woolly Monkeys

The woolly monkey is another of the South American Primates. It is covered with a thick, fleecelike fur. It is a slow-moving animal, and I have never heard of one biting anybody. It is gentle and very affectionate and is surprisingly clean. A lady whom I knew, who came through Atlanta on one occasion to discuss a plan to set up a Primate center in South America, traveled with a woolly monkey. She registered at one of the downtown motels and simply carried her woolly monkey in a basket without telling anyone in the motel what she had. I visited her at the motel, and the woolly monkey was sitting with her in the living room of her little suite and was very pleasant and affectionate to me. I asked her about the problem of soiling the furniture and the carpet in the room, and she said when the animal wished to do anything, it went to the bathroom and used either the

Woolly monkey Emma—
born in Leonard Williams'
monkey colony in Corn-
wall, England.

Courtesy of Leonard Williams

toilet or the bath. On one occasion, after she had been there for a few
days, she had taken the animal out for a walk. Since it was a two-story
motel, she got into the elevator with it to go up to her room. The
manager of the motel, as it happened, got into the elevator with her,
saw the monkey, and asked her about it, what room she was staying
in, and so on. She was frightened that she might be on the point of
being thrown out. However, this manager was smart. Instead of
throwing her out, he rang up the press and told them about the lady
with the monkey staying at his motel and invited them down to
interview her and photograph the monkey, so he got a considerable
amount of publicity for his motel and she got some publicity for her
cause.

The face of the woolly monkey is rather flat and looks somewhat
human, especially because its snout is very small. In fact, it has been
described as a "worried old man seen through the wrong end of a
telescope." It has a heavier body than the other monkeys we have
been describing. Like the capuchin, it has a highly prehensile tail with
the same patch of bare skin at the tip. Woolly monkeys vary in color

A juvenile woolly monkey entertains visitors to Leonard Williams' sanctuary in Cornwall, England.

from a sort of chocolate to a dark gray that is nearly black, and in some cases, they have a silvery-gray color and appear nearly white. They can also be pure brown. In the wild they move around the trees in small groups and appear to mingle with other monkeys during feeding without any problem. They are very sociable animals and appear to get along very well with one another.

At one time I had three woolly monkeys. I bought, first of all, a female and put her in a golden cage which stood on the conference table in my office. She became very attached to me and would call out in a very loud voice if I went out of the office, so I thought I would calm her down by buying her a very well-built, strong, active male and putting him in with her. All seemed to go well for a while, and then he seemed to be losing weight and becoming lethargic. I couldn't understand the cause of this. I thought perhaps he had some kind of virus, but the female didn't seem to catch it from him. He continued to sink, and eventually, despite all attempts to improve his weight and make him well, he died. Upon autopsy, he was found to have a perforated gastric ulcer. I thought this was an unfortunate accident,

Mother monkey Jessy with her fifth baby in the monkey colony in Cornwall, fraternizing with the colony founder Leonard Williams.

and I immediately bought another male in good health and condition and put him with the female, and again, the male began to deteriorate; finally he, too, died. On autopsy he showed multiple ulcers in the small intestine. What this lady woolly monkey did to her male mates, I just don't know. I could see no sign of her attacking them in the cage. Whatever it was, if it was due to her, it was something too difficult to be seen by normal observation.

A very interesting study of the woolly monkey has been made by Leonard Williams in his book *Man and Monkey*. Williams points out that at the time he wrote, which was 1967, more than 200 baby woolly monkeys were imported into England every year for pets and zoos. He draws attention to the fact that in the capturing of these baby woolly monkeys, the same technique is used as for most other Primates. In other words, the mothers are shot so that the infants may be taken from them and sent away to cities and towns where the animals are to live in captivity for the rest of their lives. He makes a point when he

Leonard Williams, founder of the woolly monkey colony in Cornwall, England, with his closest woolly monkey friend, Jessy. Jessy is the mother of five babies in the colony. She is also a grandmother.

Courtesy of Leonard Williams

says that "in truth, woolly monkeys are unsuitable as pets." He himself has collected a group of woolly monkeys in Murrayton, in the south of England. There he has an enclosure and separate houses, some of them indoor and some of them grassed-in outdoor enclosures. He says that the animals are free to leave the area and, in fact, free to roam the whole of Cornwall if they wish to do so, but they don't; they stay where they are and catch insects and eat grubs, and of course, they are given additional food. Great care with hygiene is taken. A keeper is on duty from 8 A.M. to 10 P.M., and feces and urine are removed every hour from the monkey houses. Each morning they are hosed down, cleaned, and disinfected.

Leonard Williams was the first to describe the ritual of chest rubbing in woolly monkeys, which is a means of identifying their territory. The procedure of chest rubbing is to wet the lips with saliva, to rub the saliva over some selected place on some selected object, and then to push the chest over the surface which has been wetted in this way. Some places in a cage or compound are used by the chest

97

rubbers, and other places only an individual monkey uses. Chest rubbing takes place several times a day and it appears to be more purposeful in mature males.

Woolly monkeys seem to be quite intelligent animals, and one of Williams' monkeys was able to unscrew a lid from a jar, turn on a tap to get water, and use its tail to reach something that it could not reach with its hands. He points out that it did not perform these operations as if they were tricks but employed its own imagination in using them out of necessity.

Woolly monkeys have a large variety of behaviors. For example, lip smacking is part of the sexual overtures of the female, and she does it only when she is in estrus. Teeth chattering often accompanies lip smacking and is thought to have the same meaning. Then there is a mixture of behavior and noise called sobbing. Williams regards this as a highly emotional response and says that it indicates submission and friendship and is usually associated with a gesture of the hand and the forearm shielding the eyes. Sometimes the animal will put its head and chest on the ground, raise its rear and curl its tail up high, at the same time using both arms to shield its face. This procedure is often carried out between two adult males and seems to be the way in which a submissive male identifies himself to the dominant male. Another activity is snuffling, which usually ends in a very intimate huddle or embrace. Williams said it is a consummation of appeasement and friendship.

The animals' communication system includes at least fifteen different calls. The sound *eeolk*, soft and sweet and short, is usually made by females, juveniles, and infants, but males may use it too. It appears to be some kind of a greeting. Then there is a variation of this: *ee-Olk*, which Williams feels includes an element of surprise in the greeting. A variation of this sound is also made by an animal lying completely relaxed that suddenly makes this noise for no obvious reason. Then there is a kind of a trill *eeoooooollk*—a strong, continuous sound, rather like a song, apparently expressing great pleasure. The same type of cry, only much louder and at first explosive then trailing away, is described as a very beautiful sound and also sometimes seems rather plaintive. Williams' interpretation of this call is that it conveys the information that "everything is all right here." Then there is a vocalization expressed as *tuff-tuff*, which appears to be a friendly communication. It is used largely by babies and juvenile monkeys. There is the *oohh-oohh*, an attention-getting noise, and *ogh-ogh-ogh*, is a kind of croak which appears to express positive response and acceptance. Adult males make this sound quite often to infants in the

98

**Indoor meeting between the visitors and the monkeys at Leonard Williams'
woolly monkey colony in Cornwall, England.** Courtesy of Leonard Williams

group. Then there is *aarrrhgk,* a distress call sometimes given by an
"aunt" monkey, who gives it because the infant she wants to mother
will not respond to her. Sometimes a *huh-huh* can be heard; this sounds
a bit like a laugh or a chuckle and is accompanied by a vigorous
sideways head shaking. This seems to be an invitation to play-fight. It
can turn into *hughh-hughh,* which is more intense than *huh-huh* and is
sounded at the point where excitement begins to turn into aggression.

When the animals are engaged in play-fighting and one of them
wants out, it will sometimes scream. This is interpreted as a signal of
submission and defeat. The scream discourages the aggressor animal
from further attacks, and it may make a squeaking response to
reassure the submissive animal. Young monkeys will often shake their
heads as they make it. When there is something exciting going on, a
play-fight or something of that sort, the noncombatants will often
produce a *nyonhk-nyonhk. Argck-argck*—short, sharp, and continuous—is
an aggressive warning sound. It is a dangerous sound because it
appears to contain both fear and aggression. Danger, fear, and distress

are expressed by *yoohk-yoohk*, usually made in response to an intruder from outside the group, rather than at an actual member of the group. There is also an explosive, short, aggressive warning sound, *ffwharff*, like a pistol shot and comes from deep within the chest. Williams said he has also heard this particular call used as a long-distance danger call. One of his animals used it on one occasion when a car that it was unfamiliar with came down the drive of the house during the evening and pulled up near the monkey house.

The female woolly monkey appears to mature at about four years of age and menstruates fairly regularly every three weeks. The young females often have a very long clitoris, and this, when it hangs down, looks somewhat like a penis. This has often led to inaccurate sexing of the younger animals. During the estrus period, if there is a male present, there is almost continuous copulation. The male obviously makes the best of a good thing, and he may keep it up until the time of the next period of menstruation.

The social groups of woolly monkeys include an alpha male, or chief male, who not only has reproductive responsibilities but also has the job of chief protector of the group. He is very protective of the mothers with young infants and seems to take a great interest in the whereabouts and the well-being of various members of his community. When a female woolly monkey reaches the stage of puberty, she acquires a certain dignity in her group; her relationship with the males "becomes one of appreciation and respect and not simply of dominance," says Leonard Williams.

Woolly monkeys are very hygienic animals, and the design of the genitalia in relation to the posture they adopt results in their waste passing into the air between the branch and the ground and not contaminating the area of the tree they are inhabiting. For example, when a female woolly monkey urinates, she projects her rear well over the branch on which she is standing, a technique Williams says is very humorous to watch. It is an accepted fact that woolly monkeys are the most hygienic in their urinary and defecating habits of all the monkeys, and this is what makes them the best monkey for a house pet.

Experiments on the vision of woolly monkeys have demonstrated that it is more acute than that of humans. Williams has seen an animal lying down and then making a sudden snatch and capturing a small insect as it flew by; in many cases, the insect was so tiny that it was almost impossible for a human to see it, certainly when flying. They have color vision, which of course, is not surprising, since this is a characteristic of all the Primates. The animals Williams worked with chose yellow as a favorite color.

Woolly monkeys can grow to quite a large size, reaching two feet or more in length plus the tail. They are typical Primates in having only one baby at a time, which they nurse for about a year.

Spider Monkeys

Yerkes Primate Center has a few spider monkeys. They belong to the genus *Ateles* and are animals with a prehensile tail developed to the ultimate extreme. It is literally a fifth hand. They are able to hang completely by it. I have seen the spider monkeys hang by the tail, with all four limbs spread-eagled, using their arms and hands to pick up food and carry it to the mouth, and then turn around and climb up the tail to grasp the branch from which they were hanging. They are great brachiators, like the gibbons, and can move very gracefully and with great speed from tree to tree, swinging from handhold to handhold. The tail can also be used to pick up food and other objects and convey it to the hands or even to the mouth. They have very bad tempers, but those at the Yerkes Center have been found to be very amiable and pleasant animals, easy to pet. The spiders vary greatly in coat color; some have black coats, and others are golden. They can be more than one color, black with colored flanks and undersides, and so on. The hair also varies in length. There are some that live some thousands of feet in the Andes Mountains; the hairs of their coats are short and close together, and their coats have a woolly appearance. Like the gibbons, the thumbs on these spider monkeys are reduced to very small digits, or they may be absent altogether, making it easier for them to swing rapidly through the trees. The thumbs, while useful for grasping, get in the way when one brachiates. They are said to be shy, but Professor Ray Carpenter, one of the pioneers in the science of primatology, has described them as pretending to be very ferocious, shaking branches, barking at him, and, in some cases, even throwing dead branches at him. Some of these branches weigh 8 to 10 pounds, and if they are dropped or thrown from 30 to 40 feet above, they can inflict serious injury.

Sanderson has given a vivid description of the unreliability of spider monkeys. With reference to one particular animal, he says that it had:

> lived for four years in a small zoo and had been given absolute freedom with numbers of children. When it was taken to a studio to rehearse for a television performance, it was at first quiet, gentle and

**Spider monkey—
Yerkes Primate Center.**

loving to its handler, whom it had known for many months. However, a member of the orchestra was sucking a lollipop, and this caught the monkey's eye. It started towards this man, leading its handler by the hand, but when the latter started to hold it back, it suddenly rounded on him without warning, seized him with both hands and both feet, and started to bite savagely. Shaken off, it bit two other men and then went straight for a woman sitting fifty feet away. It took over an hour to pacify the animal and catch it. Thereafter, it became absolutely tame and gentle again and remained so for a year. Then it erupted again, suddenly attacking a girl who refused it a piece of cake at the zoo.

We have never had anything quite as bad as that with our spider monkeys, although on one occasion when I was in the compound where our spiders were kept, together with a number of other monkey species, and was playing with my capuchin friend Benito, I was suddenly attacked by one of the golden spider monkeys called, appropriately enough, *Goldie*. It attempted to bite me; but I was able to push it off with one hand, and it only made a desultory attack and

Spider monkey (*Ateles*) inspects his hand.

South American spider monkey and baby (*Ateles*).

Courtesy of Thüringer Zoopark Erfurt, German Democratic Republic

did not inflict any damage. I suspect that in this case it was jealous of the relationship I was demonstrating with Benito.

Spider monkeys may form groups of as many as forty individuals in the evenings, but when a new day dawns, they tend to divide up into a number of subgroups which keep within calling distance of each other. The individual members of the subgroup keep within sight of each other. Some of these subgroups are unique in that they may consist of only males, although normally the subgroups contain more females than males. There is no evidence of the spider monkey's having a breeding season.

Howler Monkeys

The largest of the group of monkeys known as the ceboids, which includes the spider monkey and the capuchin, is the howler monkey. It is probably the most highly colored of the South American monkeys—the coat may be a beautiful reddish gold.

I have seen and heard howler monkeys in Barro Colorado, in the Panama Canal Zone. The noise they make is astonishing, and to me, the roaring resembled the noise of a passing train. Dr. Ray Carpenter suggested that the roaring was designed to distinguish one group of animals from another and to warn the surrounding groups that they have the rights to the particular territory which they are occupying at that time. Howler monkeys also roar defensively at a human when they see one. He estimates that they have about fifteen to twenty different vocalizations. Like the capuchin and spider monkeys, they have prehensile tails and formidable teeth. They usually are not very approachable animals, but it is possible to tame them. Like the spider monkeys, howlers throw branches and pass urine and feces at human intruders. Howlers are capable of long jumps from tree to tree, but they are not typical brachiators. They group themselves into "clans" of twenty to thirty animals, and they have a well-established hierarchy with a dominant male.

Some of the monkeys in South America are threatened species. The golden marmoset is probably one of the rarest of the nonhuman Primates in that part of the world, and some of the uakaris and some of the sakis and the woolly spider monkeys and the Goeldi's monkey are also on the endangered list. The squirrel monkey has become a very popular animal for research and for use in the pharmaceutical industry, for zoos, and also for the pet trade, largely because it is a gentle animal and is certainly cheap and easy to keep in captivity. There are 35,000 Primates exported from the area of Iquitos, Peru,

Red howler monkey.

each year, and four out of five of these are squirrel monkeys. Barranquilla and Leticia, in Colombia, are also outlets for live Primates. In fact, these three places are the principal areas of outlet. There are always significant and often very large losses of monkeys during the capture procedure, and also during the holding and the transit period, until the animal reaches its final home. In some cases, the rate has been assessed as being as low as 25 percent, but some estimates go as high as 80 percent. In other words, since about 34,627 monkeys come out each year from Iquitos alone, if the wastage is added on to that, the estimates of the legal export trade from Iquitos alone of these Primates represents 43,000 to 52,000 Primates a year. There are, of course, many more killed by the locals for food. Within most settlements and most river highways, practically all Primates have been eliminated. Hunters and trappers state that each year they have to go farther and farther into the forest to find an adequate number of monkeys.

VI. The Old World Monkeys—
The Macaques

ALTHOUGH THERE IS a relatively small number of monkey species in South America, the number of species in Africa and Asia is enormous. All are included in the big family Cercopithecidae. This family includes the many different species of macaque, among which are the Barbary apes that inhabit a limited part of North Africa and the Rock of Gibraltar, the Celebese black apes, the mangabeys, the baboons, the mandrills and the drills, the guenons, Allen's monkeys, the red hussar or patas monkeys, the langurs, the proboscis monkey, and the colobus or guereza monkeys. There are, in fact, altogether about sixteen genera in this family and probably around sixty species. They are distributed over Africa, Arabia, southern Asia, Indonesia, the Philippines, Japan, and Formosa. They live in all kinds of habitation, from relatively rocky desert to forests and savannalike areas and mangrove swamps.

With the large number of species, not a single one has a prehensile tail. Nearly all have tails but presumably use them as balancing organs of some sort and to maintain stability when they are sitting up. Another characteristic is that nearly all have bare patches of various size on their bottoms. These are known by scientists as "ischial callosities." They use these pads to sit on when they are resting and especially when they are sleeping at night in the trees, and it is said they are bare to enable the skin to "feel" the surface of the branch they are sitting on and thus help them to avoid falling off the branch.

Most of them live on vegetable food, but nearly all will eat birds, insects, and birds' eggs. Many have cheek pouches where they can store and carry away surprisingly large amounts of food, to bring back into the mouth and eat later. They usually have much more elongated muzzles than the South American monkeys, despite one less tooth on each side of the jaw. The muzzle in some of the latter, particularly in the woolly monkey and the spider monkey, is greatly reduced in size, the face being relatively flat. The extension of the muzzle in the Cercopithecidae is shown to the greatest extent in male baboons, where it is almost doglike in dimension. The members of the Cercopithecidae have thirty-two teeth, usually with very well-developed canines. In some cases, these can be elongated and appear like little tusks.

Most of the members live up in the trees—that is, they are arboreal. The baboons live mainly on the ground, and the macaques seem to be equally at home on the ground or in the trees. They all have good sight and hearing, and for tree-living animals, their sense of smell is also good. They communicate with each other not only by a variety of vocalizations—more than forty have been recorded—but also by many types of facial expression. All the females have menstrual cycles like the human, usually somewhere around thirty days, but it may be longer in some species.

Of all the animals in this group, the rhesus monkey is probably the best known because he has been widely used for many years for laboratory research; it was the work done with the rhesus monkey that eventually led to the discovery of a vaccine for poliomyelitis. It is a

From K. Lang, *Die Grusitten* (Vienna, Völkerkede, 1926)

**Fulah women make presenting salutation to
visitor, reminiscent of monkey presenting.**

member of the genus *Macaca* and its full scientific name is *Macaca
mulatta*. There are quite a number of other macaques. The Barbary
ape is one, and another is the toque monkey; still others include the
bonnet monkey, the lion-tailed macaque, which is now approaching
extinction and is on the endangered list, the pigtail macaque, and the
crab-eating monkey, which is also called the Java monkey. Then there
is the Assamese macaque, from Assam, India, the Formosan rock
macaque, the stump-tailed macaque from Southeast Asia, and the
Moor macaque, as it is sometimes called (this black-faced, black-
haired animal lives only on the Celebes Islands, north of Australia).
The macaques range from the Celebes through all Southeast Asia, the
Philippines, Borneo, Java and Sumatra, into India and Ceylon, and to
the east and north to parts of Afghanistan, Tibet, China, Japan, and
Formosa. They live in a wide variety of elevations, some at sea level
and others up in the mountains where their place of residence may be
as much as 13,000 feet above sea level. Japanese macaques seem to
survive happily through the snowy winters in parts of Japan and grow
the most beautiful, thick, bushy coats in the wintertime, which make
them look like little bears as they walk around.

Macaques have developed an anatomical structure known as a
cheek pouch, situated on either side of the face. This enables them to
rush into an area which is potentially hazardous, fill their cheek

pouches with food, rush out again and enjoy the food at leisure during a period of relaxation, without the strain of impending danger. They seem to move equally well on the ground or in the trees, and they walk or run on all four legs. They can, of course, stand on their hind legs and often do for brief periods, particularly when they are using their hands for carrying food.

The type of country that the macaques inhabit ranges from tropical rain forests to mangrove swamps and grassland. In India, they have even settled into suburbia in some towns. Religious temples seem to have a special attraction for them.

All macaques seem to be able to swim. The Formosan macaque is very much a swimmer and lives on the beaches of that island; its main food is shellfish.

Rhesus Monkeys

The rhesus (*Macaca mulatta*) as we've noted, is probably the best known of all the macaques. It is a fairly large monkey. The adult female can weigh up to 15 pounds and some of the big males may even reach as much as 25 pounds. The color of the coat is yellowish brown, sometimes lighter on the belly. The face is hairless and the snout elongated and doglike. They too have pinkish, naked areas on the buttocks (the ischial callosities again).

The rhesus appears to be a highly intelligent animal. All rhesus macaques are very aggressive and threaten humans continuously, with wide-open mouths, while in their presence. They obviously have no trust in human beings. We had some at Emory who became so tame that they relaxed and closed their eyes while we groomed them and stroked their heads and necks. The moment we took our hands away, however, they opened their eyes and threatened. Whatever you do to them seems to make no difference. They seem never to lose their aggressiveness and their threatening behavior; of course, they also threaten each other, as well as other animals. The other animals are expected to give a sign of submission, usually turning the back and presenting it for mounting. This is usually a sure way, if not the only way, to save themselves from being beaten up by the dominant animals.

Rhesus macaques are very highly motivated and thus will perform well in a number of psychological and other testing situations. I have seen a rhesus monkey at an Air Force base sitting in a mock-up of a aeroplane cockpit which could be tilted in any direction by a remote observer with a joystick; in this primitive cockpit the monkey had a similar joystick, and by using it to counteract the tilting imposed on him by the observer, he could keep his cockpit in level "flight." The animal is very quick to learn, and in no time at all is able to keep the cockpit level, despite efforts of the observer to tilt it.

A strange story, which also involves a rhesus monkey, comes from Atlanta, Georgia. It was back in 1953, when a young man named Edward Waters was driving along a highway near Atlanta with two of his friends. They claimed that they saw three small men running in the direction of a flying saucer. Waters accelerated and ran one of them down. The other two disappeared in the space ship and took off. Waters picked up what he described as a little space man and walked into the office of the Atlanta *Journal*, carrying the animal in his hands. Waters, who was an Atlanta barber, was questioned by the State Director of Intelligence and the little "man" was examined. He was twenty-one inches long and his weight was approximately four pounds. He was taken to the Georgia State Crime Laboratory. It is probably an interesting coincidence that shortly before Waters claimed that he had seen two of these little men take off in this spaceship, a bank clerk by the name of R. L. Long reported seeing a similar unidentified object in the area. The two men with Waters, Thomas Wilson, also a barber, and Arnold Paine, a butcher, faced a long session with newspapermen, radio reporters, photographers and TV crews. During the questioning, someone in the area called Waters and offered him $5,000 for the creature. In the meantime, Dr. Herman Jones, of the State Crime Laboratory, had carried the creature out to the Emory University Department of Anatomy where anatomists there recognized it immediately as a monkey which had been shaved. It was significant too that a pet shop in the neighborhood of the Waters' home reported that some monkeys had escaped around that time. The next day, Waters, in the County Superior Court, claimed that he had made a bet of $10 that by the end of that day he could be sure of getting his picture in the newspaper. He claimed that he bought the monkey and put it to sleep with ether and then shaved it, and after he had finished, he had hit it on the head with a bottle and killed it. Then he had cut off its tail to make it look more human. So that was the end of that fantastic story, except that, of course, there are some people who still believe in unidentified flying objects.

Rhesus monkeys have been well known in captivity for so many years that there are many anecdotes about them. While they have been used extensively as pets, they are unsuitable for this purpose, for when they become mature, they are unpredictable and can be dangerous.

On one occasion in London, a rhesus monkey escaped from a pet shop and scampered over the rooftops, followed by a group of men trying to capture it. This turned out to be a dangerous expedition for the humans, because not only did they have to run over the rooftops but they had to dodge heavy bricks and tiles that the monkey kept throwing at them. The animal was at liberty for several days, and all attempts to capture it were fruitless. In the meantime, it did well on foodstuffs that it stole from various houses through open windows.

Many years ago, at the famous Whipsnade Zoo outside London, there was the most beautiful oak tree, the Thailand tree, in what is now Bluebell Woods. It was surrounded by a 9-foot fence, and forty rhesus monkeys were turned loose in the enclosure. The group tore through the branches, leaping from one to another, and used the tree as a jungle for a few weeks. At the end of that time, they had devoured every leaf and bud, eaten all the bark and eliminated every branch that was not strong enough to stand their weight.

The same thing happens in the monkey compounds at our field station in Atlanta. We have twenty-four compounds which vary from 50 feet square to 125 feet square, and all of them originally had some trees. However, it only takes a few weeks for the monkeys liberated in them to do what the London monkeys did to the Thailand tree. They remove everything edible from the trees and leave them completely bare, with their barkless branches stretching out into the sky. Our compounds are surrounded by a 14-foot-high chain-link fence, with the top 8 feet covered by metal sheeting.

Once a monkey escaped from our center on the Emory University campus. He was the only animal that ever escaped. Somehow or other he got over the electrified fence with which our place is surrounded, disappeared into the woods, and turned up within a day or two in the basement of the Georgia Mental Health Institute, situated some three or four miles away.

The 9-foot palisade fence at the Whipsnade Zoo, even though it had a sort of roll-over top designed to keep the animals in, was not very effective for this purpose. The monkeys that wished to explore the world outside the compound simply climbed up the big oak tree and dropped perhaps 40 feet onto the ground outside the fence and were then able to move in any part of the zoo that they liked. They showed a devouring curiosity about everything going on, especially the repair of buildings. They stole tools from various workmen and would then carry them back into the enclosure and up the Thailand tree, where they would often lose their grip and drop them. On one occasion, a tool weighing several pounds was dropped from the summit of the tree some 60 feet above the ground and narrowly missed a workman.

Another story of monkeys refers to a court case in Caracas,

Venezuela, where a woman accused a monkey of assault. The woman said that at the time the monkey was driving a motorcycle and another four monkeys were standing on its shoulders as passengers. The police who investigated the alleged offense found that the whole thing happened at a circus and the monkey supporting the rest of the troupe lost control of the motorcycle and they all pitched into the audience, hitting the woman.

There are about 5,000,000 to 6,000,000 rhesus monkeys in India, but there is no doubt that this number is being rapidly reduced. The monkey used to be protected so that the people in India would not attack it or even stop it from eating their food. However, it is meeting with more resistance now. Rhesus monkeys thrive in many different habitats. They live in cities and in remote forests. Many of them frequent Indian villages, and most of them live in reasonably close contact with man. It is estimated that, in fact, only about 8 percent live in forests, 46 percent in villages, 17 percent in towns, 13 percent in cities, 6 percent on roadsides, 3 percent on canal banks, 2 percent in temples, and 5 percent in railway stations.

In nature, the rhesus monkey usually lives in groups of twenty to sixty, and these groups maintain their identities, often for many years. During the daytime, the group moves over a considerable area of territory, but it usually returns to a particular area at night to sleep. They do not have a very strong territorial urge, although one group is not happy about having another group near it, and the groups tend to avoid each other. Members from the various groups, when they come into actual contact with each other (for example, if they come near each other in the process of gathering food), show considerable mutual aggression. Adult males and infants and young animals have very little physical contact, which is opposite to the relationship between the infants and females, who form very strong bonds. The adult males will, however, react vigorously to any danger directed toward either infants or juveniles. Strange animals introduced into the group face aggression and fighting and the animals introduced very likely will not endure long.

Female monkeys become sexually mature at about three years, perhaps a little earlier than that, but it is not until between three and four that they actually begin breeding. At five or six years, the mating activity reaches its peak. In males, three years is the usual age for sexual maturity, but the animals keep growing until they are about seven or eight years old.

There is a definite breeding season in the wild. It is usually somewhere around the beginning of the monsoon in July and goes on for two or perhaps three months, with most of the young born in January or February. In most cases pregnancy is uncomplicated, and labor appears to be painless and brief. The mothers seem unafraid of parturition. They help deliver the baby with their own hands. They bite through the umbilical cord, eat the afterbirth, and lick the baby all over.

This latter phenomenon is extremely interesting. Ashley Montagu says the prolonged labor that occurs in great apes and in man gives sufficient stimulation to the skin of the baby to start it functioning properly when it is born; therefore, humans and apes do not lick the baby when it is born, but mammals, and monkeys in particular, in which the labor is very short, invariably lick the babies; it is believed that this licking procedure, through stimulating the nerve endings of the skin, helps get all the other functions of the body under way. However, some licking of the newborn by apes has been seen at the Yerkes Center. The licking of gorilla infants immediately after birth has been recorded but this has only been seen occasionally. There is also some evidence that the orangutans give a good deal of attention to the baby by licking it, and we have seen this in a number of instances in our own animals but not with all of them. Actually, all the great apes have very much shorter labor periods than humans do; for example, the orang, gorilla, and chimpanzee all have labor periods which may vary from twenty-five minutes to a few hours; humans usually spend many hours in labor. However, additional advantages of licking the infant have also been suggested; the licking may serve to protect against attracting predators, help free the head from birth membranes (we have seen our apes use their mouths to remove the birth membranes from around the face of the baby), and prevent the danger of chilling in colder climates. It also prevents loss of contact between the mother and offspring and facilitates recognition and perhaps the initial formation of the important mother-baby social bonds.

June Johns gives an interesting summary of the relationship between the baby and the mother:

When the baby is accepted by the mother, she nurses it to her breast, smacks her lips at it and cares for it with devotion. She grooms it several times a day and fends off advances from curious neighbors. The newborn monkey is quite helpless apart from having the ability to suck and to hold tight to its mother's fur. For some months it clings to her night and day, but once it starts to crawl, it comes in for some rough treatment. The mother's patience wears thin, and when the child offends her—by adventuring too far away or trying to taste her food—she scolds and smacks it and will even drop it off the perch. Almost immediately, she will relent, pick it up and kiss its head, but the whole cycle of punishment and forgiveness will be repeated when it offends again. As it grows older, the infant appears to get on its mother's nerves; she begins to chastise it without good cause and gradually the ties of motherhood are broken. The infant starts to find more pleasure with other youngsters and spends more and more time with them. The mother weans it by poking pieces of well-chewed food into its mouth and allowing it to sample bits of her own meal, and as lactation ceases and she comes into season, the infant is put aside while the mother takes up with a male.

Allison Jolly, in her book on the evolution of Primate behavior, discusses how mother animals, and in particular mother monkeys, are able to recognize their own babies. This is of particular interest in view of the fact that in a number of monkeys, the infants are quite different in color from the mother. For instance, the colobus monkey, which is black and white, has babies which are all white. The rhesus baby monkey, on the other hand, has a coat of dull brown which is similar to that of its mother and to all the other rhesus babies. Human children are very different in shape from adults and are regarded generally as being "cute." It seems that children, animals, or cartoon characters that have large heads and short legs in relationship to the body have a cute look, and this is accentuated by large eyes. Presumably, cuteness is thus an important factor to a human mother in recognizing her child, although many human fathers have difficulty in recognizing their own babies in the hospital nursery.

The monkey mother probably recognizes her baby by smell, as well as by sight, but the important thing is that the mother does recognize it and by permitting it to climb all over her, to suckle from her, and by cuddling and cleaning it, she demonstrates that she is reacting to it as a baby and not as a stranger.

Of course, the baby also has to recognize the animal holding it as a mother, and Dr. Harry Harlow, of the Wisconsin Primate Center, attempted to find out what were the factors of motherness stimulating to a baby monkey. He produced what he called surrogate mothers. He took soft terry cloth and put it over a rounded structure made of wire. He left another frame covered with bare wire but added bottles of milk with teats attached. The baby animal was given the opportunity of cuddling the terry-cloth mother or cuddling the wire one. The baby preferred soft terry cloth to the wire, even if the latter provided milk. So softness was the first and most important factor of motherhood.

Dr. Harlow and his colleagues were interested in studying what

Yerkes rhesus monkey
mother and child.

Feeding rhesus monkeys by hand at the Yerkes Primate Center.
Dr. Irwin Bernstein and Mr. Tom Gordon.

Bringing up a baby rhesus in an incubator at Yerkes.

happened to young monkeys who were deprived of their mothers. They found profound psychological disturbances and then studied methods of treating these disturbances because of the obvious human application of such studies. The interesting details of their investigations are contained in a recently published article called "Monkey Psychiatrists," written by Drs. S. Suomi, H. F. Harlow, and W. T. McKinney. Psychopathology is rarely seen among animals living in the natural state because in a disturbed condition they simply cannot survive; there are no hospitals for mentally disturbed monkeys in the wild. However, psychopathology can be produced experimentally in animals in captivity, particularly in monkeys and apes, and total social isolation of the baby shortly after birth is the best-known method of producing this type of pathological-psychological abnormality. This type of study has been carried out over many years by Drs. Richard Davenport and Charles Rogers at the Yerkes Center on baby chimpanzees, but Dr. Harlow and his colleagues used rhesus monkeys. The isolated animals were placed in chambers where they got no opportunity for either physical or visual contact with human beings or any other animals. In the case of monkeys, six months of isolation are sufficient to produce profound psychological disturbance; in the case of chimpanzees, Drs. Davenport and Rogers used twenty months of isolation. Animals isolated in this way moved less, had less curiosity about things, and showed little or no social behavior. They gave hardly any response to grooming, play, and other types of social interaction but sat clasping themselves, often huddled in a corner, rocking continuously and behaving very much like severely mentally deficient children or even, in some cases, like schizophrenics. None of these animals, as they got older, had any sexual responses. When monkeys which had been inseminated artificially produced their babies, they showed poor maternal behavior; they were indifferent to the baby and in some cases behaved brutally toward it, even attacking it—a thing a normal female monkey would never do. They have even been foolish enough to attack the dominant male in the group—a thing that few "socially sophisticated" monkeys will attempt. Thus, it can be seen that "total social isolation" has a devastating effect on the development of appropriate monkey behavior. Harlow and his colleagues have tried to rehabilitate these isolated monkeys by exposing them to socially normal monkeys as cage mates. However, they found this unsuccessful.

Dr. Gene Sackett describes these isolated animals as follows:

Isolation reared primates change as they mature, from rocking, digit-sucking, grimacing, self-clutching recluses, to pacing, socially aggressive, self-threatening, masturbating, self-mutilating menaces, who often make bizarre movements. When the isolate animal reaches sexual maturity, its sexual behavior is abnormal. The maternal behavior of the isolation reared female is indifferent or brutal, yet appears to improve with time. The infants of brutal mothers also show some signs of behavior pathology, so that experimentally produced pathology can be passed on to the next generation.

Psychiatrists have pointed out that these isolation-rearing studies resemble human cases of drastic neglect.

When Drs. Davenport and Rogers put their isolated chimpanzees back with the normal animals, they found that there was considerable, although not complete, return to normality. They were able to adapt socially even though they maintained a good deal of abnormal stereotyped behavior. In some cases, they showed considerable incompetence in raising babies and did not show the same care of them as the normal mothers, but they did show a considerable improvement over the years. This success has never been obtained with the rhesus monkey by Dr. Harlow and his colleagues.

The initial responses of isolated female rhesus monkeys which have become mothers to their own babies is that of almost complete rejection; however, some of the infants manage to survive in spite of their mothers, and the ones that have survived were those that continued to make the effort to maintain contact with the mother's body. If the baby was able to keep this up, still get some nourishment, and survive its mother's rejection, then after about the fourth month, the mother would gradually give up trying to reject it. Such a persistent baby had, of course, to be supplemented with additional food; it rarely received enough from its own mother. Females that were lucky enough to have a persistent baby which survived and were again impregnated then exhibited relatively normal monkey maternal behavior toward the new baby. Thus, they had secured virtual rehabilitation.

These studies led Dr. Harlow and his colleagues to study the possibility of some kind of social stimulus that might be used to rehabilitate these isolated females. If an isolated animal were put in a cage with normal animals, the normal animals would attack the isolated one and simply increase its abnormal behavior. However, if a monkey could be found which would give the isolatee what might be known as contact acceptance, it might be possible to help the isolated animal. Thus the concept of "monkey psychiatrists" or "therapist monkeys" was born. The monkey psychiatrist had to be especially trained for its job. A little earlier, I mentioned how baby monkeys would cling to a wire frame covered with terry cloth. If these structures were also heated, given a spout which would stick out and which would deliver warm milk when sucked, a lot of the essential factors of the mother were provided and a baby would cling closely to such an object. These surrogate mothers sometimes also had a crude face painted on them. The monkey psychiatrists were raised on the heated surrogates and allowed to react two hours each day with other monkeys being raised in a similar way. Some of them were raised together in pairs in their own cages and sometimes in groups of four in a social playroom. They showed a reasonably normal social development appropriate to their age. After a time, they were put to play with the isolated animals, which were three months older than they were, yet young enough not to have developed the aggressive responses characteristic of the older animals. When the monkey psychiatrists were first introduced to the isolates, the isolated animals tended to huddle in a corner and the therapist monkeys' first response was to go to the isolates and cling to them. Within a week the isolates were reciprocating by clinging to the therapist monkeys. This itself was an important advance. In another two weeks the isolated monkeys were reciprocating with the therapist monkeys in simple play. Not long after that, the isolated monkeys actually began to initiate play behavior themselves. Their abnormality began to decrease. By one year, the isolated monkeys virtually could not be distinguished from the therapist monkeys in either exploratory, locomotive, or play behavior. It was of interest that the type of play that the animals indulged in was characterisitc of the sex; for example, the males liked to have a rough-and-tumble play while the females preferred the type of play that did not involve contact.

Isolates, when they were one year, could still be distinguished from the therapist monkeys only by the fact that they occasionally would lapse back into hugging themselves in a corner. But these incidents were fairly infrequent and did not last very long. At this point, the isolated animals and the therapist animals were taken from the individual cages in which they had been receiving treatment and giving treatment and put together in a pen where there was a large group of monkeys. After about a year in such a group, by which time they were about two years old, the isolated animals showed almost complete recovery from their psychological abnormality. Their behavior was mostly socially directed and was appropriate to their age—in other words, to the age of normal monkeys. Social activities were much more mature. There was grooming and sexual mounting and no sign of disturbed behavior.

These studies, which are still going on, are of great importance since they have implications in human psychopathology. We will have to wait to see what happens to these rehabilitated monkeys when they become sexually mature, see whether they will indulge in completely normal sexual behavior and if they treat their babies in a normal fashion. These findings are awaited with great interest in the scientific world.

Dr. Harlow and his colleagues commented in their report as follows: "In conclusion, we are all aware of the existence of some therapists who seem inhuman. We find it refreshing to report the discovery of nonhumans who can be therapists."

A group of monkeys at our field station are used for breeding offspring suitable for space research. One of the reasons we did this was that practically all the wild-born rhesus monkeys we received were, upon investigation, found to have some evidence of disease processes in their muscles; therefore, it was virtually impossible to get a normal animal from this point of view. We decided to breed our own monkeys in a relatively controlled environment and so obtain animals free from the many diseases wild monkeys carry.

Many interesting things have emerged from studying the social behavior of this group. One is the tendency of females to kidnap babies. A mother will have a young baby and sometimes another mother will steal it from her. When she does, it is often very unfortunate for the baby because the animal that does the kidnapping does not have milk and the baby struggles unsuccessfully with the nipple to get a drink and finally dies, unless the mother can get the baby back. This kidnapping is very common at the stage when the mother is able to leave the baby for brief intervals to stroll a few feet away. Often, another mother will be standing nearby, trying to encourage the baby to come to her or sometimes even running forward and snatching it up. Often when the mother sees a potential kidnapping female in the neighborhood, she will chase it away. It seems that most of the animals who do the kidnapping are mothers who have lost babies recently but long enough ago for them to no longer have any milk.

The hierarchy in a colony of rhesus monkeys is extremely strict. We originally started this colony with forty-five females and five males and have lost a number of animals. When we tried to make up the numbers by adding males or females, we found it virtually impossible

A rhesus family.

Courtesy of Dr. Robert Goy of the Wisconsin Primate Center

to do so. Certainly single additions were impossible. The problem was that individual animals which came into the colony obviously had no place in the hierarchy, and before they were given a chance to find a place, they were usually set upon by almost everybody in the colony, particularly by those high in the dominance hierarchy. They could, in fact, be slaughtered. In some cases, males were severely bitten around the testicles, and in one case, the testicles were completely bitten off. We have also found on some occasions that we have had to take a female who was pregnant or who got sick out of the colony, either for treatment or, in some cases, to have a Caesarean or other birth complication. If that animal was out for more than a few days—say, in some cases we had to keep it out for a month or six weeks until it was completely well—when it was returned to the group, it received an extremely bad time and in some cases was killed. A great deal appears to depend on the animal's reaction when it gets back into the colony. If, for example, it tries to come back at a different level in the hierarchy from the one it was in when it left, it will undoubtedly receive a very bad time from the group. If it comes in at its original hierarchy level, there is a much better chance of its being accepted by the group. So there are very many problems in handling groups of rhesus monkeys such as this. The five males that we have put in this particular colony have formed a very powerful hierarchy; one of them immediately established himself as the principal, or alpha, male, and the others established themselves at various levels below him. Some of the females are chosen especially by the alpha male and are regarded by him primarily as his property. This is not a serious problem in this colony since there are nine females to each male.

Julia Hartman, who studied our monkey group in the summer of 1972, made a series of fascinating notes about some of the individual animals:

> Cassius is the largest animal in the compound. He's the number two male, and he seems to be the third-ranking animal in the group. He is a very even-tempered animal, and he often plays with the infants. On the other hand, John is the dominant animal in the compound. Cassius had been the dominant animal, but John took over from him sometime in June or July. The change in the hierarchy was not sudden, and there were no signs of a fight. One day Cassius sat back and let John eat first. John is a gentle leader most of the time and spends a lot of time grooming other animals in the group. He spends most of his time with Hazel, the dominant female, but he is also interested in Ruby. John can be distinguished from Cassius because he is not as large, by his darker color, and by his black tail. John also plays with the infants.
>
> Viejo is the third male in the group, but his overall ranking is uncertain because there are several females dominant over him. He looks like an old prizefighter with scars on his face and shoulders and the stooped walk of a beaten old man. His crossed eyes are his most striking features. He plays the role of control male with respect to outsiders. When caretakers or threatening strangers approach, Viejo is in charge of the defense of the group. Within the social group it is a completely different story. He avoids fights, mostly by moving to the opposite end of the compound. He knows from experience that John and Cassius would be quite willing to display their anger and chase him.
>
> Gypsy is one of the easiest animals to recognize. A blond color and large size make her easy to pick out. She is the second-ranking female, fourth in the group. She has a history of kidnapping and never misses an opportunity to try to get another baby. She has not been able to keep one for more than five minutes this summer, but

she keeps trying. She is a good mother most of the time, restraining her infant when it tries to visit people. On occasion, however, she leaves it with another female as a baby-sitter.

The monkey called Ruby is called by this name because her face is bright red. She is a big animal, but she is also a bit crazy. When she threatens or is being threatened, she bites her hand and screams, and her behavior in general is very erratic.

Patricia is one of the gentlest, friendliest monkeys in the compound. She is a medium size, lightly colored, with a fine coat. She and Annie are rarely separated. From the first day her baby was born and she brought it to the fence to show me, she has carried the infant on her back. He hangs on very well and remains in fine shape.

Annie is a small, skinny animal with a funny crook in her tail. She is fairly high-ranking and is pretty outgoing toward people. She is Patricia's bodyguard, and they are rarely more than ten feet from each other.

In a subgroup in the colony is Ichabod. Poor Ichabod is neurotic. He is the lowest animal in the group, and he looks awful. He will never take the most popular food being fed but will sometimes take the other. Despite all the hassles he gets into, he's first in the defense against outside threats to the group. He's the last animal in the house and the first out.

Henry is the largest animal and is presently dominant. He is very friendly towards people but can be nasty toward other monkeys. He often sucks his fingers when he is unhappy.

In the 1930's, Dr. C. R. Carpenter brought a shipload of rhesus monkeys from India and set them free on a small island, Cayo Santiago, situated off Puerto Rico. Originally, 409 monkeys were liberated, and the resulting colony and their descendants have survived nearly forty years. They are still being studied in great detail by scientists who, over the forty years, have learned an enormous amount about hierarchies and naturalistic behavior of Primates from this colony.

The interest in the macaque monkeys, of which the rhesus is one, goes back very many years. Macaques were used by Galen of Pergamon as long ago as A.D. 150 as substitutes for human beings in dissections performed for the instruction of medical students. Galen did not use the rhesus macaque but the Barbary ape (*Macaca sylvana*) which is not really an ape at all but a macaque monkey. The Barbary ape's other claim to fame is that it shares Gibraltar with the British. The first reference to the use of the rhesus monkey in science was in 1886. Dr. John Bland-Sutton in that year published an article about the animal in the *British Gynecological Journal.* He was interested especially in menstrual bleeding in macaque monkeys and was wondering whether there was an associated breakdown in the lining of the uterus at the time of menstruation. Sir Charles Sherrington, who was a professor of physiology at the University of Oxford in England, made some important studies in 1889 on the degeneration of nerve tracts in the brain of monkeys after the surface (the cerebral cortex) had been experimentally damaged. The monkeys he used were also rhesus monkeys.

In 1909, Dr. Karl Landsteiner, who revolutionized blood transfusion by his study of blood groups, used the rhesus monkey in his studies on poliomyelitis in Vienna. The story is told that some time earlier Dr. Landsteiner had gone to the director of his institute and told him that his work on poliomyelitis had been unsuccessful because he was working with rabbits and that it was impossible to give the disease to the rabbits. He believed he could transmit the disease to monkeys, making the development of a vaccine possible, and he asked if he could buy some monkeys. He was told very firmly that monkeys were much too expensive to use for experimental purposes and to go back to his rabbits. It is probable that this early rebuff to Landsteiner delayed very substantially the ultimate development of a vaccine for treatment of poliomyelitis. The final work in the production of the vaccine was, in fact, done many years later, first with rhesus monkeys and then with tissues from the rhesus cultivated in test tubes.

The behavior of rhesus monkeys was studied many years ago, the first papers appearing in 1902 and 1911. Dr. Yerkes, founder of the Yerkes Primate Center, published a paper on them in 1915. As long ago as 1938 there were more than 15,000 rhesus monkeys imported each year into the United States for various kinds of scientific experimentation. The present-day figure is not available, but we know that over 110,000 monkeys of all species were imported into the United States in 1966.

In 1940, Dr. Landsteiner, whom we mentioned earlier, was working with Dr. Alexander Wiener on the blood groups of the rhesus monkey. They found that there was a particular blood group factor which they called the RH, or rhesus, factor. In this study, the two investigators were interested simply in the blood groups of the rhesus monkey. They were motivated only by scientific curiosity and were not studying the cause or cure of any particular disease. Subsequently, however, it was shown that many humans have this factor and that humans are, in fact, divided into two blood groups—RH positive and RH negative. The presence or absence of the RH factor is an important feature in the development of certain types of disease, and in the appropriate combination of RH positive and negative factors in the parents, a

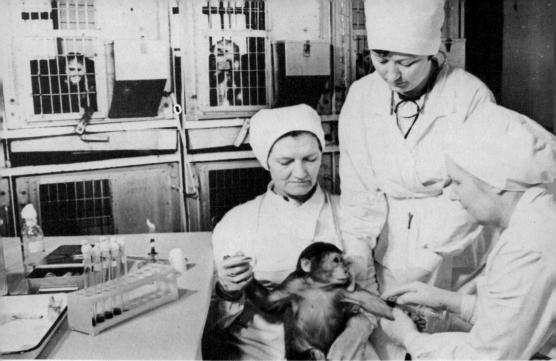

Rhesus monkey receiving an injection in the Soviet Union.

child can very likely die from a blood disease early in life. Many babies, in fact, did die from this condition before the importance of the RH factor had been established.

In most naturally occurring rhesus groups there are about twice as many females as there are adult males, in addition to juveniles and infants. Natural groups of rhesus monkeys vary in size, and there may be as many as fifty or as few as ten. The size of the group is probably related to the food and the types of shelter available in the area, and beyond a particular figure appropriate for that environment, the group will split and set up separate colonies. A group usually ranges over an area of about three square miles. They have an elaborate facial and vocal communication; in one group in Puerto Rico about forty different vocalizations were identified.

Another group of rhesus monkeys at our field station is that supervised by Dr. Irwin Bernstein and by Dr. Robert Rose of Boston University. They have been studying the hormone excretion and the hormone levels in the blood of animals at different levels of the hierarchy. They have made the intriguing discovery that animals with high aggression and animals that tend to rise to the top in status in the monkey community are those possessing a high level of testosterone, or male sex hormone, in their blood. It has not yet been decided whether this increased aggression and dominance are the result of the higher testosterone levels or whether social influences, such as dominance, result in an increased testosterone secretion. Some experiments were designed to test this. Each animal in this study spent two weeks as the only male in a compound which contained thirty females. The animal was then given a two-week period of "recovery" and then a brief fifteen-minute to two-hour period with a group of about thirty adult males that was already well established. When the male was housed with the females, he very quickly became a dominant animal and sexually active. In such cases, the testosterone level went up by 183 percent over the baseline levels, when the animal was put back in its own cage, it fell to normal. During the period when the animal was placed with the male group, it was invariably attacked by the male group and defeated and forced into submission. In such cases the testosterone level fell by as much as 80 percent. These levels remained at a very low figure for almost two months, but once the males were reintroduced into the female society and their dominant role was reassumed, the testosterone level rose again.

Dr. Rose has said about these studies that they "document the importance of sexual and social stimuli in the regulation of testosterone secretion."

Sir Alan Parkes, of the British National Institute for Medical Research, published a paper some years ago in the *Journal of Reproduction and Fertility* in which he described the strange fact that when a pregnant mouse in a cage is exposed to the odor of a strange male, it can cause a reabsorption of the young that the female is carrying and that this is apparently caused by the male's scent. Dr. Richard Michael, also in Britain, demonstrated the importance of odor in rhesus life; in this case, it was the female scent affecting the male. A male rhesus monkey knows when a female rhesus monkey will receive him sexually by smell alone without actually seeing the animal.

At our field station, Dr. Irwin Bernstein has found that a group of male rhesus monkeys who can see female rhesus monkeys 10 feet away respond by increasing the amount of male sex hormones in their blood when these females were in heat. In this case, of course, it is not possible to be sure whether this is due to a particular smell, a pheromone, or a visual stimulus. The male can, of course, tell when

127

she is in heat because of the swelling of the "sexual" skin in the anogenital area of the female.

Another group of male monkeys who were too far away actually to smell any pheromones that the female monkey may have produced did not show any changes, but they could not see the females either and they showed no increase in male sex hormones in their bloodstream.

Dr. Michael and his colleagues showed that vaginal secretion (collected from rhesus monkeys that had had their ovaries removed and were being treated with female sex hormones), when applied to the sexual skin of female monkeys who had their ovaries removed for a long time, caused the males to become sexually excited and to make many more attempts to mount the females. When the animals were able to penetrate into the vagina of the females, they ejaculated much more regularly than when control materials, which were inactive, were applied to the females instead of the secretions. In fact, Dr. Michael found that when female monkeys were given an oral contraceptive pill, the male monkeys were turned off; presumably, the pill prevented the pheromones from developing.

This brings us to the very interesting possibility that the human female may produce pheromones. Pheromones, in the sense of sexual signaling by means of smell, are part and parcel of human life and obviously represent the foundation of the perfume industry, which specializes in pleasant scents that are used to associate in the male's mind a particular type of attractive female. Pheromones have also generated the deodorant industry. There are a number of places in the body where odorous substances are produced; for example, there are large glands in the armpits, and there are glands in the head of the penis which produce an odorous substance which can produce a strong smell in uncircumcised people who do not regularly wash under the foreskin. The deodorant industry is devoted to suppressing the smell that is generated from under the arms and also, with the aid of what is delicately called a "female hygiene deodorant," the smell of the female vagina. Dr. Alex Comfort, writing in the *New Scientists,* an English publication, claims that deodorants are the worst form of pollution and quite likely foul up the silent communications that have occurred among human beings for a million years or more. There is a rather musklike scent called exaltolide, which is used by the perfume industry. It is surprisingly like what is described as a "signal scent" that the human male produces. Apparently women between the ages of fifteen and forty-five can smell this most easily, and when they are ovulating—that is, when they are at the most receptive part of their sex cycle—they can smell it much more acutely. It is very interesting, however, that hardly any males can smell this at all. Moreover, in an article in *Nature,* in 1966, Dr. R. S. Patterson pointed out that if the male is injected with estrogen—that is to say, with a female sex hormone—he can smell the material.

Schizophrenics are supposed to be able to smell hostility in a person near them. It is also claimed that human beings who are in great fear give off a smell which can be detected by animals. Schizophrenic patients are said to have a particular type of substance in their sweat which enables some people to smell them. The material is known chemically as trans-3-methyl hexanoic acid.

Some years ago, in *Nature,* Dr. M. K. McClintock described how he found that female mice that were kept together in a small confined space all tended to ovulate at the same time. Their sexual cycles just swung into rhythm. McClintock also found that if young women were living in close quarters—for example, in a dormitory—their sexual cycles also tended to become synchronized, so that they tended to menstruate around the same time, something that the campus male should bear in mind. Of course, we do not know if this is due to some odoriferous material, but it seems reasonable to suggest that it might be.

Dr. Comfort has pointed out that there are a number of anatomical structures which have little rational role except that they serve as a source of production of odorous substances, or pheromones. He says that even some of the folds in the skin may be there for the nurturing of certain bacteria in the sweat which produce an odoriferous substance.

Another item of interest in this area, though it does not necessarily relate to smell, is the story of a British scientist who had to live for long periods of time isolated on an island and amused himself by taking the dry weight of the hair cut off by an electric razor each day when he shaved. Every now and again, he returned to the mainland and had the opportunity of enjoying the company of women. He continued his hair-weighing studies and found that for a while, in the new and more pleasant situation, his beard grew much more rapidly. This suggests that the relationship with the women may have altered his sex hormone balance, affecting the hair growth. There is evidence, in fact, that the growth of the beard in males is related to the presence of male sex hormones in the bloodstream.

Whether there is, in fact, an active pheromone produced by the vaginas of human females which is more active during the period of greatest sexual urge has to be proved, but it seems likely that something of this sort does occur. If it does, then the women who use scented vaginal douches may only be complicating life for themselves and the males with whom they associate. There may be a subconscious message passed across by this pheromonic route, and the woman may be wondering why her man friend is cold just when she is feeling warm—it may just be that she has cut off her own subliminal message.

A very recent issue of *Penthouse* contained a letter from a reader (Grant H. Hendrick) which is especially relevant to this discussion. He said he was attending a church service one Sunday morning when he became aware of a strange smell. It was "not the smell of new or dry cleaned fabric, or the smell of newly polished shoe leather, or the smell of perfume, talcum powder or after shave lotion, but the smell of a woman, a woman in heat." He could not tell where the smell was coming from, so at the end of the service he stood by the door as the congregation filed out. He had no difficulty in recognizing the lady who had awakened his olfactory interest, followed her out, introduced himself, and invited her to dinner. Later that night he was able to confirm that she was, in fact, in heat, and subsequently they were married.

This seems a pretty clear-cut case of the existence of a human sexual pheromone. Perhaps this young man had an especially acute sense of smell which enabled him to cross over what would normally be a subliminal barrier or perhaps in some women the pheromone is stronger or more abundant than in other women.

The Pigtail Monkey

Scattered over Southeast Asia, in Burma, the Malay Peninsula, Indonesia, particularly Sumatra, and Thailand, are a group of macaques known as pigtail macaques, *Macaca nemestrina*. These animals are bigger and heavier than the rhesus and have a small hanging tail which is either naked or covered only lightly with hair. Their fur is yellowish, sometimes with a green tinge to it. It is glossy and, in many cases, even silky, and it lies in a smooth coat. There is no hair on the face, and some have naked ears. The hair on the top of the head is arranged so that it looks as if the animals have been given a crew cut. In some cases, this mass of crew-cut hair on the top of the head is darker than the hair on the other parts of the body.

In *The Ape People*, I discussed the studies Dr. Irwin Bernstein has been making on pigtail monkeys. He has found that these animals form very strongly organized social groups when kept in the com-

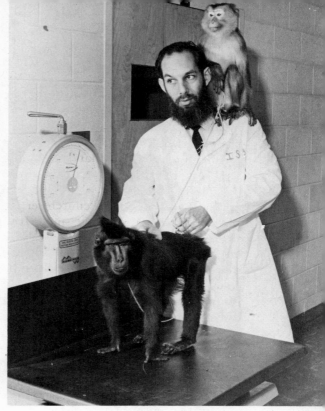

Dr. Irwin Bernstein at Yerkes Center. Pigtail monkey roosts on his shoulder, and Celebes black ape stands in front of him.

pounds at the field station, and he has also discovered that in the wild there is a similar, rigid organization. In the group that he formed at the field station, Dr. Bernstein started with only two or three of these monkeys and began adding single individuals, fully or partly grown males or females and youngsters, and studying their acceptance by the group. A lot of interesting material has accrued from his studies which we do not have the space to discuss here. One of his findings is of special interest: Sometimes animals that were introduced appeared to be totally unacceptable to any member of the group. They were treated as outsiders and were not given the opportunity to join in any activities, and they were often partly deprived of food by the others, even by the young monkeys of the group. Sometimes a young monkey would come up to the newcomer and bite its tail or pull its ear or be unpleasant in some other way. On more than one occasion, an animal introduced in this way and rejected by the group has been picked up one morning, dead, without any signs of injury—he could not accept indefinitely being treated as an outsider. As Dr. Bernstein's group

Portrait of head of Yerkes' pig-tail monkey.

grew bigger, it showed typical orientation into dominance groups. One particular animal was the most important member of the group; he was the dominant male, and he would be the animal that would come to meet a visitor who came to inspect the compound. He would look the visitor over and threaten him if necessary. In other words, not only was he the most important animal in the group and the one to whom all the others in the group deferred, but he was also responsible for the safety and the welfare of the group and had no fears about taking on this part of his duties. In addition to this animal, there were obviously number two and number three males in the dominance hierarchy. There were also a series of females, some of whom were preferred by the principal male and others preferred by the other dominant males.

In pigtail monkeys, the dominant male holds his tail curled stiffly backward over the bottom part of his back. Numbers two and three hold their tails less tightly, and in others, who do not rate, the tails hang limply down. Although the dominant female sticks with the dominant male and although the other males tend not to interfere with the female that belongs to the dominant male, occasionally, as in any well-organized society, some individuals do not wish to go along with rules. On some occasions, a female belonging to a dominant male in this group has gone over to the farthest corner away from her lord and master and has indulged in quick sexual intercourse with another male; by the time the dominant male is aware of it the act is over and the two are apart again. Even though the dominant male may chase the other male and chastise him, he is not able to prevent the act from taking place. This tendency for a little sexual nonconformity every now and then is the reason why Dr. Bernstein found, when he was working in the jungle in Malaya, that occasionally he would see a hybrid animal representing a cross between the Java (crab-eating) macaque and the pigtail macaque. He has also been able to produce

A dominant male pigtail monkey at the Yerkes Center.

such hybrid animals at the Yerkes Center by a deliberate crossing of the two species.

In a group like this, sexual intercourse, or at least a representation of sexual intercourse, is regarded as a demonstration of dominance or submission. For instance, if the dominant male approaches a particular female, she will stop, turn her back toward him, lower herself more or less onto her elbows, and stick her rear toward him. This is known as presenting, and usually the male will mount her, make a few thrusts with his pelvis, and get off. It is not sexual intercourse; but the female has shown that she regards him as a superior being, and he has demonstrated the acceptance of the submission. In fact, even males will use this method of presentation to demonstrate to a dominant male animal that they are subservient to him. This, in the past, has been misinterpreted by some observers as an indication of homosexuality among the males, but it is not so. If an animal wishes to demonstrate to a superior male that he accepts the superiority, he simply turns his rear end to the animal, and the dominant animal will mount him and give a few thrusts with his pelvis as if he had mounted a female. In this way the dominant male accepts the other male's acknowledgment of his own superiority.

It is startling to realize that this presentation of the rear end has been used by humans in the recent past and may still exist in some remote parts of the world. Dr. K. Lang described in 1926 how women of the Fulah tribe in Africa turn their backs to a person being presented to them and then bend forward so that their bottoms are presented to the visitor—a gesture almost identical with the pigtail and other macaques. At the gateways of some old European cities and forts, bare buttocks are depicted presumably as a sort of gesture of greeting and submission to a visitor. In some Germanic tribes in the past if there was thunder, both men and women would bare their

133

bottoms and stick them out of their doors to appease the god Wotan.

Many Siamese farmers keep pigtail monkeys for the purpose of collecting coconuts. The farmer puts a rope around the animal's waist and lets it run up a coconut tree; the monkey then takes the coconut in both hands and twists the nut until it falls to the ground and is collected by its master. Several of the pigtail monkeys we imported for studies at the Yerkes Center from Bangkok, Thailand, had been trained as coconut pickers. The early Chinese chroniclers, writing about Siam, mentioned this custom and advocated the introduction of the use of such apes in China. This apparently is one of the reasons why these animals were later used for picking tea and also for picking peppers. In Malaya, pigtail monkeys have been used by botanists to collect species of plants and one has been known to have collected specimens of 350 species in a six-month period. These monkeys run up trees with a rope attached to them; sometimes this rope becomes entangled, but they are extremely good at untangling them and in this way are able to get down safely. Some Thai banks have used monkeys to detect false coins and to sort them from good coins. It was said to take about two years to train a bank ape for this purpose, and they are then valued at $5,000 each.

Japanese Macaque

Japanese monkeys, *Macaca fuscata*, are found on the main Japanese and other islands. They are very strong animals and are able to survive the cold winter and snow. A few years ago I visited Japan and saw the Japan Monkey Center, near the town of Inuyama. My Japanese colleagues took me to lunch at a nearby inn, and afterward, we went over to a park where a number of Japanese macaques lived. There was still snow on the ground, and when we arrived a troupe of monkeys came over the hill. A number of other people in the park began throwing nuts around on the ground for the monkeys. The animals walked in and around without molesting us—they made no effort either to avoid the humans or to make contact with them. Having nothing to give the animals, I bent down to pick up a couple of nuts to hand to one of them; one of the big males immediately thought I was stealing his food, and he rushed close up to me and stood a few feet away, growling and making threatening gestures. I threw the nuts at him and decided I had learned my lesson.

We noted that these animals were particularly well covered with long, shaggy fur; in fact, waddling along the ground through the snow, they looked like a group of little bears. There is a lot of interesting

Portrait of male Japanese macaque.

Courtesy of Mr. Harry Wohlsein
and Dr. William Montagna,
Oregon Primate Center

Japanese monkeys at the Oregon Primate Center.

Courtesy of Mr. Harry Wohlsein and Dr. William Montagna, Oregon Primate Center

Courtesy of Mr. Harry Wohlsein and Dr. William Montagna, Oregon Primate Center

Japanese monkey and baby.

information available about the Japanese macaque. There have been studies in some detail by the Japanese and also by American and European scientists. The animals depicted on the originals of the Oriental statuettes of three monkeys with their hands over their eyes, their ears, and their mouths, expressing the teaching of Buddha, "See no evil, hear no evil, speak no evil," were Japanese macaques.

About thirty troupes of Japanese macaques have been well studied, the size of the troupes varying between thirteen and several hundred, and it has been noted that some colonies develop special behaviors like those described below.

The monkeys in one of the colonies on the island of Koshima, off the southeastern coast of Kyushu, had been provisioned for some time. The island was uninhabited, except for a family of fishermen who lived there, and the animals had been fed for eleven years but were not bothered by tourists. The monkeys at Koshima rarely went into the sea until about four years ago. Then, one day, some of them went into the water and began to swim. It is possible that they went there looking for sweet potatoes that had been thrown into the sea during the provisioning. Three years after that, only thirteen of the fifty members were not swimmers; the others had developed the habit by imitating the animal that had first taken to the sea.

Sweet potatoes were a regular part of the food left at a feeding station on the beach, and on one occasion it occurred to one of the young females to wash the sand from the sweet potatoes by dipping them into a little stream that was running through the beach and into the ocean. This so-called washing behavior gradually spread through the entire group. The younger animals, interestingly enough, were the first to take it up; prior to that, the method of removing the sand had been to rub the potatoes with the hands. It took about four years for the new habit to spread to 50 percent of the animals and nine years for it to spread to 71 percent. After about ten years, nearly 90 percent of the group was washing the potatoes.

A modification of the cleaning procedure developed when some monkeys started to wash the sweet potatoes in sea water. Some appeared to like this method better, perhaps because the salty water gave a nicer taste to the sweet potatoes. Some of the monkeys, even when the potatoes were near the little stream originally used to wash them, would carry the potatoes all the way to the sea to wash them. Sometimes when they did this, they carried potatoes in both arms and hands and walked on their hind legs. As the custom grew the monkeys entered the sea more often and walking erect in the water with a

Baby Japanese macaque sucks its thumb.

Courtesy of Mr. Harry Wohlsein and Dr. William Montagna, Oregon Primate Center

couple of handfuls of potatoes became a habit rather than a novelty. Following this, they began to walk in the bipedal way more frequently on land. Some have been known to walk bipedally for as much as 175 feet.

The monkeys on this island also invented a way to separate wheat from sand. Wheat is one of the foods given to the animals; it is thrown on the sand, and the monkeys apparently find the process of picking the individual wheat grains out of the sand very tiresome. So some of the monkeys have solved this problem by carrying handfuls of wheat and sand to the sea and plunging them into the water. The sand falls to the bottom, and the wheat floats on the top and can be scooped up and eaten. Apparently this procedure has never been taught or shown to these animals, and there seems little doubt that it can justifiably be called an invention and is, in fact, a very important invention. This ability shows an extremely significant advance in mental behavior—

Japanese macaque baby gets a free ride on mama.

Courtesy of Mr. Harry Wohlsein and Dr. William Montagna, Oregon Primate Center

that of spontaneous problem solving. The ability to solve this type of problem is characteristically found among Primates and is rarely found among lower animals.

Among the Japanese macaques—and this probably applies to many other monkeys—basic activities such as the selection of foods, the right kinds of food to eat, the use of cheek pouches, and the feeding by the female of its newborn baby apparently all have to be learned. One troupe that was transferred from Koshima Island to Inuyama took quite a while to identify the various edible plants and fruits in the new area. Dr. S. Kawamura reports than an infant Japanese macaque, raised with its mother, had never learned to use its cheek pouches. Later it was placed in a cage with another Japanese macaque. In the beginning, it was never able to eat because the other monkey stuffed its cheek pouches with practically all the available food as quickly as it could. Apparently, the sight of the other animal's eating behavior eventually showed the youngster how to use its own cheek pouches.

Sometimes the large troupes of animals break up into subsidiary groups. Dr. J. Itani, the Japanese primatologist, has described how sometimes a male Japanese macaque living in solitude will approach a troupe and start recruiting female animals and the younger males present on the periphery of the group. By these means, he gradually builds up a new troupe, which becomes completely distinct from the original troupe.

The Oregon Primate Center has a group of Japanese macaques on which it has been carrying out a number of studies. One of these studies showed that when the animals' cage becomes too crowded, the amount of aggression in the group is greatly increased. This study supports those by Dr. Charles Southwick, of Johns Hopkins Univer-

Assam macaque at the Yerkes Primate Center. His name is Red, and his exploits are described in the author's *The Ape People.*

sity, who made a fine study of the rhesus monkey in India. Dr. Southwick found that when rhesus monkeys are put in a large outdoor cage, the aggression in the group increases sharply when the space is suddenly cut in half.

Recently, under the supervision of Dr. John T. Emlen, of the University of Wisconsin, 115 Japanese macaques arrived at Laredo, Texas. They were received into a newly established 100-acre ranch specially designed for them—the Arashiyama West Primate Research Ranch. The provision of a home for these animals was made by a committee of Japanese and American scientists. These unusual immigrants have taken up residence in an area very different from their home, and it will be interesting to see how well they do, particularly how they cope with rattlesnakes, scorpions, and coyotes. The monkeys were a present from Sonosuka Iwata, who was the president of the Iwatayama Monkey Sanctuary. Edward Gardener, a Texan, made available the facilities to permit the ranch to be established.

Bonnet macaque and baby.

Courtesy of Dr. Rahaman
and Dr. Parthasararthy of
Bangalore University, India

Assam Macaque

The Assam monkeys, which are closely related to the rhesus, live in
the hilly parts of north of India, known as Assam, where the largest
industry is tea growing. The Assam monkey, *Macaca assamensis*, like the
rhesus monkey, is a yellowish brown and sometimes a little darker in
the head. The Assam monkey is said to roll rocks down the mountain
on people passing below. Whether this is actually deliberate or
whether the rocks are dislodged when the animals run about the sides
of the mountains after getting excited at seeing people is very difficult
to say.

In my book *The Ape People*, I referred to an animal known as Red, a
large, rhesuslike animal that we had believed to be a hybrid between
the rhesus monkey and some other species of macaque. This was the
animal which, in a mixed species of monkeys, formed an alliance with
a red stump-tail macaque to defeat the dominant animal in the group
which was a rhesus monkey. The story of Red's take-over and his

Bonnet macaques grooming.

Courtesy of Dr. H. Rahaman and Dr. M. D. Parthasararthy of Bangalore University, India

Bonnet macaque. The arrangement of hair on the head gives the name to the group.

Bonnet macaque presenting.

Courtesy of Dr. Rahaman and Dr. Parthasararthy of Bangalore University, India

subsequent deposition by the red stump-tail macaque was told in detail in *The Ape People*.

Red came to us on November 5, 1963, when we imported a number of pigtail macaques. In one of the groups was an animal that looked very unusual and seemed to be around the age of three. He differed from the normal rhesus macaque and, among other things, had golden yellow hair. Because of the various differences between him and the rhesus monkeys, we decided that he must be a hybrid. As he grew older, the yellow color changed to a deeper olive color, and while he had been very docile and friendly during his early period, his behavior became much less cooperative in later life. Following his defeat in the mixed group, he was put in a cage of his own and during a period of regular TB testing, was found to give a positive reaction for TB and, because of this, was euthanized. Unfortunately, he showed no indication of infection at the autopsy. The availability of his body and, later on, of his skull made it possible for Dr. Bernstein and Dr. Osman Hill to establish that he was not a hybrid animal but that he actually represented a specimen of *Macaca assamensis*, the Assam macaque. When he was young, he had rather a gracile form, but as he became older, he became much stockier and squarer in appearance.

Formosan Macaque

Another type of monkey called *Macaca cyclopis* is also found in Formosa. Its tail is shorter than the rhesus monkey's, and it lives in caves on the coast. It exists on shellfish, and a dip and swim in the sea are a common part of its activities. It is a beachcomber and scavenges anything that looks, tastes, or feels edible.

143

Bonnet macaque showing his cheek pouches stuffed with food.

Courtesy of Dr. Rahaman and Dr. Parthasararthy of Bangalore University, India

Bonnet Macaque

The bonnet monkey is another macaque, known scientifically as *Macaca radiata.* These animals are found in many parts of India. They live in very well-organized societies and tend to locate themselves very close to human habitation since they eat cultivated as well as wild food. Bonnet monkeys run about in troupes varying from about ten to fifty members and including males and females of all ages. They have well-defined home ranges which overlap with those of neighboring troupes so that occasionally fights or aggressive demonstrations between two groups occur when one group feels that another is intruding into its territory.

Bonnet monkeys prefer to roost at night in tall, well-spread trees, although they are mostly ground dwellers during the day. Their tendency to sleep in the trees has been developed for safety purposes since it protects them from predators. When the troupe retreats and runs away from a threat, the adult males drop to the rear so that they form a defensive band between the danger and the rest of the troupe. The big males then see that the rest of the troupe gets up into the treetops before they themselves come up. They have a fairly extensive system of communication which uses gestures, facial expressions, including various types of grimaces, and quite a variety of calls.

Dominance hierarchies are not as rigid among the bonnets as in the rhesus monkey groups. Apparently shifts in dominance take place among the members of the troupe, depending on the situation or the activity.

A study of the bonnet monkey has recently been published by Drs. H. Rahaman and M. D. Parthasarathy who work in the Department of Neurology, Central College, Bangalore University, in India.

Infant bonnet monkeys are allowed by their mothers to stray farther than any other macaque babies, and they are allowed to stay away much longer than the pigtails. Allison Jolly points out that the pigtail monkeys restrain their infants and punish them much more than bonnet monkeys do. As adults the pigtail monkeys do not tend to cuddle with each other as do other macaques, especially the bonnets. The bonnets clump together in groups when they are sleeping; the pigtails tend to keep a distance from each other. She also points out that the punishing of the infants by pigtail mothers does not necessarily mean that the infants are less attached to the mothers. "On the contrary," she says, "laboratory studies seem to show that punishing or even quite brutal treatment of infants leads the infants to cling even closer not only in the early stages but later in life." Studies of the dedication of a baby macaque to its mother were carried out by Dr. Harry Harlow, who arranged one of his surrogate mothers so that when the infant clung to the surrogate, it could be blasted in the face with air, which rhesus monkeys hate. It could turn the air off by ceasing to cling, but of course, in most cases it hung on and got even more blasted. Dr. Gene Sackett did experiments with rhesus monkeys in which he permitted the infants to choose among different females. In most cases, they chose their own mothers if they had had anything like normal mothering but chose their mothers even more often if the mothers had been brutal and punishing. The pigtail mothers who were punishers also had their infants clinging and staying closer to them. This was not only on the infants' initiative but on the mothers' as well.

Perhaps this has possible application to the rearing of human children, not in the sense that they should be treated brutally or be unnecessarily punished, but tightly structured behavior in the home where a child knows precisely what it can do and what it cannot do gives it a secure background that will make it a more reliable adult when it grows up. It is probable that many of our "dropout" children are in many cases the product of permissive parents who never gave their children the security of a well-structured home environment.

Toque Monkey

There is an animal related to the bonnet monkey called the toque monkey (*Macaca sinica*) which lives almost exclusively on the island of Ceylon. The toque is a small monkey, with a long tail and a face without hair; it can wrinkle its forehead, making it look like a worried human adult. The animal has a rather long topknot of hair which is darker in color and which falls down over his ears and brow. Toque monkeys are agile and a nuisance to farmers since they eat cultivated food, particularly fruits and grains but also vegetables, and chicken eggs.

They are rather shy, even though they are curious and seem to be frightened of people especially if they are cornered. On such occasions they squeeze together in a bunch and scream and chatter as if their lives were threatened.

Java Macaque

One of the most widely distributed monkeys in Asia is the Java monkey, technically known as *Macaca fascicularis*. This species ranges from Burma and Thailand all over Indochina, the Malay Peninsula, and all the Indonesian islands, and it must occur in many millions in these areas. These animals are known popularly as crab-eating macaques, and many of them do, in fact, inhabit the jungle close to the sea and fish and swim among the creeks in that area, eating crabs or any other crustacean they can lay their hands on. They will, however, live on anything else they can find, particularly fruits and grains and various cereals, and they occur in great droves and have been known to invade a plantation and eat everything they could find. For some reason, on the island of Bali they are treated as if they had some religious significance and are rarely hurt or killed.

Red Stump-Tail Macaque

Perhaps the ugliest monkeys are the red stump-tail macaques, *Macaca speciosa*. The adults are quite large, muscular animals and their faces are colored pink or red and often have uneven and irregularly shaped blackish or brownish patches on the red, making them look particularly ugly. They are also red around the rump. I have found, in the group that we have out at the field station, that the adults are rather vicious and certainly untrustworthy. The young are more gentle and gambol around and play rather attractively.

The stump-tail is so called because it has a small, practically naked tail a few centimeters in length. It is also called the bear macaque because it is large and powerfully built, moves slowly, and is more of a terrestrial than an arboreal monkey. There are three subspecies of the animal, which is found in Assam, upper Burma, Laos, Indonesia, and extends into southern China. The adult is usually about 2½ feet in length and may weigh as much as 15 kilograms (around 40 pounds).

146

A red stump-tail macaque at the
Yerkes Primate Center.

The newborn animal has a creamy white fur, and its face, its hands
and feet, and its buttocks and genitalia have a flesh pink color. As it
gets older, it develops the facial pigmentation already mentioned, as
well as red buttocks and genitalia and a dirty brown coat, sometimes
with a little touch of red in it. It is one of the Primates that are subject
to baldness, and after adolescence, the forehead and cheeks tend to
lose their hair progressively.

In the wild, the stump-tail monkeys come down from the trees,
where they sleep, at dawn and feed until about ten o'clock, then rest
and groom each other until early afternoon and start feeding again in
a desultory way. From midafternoon they feed pretty actively until
about six o'clock and then go back in their trees, up to three-quarters
of an hour before dusk. They eat small fruits, particularly of the ficus
tree, small berries but only a little flesh. It takes a fairly long time to
collect this type of food, and they may walk as far as two miles in
search of food during the day. When the stump-tail monkeys come
into what would be described as human territory, where they

normally expect to see humans, according to Mireille Bertrand, they are very cautious. They do not exhibit any fear of humans or excitement if they see them, and they have very little vocalization. They are likely to eat from bushes by the side of the road in full view of people from a nearby village, but they do not react to them at all as they pass by. Very often, they are so quiet that even the villagers do not know where they are. However, when they are up in the hills where human contact is rare, they get very excited, and there is a great deal of calling and vocalization to each other if a human appears.

When these animals are in captivity, there is an interesting change in the behavior. For example, after a few weeks, they do not react anymore to alarm calls from other stump-tail macaques. Instead of this watchful kind of behavior, they become more interested in the investigation of specific objects and of minor events, both inside and outside the enclosure. This shows how careful we have to be in interpreting wild behavior from observing animals even under semi-caged conditions.

It has been shown by Professor Itani, of Japan, that most wild Japanese macaques will accept new food very reluctantly and only over a very long period of time. The younger animals, however, will take their new food much faster than the older animals. Bertrand found that infant and juvenile stump-tail macaques would eat quite a variety of foodstuffs, even including those as diverse as cooked fish, cough drops, and pomegranates. However, while they would smell smoked ham and cheeses, they would reject them as foods. Bertrand notes that these foods were rejected by five other species of macaques that he tried them on and also by some guenon monkeys.

The red stump-tail macaques drink like human beings—that is, by sucking with the mouth. But sometimes they dip their fingers into liquid and lick them. This happens when the container of fluid is too small for the head to be inserted into, when the liquid is too hot, and when it is a new drink they have not met with before. When they were given something to drink in a cup, two of the younger animals studied would actually carry the cup to the mouth and, after having drained it of what it contained, would put it back on the ground with great care. They had not been taught this type of behavior, and Bertrand wonders whether it might be an example of imitated learning from humans. They all, of course, drink water and in captivity all of them drink cow's milk. When they were offered tea, eight out of ten drank it, obviously a very civilized behavior.

Bertrand has some interesting observations about monkeys' swimming. He points out that he did not see either wild or captive stump-tail macaques swimming, but some of the captive animals which were provided with pools would jump repeatedly into the shallow part of the pool, mainly in the form of play. The lion-tailed macaques, however, would only dip their limbs or wade gently in the water. We know that rhesus monkeys swim without difficulty. The Yerkes Center had a group of rhesus monkeys which were kept on an island in a lake outside the house of the president of Emory University On one occasion, when food was late in arriving and they got hungry, the entire troupe took to the water and swam to the shore and began to run around the outside of the president's house, peering into the windows—a procedure which led to their rapid banishment. My capuchin, Benito, was also a swimmer. He would not leap into the water voluntarily from the shore, but I would take him out on a rubber air mattress, and as he would see the shore of the lake getting farther and farther away, there would come a time when he could not stand it any longer and he would jump into the lake and swim actively for the shore. He could swim so fast that if I had not had frogmen's flippers on, I would have had difficulty in keeping up with him. The Japanese macaques also swim. Originally, in the area of Koshima, they learned to beg for and swim after peanuts that had been thrown into the sea. Later on, some of the juveniles in play began to swim and also to dive when the weather got too hot. They would do this purely as play, and eventually, as Bertrand points out, this cultural pattern spread to older animals. In what is known as the Shiga troupe of Japanese macaques, some of the young animals bathed and swam in hot springs but only during the cold season. It is well known, of course, that the crab-eating macaques (Java macaques) can swim and dive with great facility. Pigtail macaques are also able to swim and dive, and Cuma reports having seen hamadryas baboons swim. The rhesus monkeys of Cayo Santiago, off the coast of Puerto Rico, can also swim, and there is an interesting film showing these animals jumping off a tall tree and landing into water with a kind of belly-flop dive. C. R. Carpenter has recorded howler monkeys swimming, and proboscis monkeys and talapoin monkeys are also described as strong swimmers. It is strange, in view of this, that the great apes are unable to swim.

The walk of dominant Primates is of great interest. Bertrand found, for example, that when the dominant stump-tail macaque walks, he uses a display walk: He will walk very deliberately, looking straight ahead and usually holding his tail up. Despite the fact that this is characteristic of the dominant adult males, sometimes some young juvenile males will use it. Often, when a powerful newcomer is introduced to the group, a number of the animals will use this walk. If, for example, an alpha male wants to walk toward the scene of a quarrel or something desirable, he walks very slowly. Bertrand also says that if he played too roughly with the alpha male of a group of pig-tail macaques he was working with, the animal would put up with it for a short period, but then it would walk slowly away. On the other hand, if he was playing with the juveniles, as soon as he released them, they would run away. This slow and measured walk seems to be the characteristic of the dominant males. He points out that dominant rhesus monkeys and Japanese macaques also walk in a particular way. Allison Jolly says that *Lemur catta* dominant males have a "swaggering walk." A "confident gait" was described in olive baboons by Irvin DeVore in 1962, and T. T. Struhsaker noted a "confident walk" in vervets. Gorillas have been described as having a "strutting walk" by George B. Schaller. Suzanne Ripley made this observation about langurs: "A dominant male has an easy, powerful, measured gait and never appears to be pressed or in a hurry, even when he is in full retreat from human, dog or other langur males."

Stump-tail macaques, particularly the juveniles, are said to be very docile with humans and to be very good for laboratory purposes because they are friendly and are easy to handle. I have noted, however, that the stump-tail macaques at the Yerkes Field Station are the least pleasant of the many groups of monkeys that we have. I find that if I get too close to the fence, there is a very vicious snatching and grabbing and snapping at me, particularly on the part of the older animals. The younger ones simply keep out of the way and make very little effort to make contact. Bertrand quotes a couple of reports which also suggest that these animals are quite vicious. For instance, he quotes S. McCann, who in 1933 said of the stump-tail macaques of Assam: "They are extremely noisy and appear to fear nothing, at times running from man. . . . According to the Nagas, when chased from the fields they frequently show fight. They are apparently very pugnacious when disturbed. The Nagas are somewhat timid of them at times, on account of their vicious habits. Possibly they may attack a lone man or woman."

He also quotes a letter from Gordon Young of the Chieng Mai Zoo in Thailand:

> The attack on humans is exceptional. And yet, it surely occurs. I've seen it at least three times, and I have many, many hunter friends who have had attack experiences, some of them never came back alive. . . . I know of native hunters of the Lahu, Lisu and Atacha

tribes who have been killed by stump-tails. One hunter, a Lahu I knew personally, approached a troupe that was preparing to bed down and sleep on a sort of rocky mesa top. He shot a female, it screamed and the attack was on. The man was ripped completely to pieces by the troupe.

Bertrand tried giving mirrors to some of his stump-tail macaques in captivity. He found that most of them would go behind the mirror as if they were looking for the monkey behind it; others, holding the mirror with one hand, would swipe behind it with the other hand as if they were trying to contact the animal they thought lurked there. They would also sniff and touch the reflection, and they would bite and lick the mirror and also try to tear it apart. He says that when he tried giving mirrors to other monkeys, they all did the same thing. There was no evidence that the animals ever recognized what they saw in the mirror as their own image. He describes one animal that always greeted her image with a kind of pout; she even presented to the image twice—in other words, turned her bottom toward it, stuck it in the air, offering it for mounting. Other animals would threaten the mirror with open mouth, canine teeth showing. I have seen this on a number of occasions with some of our monkeys. One animal would even slap the reflection in the mirror. A stump-tail has been known to drink from a pool in the zoo and, after having had a drink, threaten its reflection in the water. I have also seen this with rhesus monkeys. It is probable that all monkeys do not react to their image in water in the same way because some of them have got used to the other monkey in the water.

Apes behave toward mirrors much more as if they recognized themselves and they also react quite well to pictures and drawings. In *The Ape People*, I described how one of our chimpanzees, Vicki, was able to select her own picture from a pile of monkey and human pictures and placed it on the human pile, not on the ape pile.

About seventeen types of vocalization have been established for the stump-tail macaque, but its vocal repertoire is undoubtedly greater than that. Many of the sounds are not clear and stereotyped but are graded into each other with many intermediate sounds. Many of the vocalizations that can be heard in animals in the wild are heard only very rarely or not at all when the animals are in captivity. It is possible to persuade the animal to make some sounds more easily than others, and some sounds are produced only in response to particular situations such as excitement or some other emotional state or in response to calls of other animals. The vocalizations resemble those of

The Moor macaque.

Courtesy of the Cheyenne Mountain Park Zoo

the Japanese macaques more than they do any other species of macaque.

Moor Macaque and Celebes Black Ape

In the Celebes islands, just north of Australia, there are two species of monkey. One is called *Macaca maura*, or the Moor macaque, which is a dark gray or blackish furred animal with a very attractive face, and the other is a very unusual animal, the Celebes black ape. This last animal has been known scientifically for many years as *Cynopithecus niger*, indicating that it was probably related to the baboons. However, the experts have recently reclassified this animal and called it *Macaca nigra*, which suggests that they now believe it to be a macaque and to be less related to the baboon than was suspected. For some time it has been called the Celebes crested macaque. This animal has practically no tail, but it walks, stands and looks rather

152

like a baboon, and its skull shows prominent bony ridges above the eyes. The face is rather long, with peculiar elongated nasal ridges on the top of the face, and the animal has black skin and hair. On the top of the head is an elongated tuft of hair which can be erected into a crest. The coastal tribes in the north of the Celebes Islands worship the black ape as a sort of god. Every so often they set a raft laden with food adrift in the river. The food is intended as an offering to the ape gods.

At the Yerkes Center we have both the Moor macaque and Celebes black ape, and we have been successful in crossing them and producing hybrid animals. Scientists at the Oregon Primate Center have recently made some interesting discoveries in the forty Celebes apes they have. They had noticed that in this particular colony the loss of infants was greater than it was with other species of monkeys at the center. The reason for this was the cause of some concern, and a number of possibilities were considered; eventually, they decided that the Celebes ape mothers seemed to have a lesser maternal instinct than the other monkey species at the Oregon Center and that a number of the infant deaths were really due to neglect by the mother. By taking the babies away from the mothers, bringing them up in an incubator, and then caging them with other young monkeys, they found that the infant Celebes apes developed perfectly well and could be returned to the colony. Then an interesting thing happened; they found that in this group there was a surprisingly high incidence of diabetes. They believe that this diabetic condition has existed in the group undetected for years. Most of the group of Celebes apes at the Yerkes Center originally came from this same group at the Oregon Center, and when scientists from Oregon came to study the Yerkes animals, they also found diabetes in this group. Diabetic signs were found among both sexes of all ages. The scientists at the Oregon Center are now studying Celebes black apes at zoos throughout the country in an attempt to see if the tendency toward diabetes is a characteristic defect of the species. If this is so and even if it is not so, at least the Oregon group can be used as a model for the study of human diabetes, and this should make an important contribution to our knowledge of this disease.

Celebes apes are extremely pleasant to associate with, since they appear to be very much people-oriented and are very nonaggressive toward humans; it is possible to go into a compound with them and not be attacked. However, I have seen a woman badly slashed by the dominant male of a group when she made the mistake of giving food to a junior member before giving it to him. We have used one or two

Celebes black ape at Yerkes Primate Center.

The author with Jill—
Celebes black ape.

Barbary ape.

Courtesy of Barbara and Michael MacRoberts

Family group of Barbary apes.

Courtesy of Barbara and Michael MacRoberts

of these animals at the Yerkes Center as public relations animals. One of them, called Jill, was a very friendly creature. She could be taken out of her cage and carried around and would perform a lot of friendly lip smacking toward whoever made contact with her.

It is an intriguing fact that in the Celebes islands there are only three species of Primates: the Celebes black ape, the Moor macaque, and the little tarsier. The scientific name of the Moor macaque means "dark color." The Moor macaques are both forest roamers and beachcombers. They move around in very large groups, eating almost anything, but they have a predilection for animal food.

The presence of these animals in the Celebes islands is surprising since presumably they could only have come from the Philippine Islands, and there is nothing in the Philippines that remotely resembles them.

Barbary Ape

The scientific name of the Barbary ape is *Macaca sylvana*. It is the only species of monkey which is truly a native of Europe. It also occurs in North Africa, in the Atlas Mountains, and was said to have been introduced into Gibraltar by the Moors. There is a legend that there is an underwater tunnel between Gibraltar and North Africa and that the Barbary apes got to Gibraltar by transversing it.

The Barbary ape is a large animal, and it is not really an ape but a macaque monkey. It is very strong, with powerful jaws, and is greenish brown.

Gibraltar, an isthmus of Spain, was known as Jebel-al-Tarik, the Mount of Tarik, and the modern name is a corruption of the old Moorish name. It is three miles long and three-quarters of a mile wide and extends into the Mediterranean toward North Africa. To the east, across the bay, is the old Moorish city of Algeciras, which from Gibraltar looks like an attractive collection of white buildings with red roofs that shimmer in Mediterranean sun. The rock was seized by the British in 1704 and remains a garrison town to this day. I have visited it both before and after the last world war and during the war landed in Gibraltar Bay in a flying boat. I found it a fascinating collection of narrow streets, old walls, residential houses, military offices, and a Moorish fort. I also saw and talked at length to the apes, who regarded me sagely for a while and then looked for something more interesting to eat than the few grapes that I had brought with me to capture their attention. They have always been a tourist attraction and were recently studied in some depth by Barbara and Michael MacRoberts.

In their account these two scientists introduce their subject with a beautiful little paragraph:

> But above the pepper trees and wild olives of the town, well beyond reach of the staccato cries of fish vendors and the declamations of morning shoppers, the only feral population of monkeys in Europe roams the military land that stretches the length of the Upper Rock. Here, among the rusting cannon and grass-choked trenches, monkeys have lived for centuries; like the people of Gibraltar, they too have had a long and tumultuous history.

Tumultuous it has been. How they first got there is not known, but there are two suggestions. One of these is that they are representative of monkeys that used to live more widely in Europe during Pliocene and Pleistocene times, 12,000,000 to 1,000,000 years ago. However, no fossil macaques have ever been found in Gibraltar, in which there are a number of caves where they might very well have been preserved. The second view is that the monkeys were exported from North Africa centuries ago; in fact, it is known that in the Moorish markets in Gibraltar, after the British seized it, monkeys were being sold. They were undoubtedly sold there before, and they are recorded as existing in the wild state in Gibraltar in Spanish writings of the seventeenth century. It is known that by the beginning of the nineteenth century there were 130 animals on the Rock.

Today they occupy the two high-up areas in Gibraltar, in two separate groups. Lower down the hill (the Rock of Gibraltar rises to 1,400 feet above sea level), the humans live, and this gives the monkeys a chance to scavenge and steal as a supplement to their natural diet. Their thieving propensities led to considerable antagonism between the monkeys and the humans, so much so that in 1856 the commander of the garrison had to issue an order which forbade the killing of the "apes" and a signal master was put in charge of them and told to keep a census. Subsequently, the townsfolk became very incensed with the animals and, in 1863, would have virtually exterminated them (only three were left) had not the governor, W. Carrington, issued another order and even imported four more animals from North Africa. After the new imports had been added and the group grew in number, it eventually split into two groups. However, because of their continuous assaults, people began to shoot and trap them, and their numbers declined again.

Because they were immune from any kind of control, they terrorized certain of the population of the Rock, including dogs, cats and

Baby Barbary ape.

Courtesy of Barbara and Michael MacRoberts

poultry. Any shop or house was likely to be invaded by them and damaged, or articles, particularly food, stolen. Sometimes, it is said, they had jumped on high officials, particularly those traveling in open carriages, and carried away the handkerchiefs or sometimes the hats, and they have even been known to steal wigs.

Sergeant Brown, who was in charge of the animals at one time, kept a detailed diary of the activities. He also acted as a part-time veterinarian. In 1872 a young officer shot two of the monkeys because they tore up the clothes in his wardrobe. There was an official inquiry, and the officer had to make a public apology and was forced to buy two monkeys to replace those that he had killed. But because they were new animals and were not part of the established hierarchy, the two new animals were thrown off the Rock by the other monkeys and drowned in the sea.

In 1873 one of the monkeys, a large male, became so impossible to the human beings on Gibraltar that Sergeant Brown was ordered to place the animal under arrest. The animal was chased all over the Rock, sat up on a rooftop, and kept up a pretty accurate fusillade of tiles at the troops trying to capture him. When his ammunition was exhausted, he made a bolt for the open fields; eventually he was caught by the sergeant and three other soldiers, who threw cloaks over him and finally threw him into a specially prepared cage.

After 1880 the monkeys had to have more and more restraints put on them because they had become a great nuisance. In 1921 one of the old males began attacking and killing female monkeys; he had to be removed and finally ended up in the London Zoo.

Eventually, the British Colonial Office put an item into the Gibraltar budget for feeding the apes. In 1913, this amount was one

**Baby Barbary ape
takes a walk.**

Courtesy of Barbara and Michael MacRoberts

pound sterling per year, but it rose rapidly to six pounds a month by 1921. In that year, an "Officer-in-charge-of-Apes" was appointed from the Royal Garrison Artillery. The commanding officer of the Gibraltar regiment now holds this prestigious office. In 1930 the individual monkeys were given names, and new animals were brought in from North Africa to replenish their dwindling numbers.

During all this time, there had grown up a myth that when the apes left the Rock of Gibraltar, the British would have to go too. This myth had gained wide currency, and when, during World War II, it was reported to the Prime Minister of England, Winston Churchill, that there were only two of these animals left, he sent, on August 25, 1944, a message to the authorities at Gibraltar expressing his anxiety over the rumors which were being disseminated concerning the welfare of the Barbary apes on Gibraltar. He expressed the wish that they should not be allowed to die out. On September 8 a directive was issued from the Prime Minister's office indicating that every effort should be

159

Baby Barbary ape and mother.

Courtesy of Barbara and Michael MacRoberts

made to bring the establishment of the apes back to the number of twenty-four and that this should be maintained thereafter.

Paul Gallico's *Scruffy*, a magnificent and wildly entertaining book, based partly on fact and partly on a vivid imagination, deals with a large recalcitrant male ape on the Rock that gets involved in a number of imaginary incidents. It is obvious, however, that there are many events, described in this book as fiction, which have some kind of a factual basis, dating back to perhaps some incidents in the nineteenth century.

The monkeys that now live in Gibraltar all are descended from the Barbary apes brought there about thirty years ago. They are still fed regularly, and sick and injured animals are cared for. There is a total of about thirty animals, divided into two troupes, and if the number gets much above this, some are trapped and dispatched to zoos in various parts of the world.

The Barbary ape must have been much more common in Greek and Roman times; he was certainly well known during that era. The father of anatomy, Galen, who worked in Rome in the second century A.D., used the Barbary apes for his dissections since he was barred by the church from using humans.

The MacRoberts' give a very colorful description of the activities of one of the troupes. The animals spend the night on the ledges of sheer cliffs, and at sunrise, they begin to stir. Then they climb the cliffs to sit on the warm concrete of some of the gun emplacements. This is a period of social activity. The mothers suckle the babies, the adults groom each other, and young animals lark about. When the sun gets a little higher, they proceed down the western slope of Gibraltar, headed for the city. On the way, they pass over the concrete water catchment area where they stop for a drink of water from a pipe that is leaking. The first buildings of the city they come in contact with are the high-rise apartments, and here they start on the trash cans, throwing the lids on the pavement and frequently overturning them and scattering the contents. The housewives know to keep a close eye on these rascals, especially because they have sometimes helped themselves to a sheet or article of apparel from the clotheslines. Closer to town are a number of fruit trees, including figs and pomegranates; if these are in fruit, the troupe may move down there. If they do, the cry *"Mira, mira, los monos"* rises, and everyone is on the watch for them.

The Barbary ape troupes have a dominance hierarchy, like most other monkeys, with alpha and beta males at the top of the scale. In the Middle Hill troupe in Gibraltar, four adult females come next, juveniles and infants ranked below these six, and among the young animals, age, not sex, was the determining factor of their rank in the hierarchy.

The animals have a long period of rest during the middle of the day on housetops and walls, and when that is drawing to a close, some feeding on natural vegetation and raiding of the same trash cans they had searched in the morning occur, and the group then starts to move slowly back up the hill.

Here is the MacRoberts' account of the final part of the monkey's day:

> The trip back up the Rock is more rapid than the leisurely descent in the morning, and within thirty minutes, the troop once more spreads out across Princess Caroline battery. One of the adult males wanders off, out of sight, an infant still riding on his shoulders. Females sit and doze. Once again, the juveniles play. As the sun

Lion-tailed macaque at the Yerkes Primate Center.

begins to set, the monkeys descend to their sleeping cliff where they will remain until morning. From Alamed Gardens in the southern part of the city, north past the cemetery where some of Trafalgar's dead are buried, dusk turns the town slowly gray, and the lights of the harbor begin to wink. The Mount of Tarik stands black against the sky.

Lion-Tailed Macaque

Known scientifically as *Macaca silenus*, this shy animal lives in the monsoon forests which cover the Western Ghats, a mountain range in India extending from the former Portuguese colony of Goa southward. The animal is also called a wandaroo monkey, which is misleading because this name really belongs to certain langur monkeys in Ceylon and which is often incorrectly translated into English as "wanderer."

The lion-tailed monkey is extremely thick-haired, and the hair itself is very dark chocolate to black. The face is framed in light-colored, very long hair which extends onto the neck and shoulders and looks a bit like a hairy, middle-aged ruff. The tail is quite long and may be 24 inches in length; at one end it has a tuft of long hair which gives it its name of "lion-tail."

The older males are large, imposing animals. The animal is very fierce and menaces anyone who penetrates into its forest abode. There are several reports that native children left in clearings at the edges of the forest were killed by the lion-tailed macaques. They are very vicious if caught and caged when they are grown, but Sanderson says that they can make very gentle pets if they are caught young, hand raised, and kept away from other monkeys. He had one as a pet; his description of it was "gentle as a spaniel, as polite as a well brought-up human child and as intelligent as a three year old chimp." It was

162

Wanderoo (lion-tailed) monkey and baby.

Courtesy of the Thüringer Zoopark, Erfurt, German Democratic Republic

Wanderoo (lion-tailed) monkey suckling her baby and looking down at it with maternal expression.

caged in the house but could and did get out and wander around the house but did not break anything. It liked the kitchen best. It would sit on its stool and watch, with fascinated interest, all the activities that went into preparing a meal. It would only answer to its complicated name of Ootacamond—a hill town in India from which it came—and, according to Sanderson, "Such forms of address as 'Hi, you' and other vulgarities it ignored with obvious and pointed distaste."

Despite its present rarity—it is on the list of endangered species—it must have been much better known some centuries ago. Breydenbach, in 1486, illustrated one of his publications with a page of drawings of various animals, one of which was a monkeylike animal with a ruff around its neck, just like a lion-tailed macaque, and in 1216, Aelian, a Roman who wrote his scientific works in Greek, mentions the lion-tailed macaque in his *De Natura animalium*.

It is unfortunate that we know practically nothing of this animal in its natural state, and we wait for some enthusiastic behaviorist to move into the Western Ghats and tell us all about the private life of the lion-tailed macaque, if enough of them are left to study.

VII. The Old World Monkeys— Mangabeys, Baboons, Drills, and Mandrills

THE MACAQUES which have occupied quite a bit of our story so far belong, with the mangabeys and baboons, to a group called the Cynopithecoids, known also as dog monkeys because of their long muzzles. The Cynopithecoids occur in uncountable numbers, especially in Asia, probably exceeding the human population of the world. We have now to talk about the rest of this group which includes the mangabeys, mandrills, drills, and baboons.

Mangabeys

The mangabeys are called by the Latin name of *Cercocebus*. They are found in the tropical forest area of Africa, and they extend from Guinea to Uganda and also into the basins of the Congo and Niger rivers. There are about eleven species of mangabeys. Their bodies are slender, their heads oval; they have large cheek pouches, large bare patches on their bottoms (the ischial callosities), and long tails.

Mangabeys have shorter muzzles than macaques. The eyelids of most mangabeys are white, but they do not have the same brilliant colors as the guenons. Nearly all are arboreal, and they are great grimacers, frequently displaying their teeth. The coloration and swelling of sexual skin around the female anus and the genitals during the heat, or estrus period, is very well developed in the mangabey. They have a menstrual cycle, like the human, which is about thirty days, and they also have menstrual bleeding. It is interesting that the sexual swelling is a phenomenon that is characteristic of Old World monkeys; it does not occur in the New World monkeys. The estrus period occurs when the ovum drops from the ovary into the Fallopian tubes which carry it to the uterus, and it is a period when the female is receptive to the male. This is the time the bare skin surrounding the female external genitalia and extending to the anus becomes very swollen and often highly colored.

All the macaques, mangabeys, and baboons show this sexual swelling; it is also characteristic of the chimpanzee. It is also found in the little talapoin monkey and in Allen's swamp monkey, which we will mention later.

The mangabeys do not have an extensive or loud vocalization, but they seem to have developed another way of communicating with each other. Either they have pure white eyelids or they have very light-colored skin around the eyes, which contrasts very strongly with the dark skin of their faces. Even in very deep forest where there is no twilight, these white marks can be seen very, very clearly. Ivan Sanderson has noted that when one mangabey wants to communicate with another mangabey, the first one will stare at the second monkey and then blink its eyelids very rapidly. They may even open their mouths and wiggle their tongues up and down at great speed. "It is believed," says Sanderson, "that the patches on their eyebrows really act as a type of beacon to attract the attention of another animal." The flicking of the eyelids then is a method of semaphoring some kind of concepts or emotions.

166

Mangabey at the Yerkes Primate
Center.

The mangabey has a rather generalized or primitive structure, and
it is probably close to the ancestral animals from which the drills,
mandrills, and baboons developed and which lived 10,000,000 or
12,000,000 years ago in Pliocene times.

There are five species of mangabey. There is the great cheeked
mangabey, the black mangabey, the white-collared mangabey, the
sooty mangabey, and the agile mangabey. They all tend to be black or
gray except where relieved by white.

They are forest-living animals, and tropical rain forests are their
home. They live fairly high in the trees. One group prefers low-lying
forest near swamps, and they keep high up in those trees. The sooty
mangabey is usually found in secondary forest—that is to say, forest
that has been cut out once and has regrown. It usually sticks to the
lower realm of the trees. It is particularly common along the banks of
the rivers and can be seen in the branches about 10 or 12 feet above
the ground in that region.

The white-collared mangabey prefers to live on the ground but does
take to the trees on occasion, mainly for eating and sleeping.
Mangabeys eat seeds and palm nuts, all kinds of fruits, and various
leaves. The large monkey-eating eagle is one of their enemies.

Mangabeys are fairly large monkeys, and together with the head,
the body length may reach as much as 36 inches and the long tail may
extend for almost another 30 inches. The limbs are very long and
slender, and so is the body.

Generally speaking, mangabeys are very agreeable animals in
captivity and very rapidly learn to respond to handling; it is possible
to put a hand in the cage and groom them without any real danger.
However, we had one mangabey, called Mr. Magoo, who gave me

167

Crested mangabey.
Courtesy of the Cheyenne Mountain Park Zoo

quite a surprise. I had been petting and grooming him for more than a year, with the animal responding very well and obviously showing a great liking and affection for me. On one occasion, however, I bent down low to look at something quite close to his cage and apparently bent below the lowest level at which he could, even with squatting, bring his eyes. Seeing me in this position and presumably in what appeared to him to be a submissive position, he immediately and viciously grabbed my face, hair, and glasses through the wire, and pulled me up against the cage, and it was only by the greatest piece of luck that he did not sink his canine teeth into me. This experience taught me quite a lesson.

The Baboon

Back in Stone Age times, when man first began to practice the primitive art which blossomed rapidly into the beautiful cave paintings of Lascaux and other places, the monkey took his place among the animals represented by the artists. As we move into early civilized time, the monkey and especially the baboon appear not only to have become more important to the men of those times, but even to have been worshiped.

In Asia, Africa, and Europe, great importance seems to have been attached to the baboon and other monkeys. In Tibetan mythology, there is even an ingenious account of the evolution of man in which he is derived from a copulation between a monkey and an ogress that lived in the mountains. The way man behaves today certainly makes this seem more likely than conventional theories of man's origin.

In ancient Egypt, the baboon became a very important character; it lived as a pet in many Egyptian households and was worshiped *in vivo*

168

and in effigy in the temples; and when it died, it was mummified and buried with as many honors as if it were a priest. In the earlier periods of the Egyptian civilization, the kings, the members of the court, and the priesthood often wore baboon tails attached to their costumes on ceremonial occasions.

According to some of the authors in the early part of this millennium, the moon was supposed to affect the behavior of baboons:

> At that time of the hours when the moon joining forces with the sun becomes dark, then the male cynocephalus neither looks up nor eats; but is bowed down to the ground in grief, just as though sorrowing at the seizure of the moon. The female cynocephalus, in addition to not looking up and suffering the same things as the male, also bleeds from the private parts. For this reason even now cynocephali are fit in the temple so that from them the time of the conjunction of the sun and the moon may be ascertained.

When a sacred baboon died, it was embalmed and then wound around and around with bands of clothlike material with its name written on it. It took about two and a half months for the whole procedure to be completed and another two days for the funeral ceremony. Until that was completed, the baboon was considered not really dead. Explorations at the site of the old town of Thebes in the Valley of the Kings have shown that there was an enormous monkey cemetery there. Examination of the remains showed that many of the baboons had bony deformations indicating that they suffered from rickets or other diseases and also indicating that they had been poorly taken care of and certainly improperly fed.

Egyptian art often depicts baboons in a position that appears as if they are worshiping the sun. There is some evidence that they were actually trained to do this, but of course, we know that all Primates like to sit and bathe in the sun, particularly in the early morning. Also, in the evening, when the sun is setting, the baboon tilts his head upward, probably looking for a tree to roost in for the night—he has to tilt his head back because his large bony ridges make it impossible to see upward any other way. This act has been interpreted, especially by the ancient Egyptians, as an act of sun worship.

A curious description of baboons' making preparations for the night was noted by Theodore Knattnerus-Nerer, director of the Rome Zoo. Toward evening, these animals, which were hamadryad baboons, used an almost formal procession, headed by the chieftain who was followed by the older males, then the females with their young, then by the younger males and females, the rear being brought up by an old male. In this fashion, they would make several circuits of their

Portrait of hamadryad baboon family.

cage to the sound of a kind of rejoicing led by the chieftain. The troupe joined in. Dr. Nerer said that he was reminded of the baboons praying to the sun as represented in the sculptures of those keen observers, the ancient Egyptians.

Greek literature refers to the licentiousness of baboons and other monkeys. They are said to attack women and children with the object of copulating with them. There is a story in *A Thousand and One Nights*, as translated by Sir Richard Burton, in which an Abyssinian baboon is said to have attacked a woman in a Cairo street and tried to rape her. It seems that he was then killed by being bayoneted by a sentry. If such an incident did occur, it was probably a straight-out attack by the animal on the woman which was interpreted as a sexual attack. In the book *Those About to Die* by Daniel P. Mannix, the use of baboons in the Roman arena to rape little girls and even to kill them is reported.

In *A Thousand and One Nights*, Scheherazade tells a tale called "The King's Daughter and the Ape" and in this, she described the daughter of a sultan who was obviously very much of a nymphomaniac—she

Hamadryad baboons at the Russian Primate Center. Animal on post shows swollen sexual skin indicating she is in heat.

Courtesy of Dr. Boris Lapin

felt the need for copulation every hour of the day. She was recommended by her servant to take a baboon as a lover. He lived with her, doing nothing but eating, drinking, and copulating. Eventually, the young lady had to flee to Cairo with her baboon. There she was observed by a Cairo butcher to wine and dine with the ape and then carry out repeated, passionate copulations until she swooned away. The butcher killed the ape and offered to substitute in the ape's place, but he found the girl's sexual appetite completely exhausted him. Finally, with the aid of an old woman, he produced some extracts which he gave the girl. Two worms issued from her body, and after this her nymphomania was cured.

In the Middle East, exhibitions have regularly been staged in which tame monkeys, some of them baboons, were made to copulate with women on the stage. Bernard Heuvelmans described his experiences of seeing a tame young baboon assaulting women who were sunbathing on a Mediterranean beach and attempting to copulate with them.

Newborn baboon at the Southwest Foundation for Research and Education.

There are records from Ethiopia of gelada baboons that have been used to perform a variety of jobs. They have been bearers of torches at banquets and have been taught other exotic tasks. The Egyptians actually imported many thousands of baboons from Ethiopia and from the Sudan. The worship of the baboon and of the god Thoth came to its peak at the time when Heliopolis was Egypt's capital. This was during the dynasty of Ikhnaton. This was also a period when the sun was worshiped more than any other god. Ra was the sun-god, and Thoth was his scribe. It is interesting that the hamadryad baboons normally bark at dawn and when the sun comes out from behind the clouds. This is sometimes seen in present-day baboons, and it fitted in very well with their role in ancient Egypt.

Probably the oldest carving which was produced by the culture of the Nile was that of a baboon, and it bore the name of King Narmer of the First Dynasty. When Tutankhamen died, the western wall of his burial chamber depicted twelve baboons apparently placed there so that he would be guarded through the hours which constituted his first

Portrait of a young baboon at the Southwest Foundation for Research and Education.

Courtesy of Dr. Sy Kalter

night in the nether world. A picture of Queen Hatshepsut, of the Eighteenth Dynasty, coming back in her barge from the land of Punt (Somaliland) shows a number of hamadryad baboons sitting on the deck of her ship. It is very interesting that the Egyptian hieroglyph "to be angry" has a pictograph following the sign which demonstrates a baboon in a threatening posture, with its teeth bared and its tail arched. The hamadryad baboon was supposed to have been trained by the Egyptians for many roles: as a fisherman, as a servant, and in some cases as a helper in the fields. It was also supposed to serve wine and food during coronations. It was undoubtedly used as an entertainer. It is very rare to see trained baboons performing in modern times.

The chacma baboon was once trained in the eastern province of the Cape of Good Hope to move the signals on a railway track. His master, who had been a railway man, had had his legs cut off in an accident, and as a result, he taught the baboon to pull the levers that worked the signals. The animal was taught the number of each lever and then on command pulled the one which his master ordered. The two went back and forth between the station and home each morning and evening on a railway trolley. When the tracks were level or were graded upward, the baboon got off to push the trolley.

Modern Africans are supposed to use baboons as water diviners. When they come to an area where they either smell or sense the presence of water, they are said to dig until they come to it. Professor Raymond Dart has recorded in one of his scientific papers the story of a chacma baboon called Ahla that had been used as a goatherd. It had been trained to escort the animals to pasture at the beginning of

Courtesy of Dr. Boris Lapin

The Russian Primate Center in Sukhumi (Abkhazia).

each day, to stay with them, and to keep away the wild animals which could have hurt them. It would round them up in the evening and escort the herd back home, all of this without any human supervision. I have seen kelpie dogs do exactly the same thing with sheep in Australia. Another farmer in Africa has used baboons to pull the weeds out of his orchards.

The Egyptians are also supposed to have taught their baboons to carry loads, sometimes on a pole with a baboon supporting each end. They are also supposed to have been used to catch birds and even to pile wood.

For some reason or other, the Greeks, Hebrews, Romans, and Christians were concerned about the baboon idolatry of the Egyptians. Although there are pictures and sculptures of the baboon all over the Mediterranean, carried there by the Phoenicians, the same respect for the monkey did not go along with the artwork. The Koran, the Bible of the Mohammedans, has claimed that Jews who broke the Sabbath were turned into monkeys, and by monkeys they probably meant baboons. King Solomon was given baboons by his traders. Both the Hebrews and the Romans thought that the baboon was very ugly and, because of this, unlucky.

Theophilus, A.D. 391, attacked and destroyed the temples of Thoth which had been built in Alexandria. He kept one statue of a sacred baboon so that he could show it to his followers as a horrible example of the type of creature that was worshiped by the pagans. The Roman author Cicero had an acquaintance who shocked him by keeping a pet baboon. It was mentioned earlier that the Romans used baboons in their games for a variety of sadistic purposes.

In the Middle Ages the baboon took the form of a religious symbol,

174

Feeding hamadryad baboons by hand at the Russian Primate Center.

Courtesy of Dr. Boris Lapin

but it was a bad religious symbol because it was often depicted as Satan. Eventually, the goat surpassed the baboon and the monkey as a symbol of the devil.

Titian was one of those in the Middle Ages who used the monkeys and baboons for purposes of satire. For example, he did a woodcut of Laocoön and his two sons who were seen writhing with snakes in the original. Titian's woodcut showed them replaced by monkeylike creatures which may have been baboons, but this cannot be determined for certain. Michelangelo made some carvings of the "Dying Captive" and "Heroic Captive" with a baboon on the side of each, presumably to suggest that the baboons represented man's lower soul, the bestial part of his existence.

Pieter Breughel depicts monkeys in some of his paintings, and a number of paintings of other artists in the twelfth and thirteenth centuries have included monkeys. There are even paintings of Adam and Eve in the Garden of Eden with a monkey sitting alongside them eating apples.

A painting by Teniers the Younger, which is in El Prado in Madrid, shows a sculptor in the form of a baboon making a model of the devil. Both Toulouse-Lautrec and Picasso have figured baboons in their paintings. It is thought that Jonathan Swift, in *Gulliver's Travels*, intended the Yahoos to be Primates of some sort, possibly gelada baboons.

Aldous Huxley in his satire on death, funerals, and cemeteries, *After Many a Summer Dies the Swan*, describes a frantic search by his hero to find the secret of eternal life; he then discovers that the human, when preserved long past normal old age, regresses to ape form and

175

behavior. Huxley opens another of his books, *Ape and Essence*—a projected look at life after World War III—with a movie audience of baboons watching a film of baboon people: financiers, housewives, technicians, bishops, and soldiers, the latter leading Einsteins and Pasteurs on chains. A baboon singer seductively chants, "Give me, give me, give me, give me detumescence." The narrator speaks:

> Surely it is obvious
> Doesn't every schoolboy know it?
> The ends are ape-chosen; only the means are mans.
> Papio's procurer, bursar to baboons,
> Reason comes running, eager to ratify;
> Comes, a catch-fart, with Philosophy, truckling to tyrants;
> Comes, a pimp for Prussia, with Hegel's Patent History;
> Comes, with medicine to adminster the Ape King's aphrodisiac;
> Comes, rhyming and with Rhetoric, to write his orations;
> Comes, with a Calculus to aim his rockets
> Accurately at the orphanage across the ocean;
> Comes, having aimed, with incense to impenetrate
> Our Lady devoutly for a direct hit.

Jean Giraudoux, the French playwright, in *Tiger at the Gate* has one of his characters give a description of war as "like the bottom of a baboon. When the baboon is up in the tree with its hind end facing us, there is the face of war exactly; scarlet, scaley, glazed, framed in a clotted filthy wig."

There is no doubt that the swollen bottom of the female baboon and also of the female rhesus monkey at the ape's time of heat, or estrus period, is an incredible and an uncomfortable sight for humans.

Africa contains an enormous number of what are called the true baboons. These animals are not usually found in the forest as such; they are found especially in the grassland areas where there are a few trees. There are six types of baboon: the olive baboon, the yellow baboon, the Guinea baboon, the chacma baboon, the gelada baboon and the sacred baboon. Some of these were given different generic names but they are all now grouped together under the generic name of *Papio*. *Papio anubis* is the olive baboon; *Papio cynocephalus* is the yellow baboon; *Papio papio* is the Guinea baboon; *Papio ursinus* is the chacma baboon; *Theropithecus* is the gelada baboon and *Papio hamadryas* is the sacred baboon.

The chacma baboon occurs over most of south Africa, and the anubis (olive) baboon, extends right across northern Africa below the Sahara from the Red Sea and the Indian Ocean on the east to the

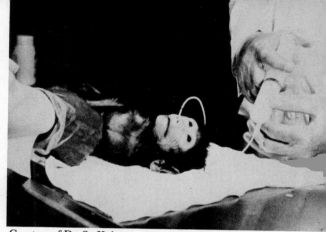

Sick baby baboon gets medical attention at the Southwest Foundation for Research and Education.

Courtesy of Dr. Sy Kalter

Atlantic Ocean on the west. At either end of their range, the hamadryad, or sacred, baboon is found. In the west, in the region of Guinea, the Guinea baboon is found. Between the chacma baboon of South Africa and the lower ranges of the olive baboon is the yellow baboon.

In these vast areas, a whole range of geographical habitats are found from open forest and savanna to scrub and semidesert and even bare outcrops of mountains. In all these areas, the baboon thrives. Ivan Sanderson makes this comment about them:

> They are one of the few forms of life on this planet that are not much unsettled by our activities and that neither retreat before them nor make any attempt to comply with them. Baboons just carry on as if we were not around, making what use they choose of our agricultural and other efforts and often feeding and multiplying at our expense. In some parts of Africa, and notably in the Union, they have become not only a pest but a menace.

The baboon has a body rather like a dog, with longer arms than legs, so that its back slopes back when it walks just as it does on the drills and mandrills, but the baboon's tail is much longer. (The very small tail carried by the drill and the mandrill is actually held out with the top curved upward.) Baboons have a large head with a muzzle rather like a dog's (the reason they are called Cynocephalidae), and there is a pair of very small eyes which are set close together. They all have a similar sort of dull coloring; it will be olive, yellow or brown, with a reddish type of hue to it. The face is sometimes black and sometimes pink. All these animals have large ischial callosities. They are highly gregarious animals and move in very large parties which are very highly organized.

Moving in such groups, they feel very secure and will attack property and crops to a very destructive degree. They ignore animals such as the ostriches and the antelopes; but they enjoy the eggs of ostriches, and they also eat honey and insects. Their paramount enemies are the carnivores—the lions, leopards, and cheetahs.

Chimpanzees will often kill young baboons. Jane Van Lawick-Goodall has described in her book *My Friends, the Wild Chimpanzees* how one of the large male chimpanzees in the group she was observing grabbed a young baboon by one leg and killed it by bashing its head against the ground and subsequently eating it. However, she found that in general, troupes of baboons and chimpanzees ignore each other, although sometimes young baboons and young chimpanzees will actually play together. Jane Goodall has described a very interesting relationship between a young baboon which she called Goblina and a young chimpanzee called Gilka. On one occasion Gilka had pulled herself away from a chimpanzee grooming session and had climbed a tree; from this vantage point, she peered down at a group of baboons that was passing 400 yards away. Then she saw a little baboon detach itself from the group and come toward her tree; this was Goblina. Gilka slid down the tree and chased after the little baboon. They embraced each other and pressed their faces together, and Jane Goodall said it was probably a playful overture but that it looked to her more like a greeting between a couple of old friends. She mentioned that after this, the animals played together very actively; they chased and startled each other and rolled about on the ground, and they wrestled and tickled. This relationship lasted for several months, but eventually they drifted apart, particularly when Gilka was in her fourth year. She was bigger than the baboon, and she then played so roughly that the baboon got frightened. Also, the baboon had matured faster than the chimpanzee, which is usual, and by four years was practically full-grown. So eventually Goblina left and rejoined her group of baboons.

Julie MacDonald has written a very interesting book, *Almost Human*, about her life with a young hamadryad baboon. She found it in the possession of a New York animal dealer, one of six baboons that had been imported from Arabia. It had been in his shop for eight months and was developing what is called cage paralysis. This is thought to be caused by incarceration in a small cage, and it is probably a combination of dietary deficiency, including vitamin D deficiency, cramped quarters, lack of exercise, and in the case of baboons, considerable stress from social deprivation, since they are fundamentally social animals, used to very highly structured baboon society. Boris Lapin, the director of the Russian Primate Center, has stated that you seriously stress a baboon simply by isolating him from his social group.

Julie MacDonald refers to what she describes as "the worrisome guilt" of ordering a Primate directly from the wild, knowing that as a result of its delivery, its mother and possibly its family group may have been killed. She quotes Karl Hagenbeck, who described how baboons were caught at the turn of the century. The group in this instance was driven into a trap set near some water at which animals had been drinking, and "the astounded prisoners sit benumbed with terror and unable to move; then they anxiously begin to seek an exit. The herd outside, no less surprised, flees at the first alarm; but they soon return and congregate around the trap, urging the captives with earsplitting yells and grunts to find their way out. Some of the boldest jump right onto the top of the trap and appear to carry on an excited conversation with their friends inside. . . ."

She describes how the captors used long sticks with forks on the end to pin the animals to the ground by the neck so that they could make them secure. Many of the older creatures were shot, while the younger animals were driven into small cages. She also describes how Africans often caught baboons by hunting them continuously until the younger animals began to lag and were captured in a virtually exhausted state. But, she points out, and rightly, this was very dangerous since the dominant males often hung back to protect the young under such circumstances. She also mentions one of the Roman methods of capture which seems a little more desirable than those described, and that was to put out a number of containers of wine which the baboons drank and became intoxicated on; then all that the hunter had to do was pick up the drunken animals.

Although Julie MacDonald's description referred to the turn of the century, a tremendous amount of cruelty still goes on in the capture of animals in the wild. It must be a very terrifying thing for a group of baboons or a group of monkeys to be caught, their social group decimated, and the isolated animals sent all over the world. Not the least of this problem is the trouble that the animals get into and the cruelty with which they meet during the shipping process. I remember a few years ago, we ordered some apes from Africa (we do not order apes anymore, we breed our own), and on one occasion I heard from the dealer that he was unable to deliver the animals because the plane had been grounded at Orly Airport in France for some time and that

Young baboon gets medical checkup at the Southwest Foundation for Research and Education.

Courtesy of Dr. Sy Kalter

Baboon born under special hygienic conditions so it is free of germs gets bottle of germ-free milk.

Courtesy of Dr. Sy Kalter

the animals had been left inside the plane in the heat without attention and without food or water until practically the entire planeload of animals had died.

In October, 1972, an explicit article by Lionel Kent was published in *Male* magazine describing the shipping of animals from South America. Kent says that many "pet merchants and shippers have found a gold mine peddling animals in America, so they jam them in and ship them out under the most inhumane conditions. So what if nine out of ten animals die in transit—there are still more to be caught and sent. . . ." He then describes in detail the state of animals arriving on a plane at Miami Airport from Peru and goes on to say, "Peru is hardly the only nation to deal in the flesh of jungle animals. Pet merchants and shippers are active in Africa, India and anywhere else they can make a profit. In 1971 alone, according to U.S. Customs records, 330,000 animals reached our shores alive. This means, according to experts in the field, that nearly a million died on route—because the ratio of death to survival is three to one."

We know that in the legal trade in orangutans from Southeast Asia whole families were shot in order to obtain one baby. For the one baby that arrived safe in Europe or in America, probably four or five babies died on the way. Such a procedure is a very effective means of destroying this species altogether. Fortunately, the import of orangs into the United States has now been banned, and although export from their country of origin has been banned for a long time, there are smugglers at work to get the orang into places where it will command high prices. Even today, primates—at least baby chimpanzees and gorillas—are frequently obtained by shooting the adults in the group and capturing the babies. The adults are scarcely ever captured and sent abroad because the caging has to be immensely strong and extremely heavy, so that air transport of the animal is prohibitively expensive. It is also very expensive to send an ape on long journeys by boat, and its chances of surviving a boat trip are small because of the poor conditions under which it is kept. Typical of this is an incident Lionel Kent described in which a ship arrived in New York after a voyage of thirty-three days from the port of Mombasa in Kenya. There were $50,000 worth of animals in the cargo; some of the animals were to be sold to zoos in different parts of the United States, and some of them were going to animal dealers. The animals came from a tropical climate, but were all put on the top deck of the ship without any provisions for keeping them warm. This trip was made in the dead of winter. There were eighty animals altogether, and a giraffe, two gazelles, and eighteen zebras died as a result of the cold. Many of the others were in a very bad state when the ship docked in New York. Some were too weak to stand, and those that could not stand did not have room in their small cages even to lie down. Many of the animals had open sores and wounds. These apparently had been caused when the ship was on the high seas and the cages broke loose and crashed about the ship. The feed given these animals was, according to Lionel Kent, inedible. A number of the animals had died at sea and had been thrown overboard, and apparently the animals had howled with pain and suffering through most of the journey across. Not only were the animals exposed to the cold, but they had no protection from heavy rain. Although the people responsible were indicted when suit was brought against them, the indictment was dismissed by a federal judge.

In recent years, there has been a great improvement in shipping animals. Most shippers make every effort to see that the animals are well treated and that they arrive in the United States in good condition. The good animal dealer also makes a special point of knowing the condition of the animals before they are put on sale to scientists or to the pet trade.

The use of Primates as pets is not to be encouraged. Certain monkeys carry viruses which might be responsible for causing lymphoid cancer, and the possessor of these animals could very well catch such a virus. Monkeys can carry a variety of viruses, including the deadly B virus. They all carry a variety of worm infestations which can be communicated to the pet owner or his children. However, there are some monkey pet owners who have really been dedicated to the welfare of their pets, and because of their insight and their ability to record and describe their experiences, they have materially helped us understand the minds of these animals. It is the casual monkey pet owner that is a menace to himself and his monkey. Julie MacDonald belongs to the first category, and what she has to tell us about the mind and behavior of her pet baboon, Abu, is valuable in helping us understand these animals. Abu was a young female hamadryad baboon, and she cost $50. Julie MacDonald took the pain and time to nurse her from her cage paralysis back to perfect health.

One of the problems of keeping Primates—and one that is very important in maintaining the large Primates such as those at the Yerkes Primate Center—is that these animals are very susceptible to virus infections. At the Yerkes Center, we take extreme precautions against transferring human viruses to our animals, including the

Abu—Julie Macdonald's hamadryad baboon. From *Almost Human* by Julie Macdonald (Philadelphia, Chilton Books).

banning of casual visitors who could bring in a strange virus that could be disastrous. Our staff wear face masks when they are handling young animals that might react badly to a human virus, and animal caretakers with colds or influenza are usually sent home. Despite all these precautions, the animals do develop respiratory infections which require expert veterinary treatment.

Julie MacDonald met this problem with her baboon, Abu. She found her one morning huddled back into her sleeping box, hardly able to move and breathing in short gasps. She had difficulty in getting a veterinarian to visit the house. She managed to get one to write her a prescription for penicillin, which she injected into the animal herself, and following the injection, the animal turned a grayish color, her pupils dilated, and she looked as though she had penicillin shock. An appeal to another veterinarian resulted in a house visit and the prescription of a small dose of cortisone every half hour. The same night it was thought that Abu was likely to die, and finally the new doctor recommended that she be given aspirin every few hours. This was administered throughout the night, and shortly after the first tablet, the fever began to subside and Abu slept. In the morning she appeared out of danger and, though weak, was able to sit up and eat some breakfast. Eventually she became well and restored to complete health.

One night, about a week after this, Abu managed to break her right leg and had to be taken to the animal hospital and have the leg set. She adapted very well to the restrictions of the cast and did not make any effort to break it or tear it away.

These two events seemed to make all the difference in the reaction of the animal to human kindness and also toward her animal companions, which included a dog and a cat. Abu began to show affection to all of them. After the cast was removed, it took her about a week to regain use of the leg, and then it became apparent to everybody that she had accepted the family as part of her troupe. Being in a troupe is part of a baboon's normal life, and Abu's recognition of the humans and animals around her in *her* troupe was the key adjustment she had to make to fit happily into this new environment. As a result of this adaptation, she wanted to be constantly with the group, and it became devastating to her to be left alone for any length of time.

The terror that baboons feel at being left alone is probably related to the fact that if this happens in the wild, it is pretty much certain death from predators such as leopards.

Abu also developed grooming as a serious business, and she would sit on a human lap or shoulder and pick through the clothing and hair with a great deal of care. She was very interested in buttons and zippers and also in eyelashes and eyebrows. Grooming can begin in baboons as early as four to six days after birth.

Abu also developed the habit of eating at the dinner table and sitting in a high chair to do so. She was not trained to do this, but on one occasion when she approached the table when her owner was eating, her owner pointed to the high chair and Abu immediately climbed up onto it and sat. After that, when the family was eating and the high chair was at the table, she would automatically sit in it. It is also interesting that she would show no antagonism to being put on a leash. From the beginning, she had a collar slipped around her waist and a leash placed on it, and she accepted this situation simply as a part of the new life. Her owner records that subsequently, as soon as the collar was put around her waist, she would turn around and stand to have it fastened.

She was very catholic in her food tastes and especially enjoyed spaghetti; she would place one end of it in her mouth and suck it up into her cheek pouches. Among the things that appealed to her apparently sophisticated taste were Caesar salad, spinach, oatmeal, cooked chicken, beef, eggs, cheese, and all kinds of fruits, vegetables, and desserts. In the wild, baboons eat grain most of the time if they can get it. They also eat birds, if they catch them, and birds' eggs. They have also been said to eat slugs and snakes.

A significant event, particularly to the owner of a baboon, especially a female baboon, is its attainment of sexual maturity. For some reason or other, baboons have a reputation for sex, and early writers claimed that baboons liked to satisfy their strong desires in public and that they were particularly impudent in the presence of women. As she became sexually mature, Abu started presenting to the owners and also to the dogs and cats. Her sexual presentations caused some problems in the household because when she presented to the two dogs, both of which were males, her "flirtations were entirely too provocative," and this led to fights between the two dogs. It was probably the readiness on the part of a female baboon to present that led the baboon to be labeled as a sexy animal.

Julie MacDonald had a number of exotic animals in her house, apart from her baboon and her dogs and cat. She had a pygmy marmoset and a saki monkey from South America and also a kinkajou and a jaguarundi. The pygmy marmoset and the saki monkey were Primates, and Abu got along extremely well with both of them and played with them, groomed them, and was very protective of them. The kinkajou is a bearlike creature which is not a Primate and the animal which Julie MacDonald had was a fairly solitary animal, a characteristic also found in our own kikajou.

I have not had any personal experience with the jaguarundi, a catlike creature, also from South America. It does not meow like a cat but simply emits a high-pitched sigh. It has a black and white cape and is capable of moving at great speed. Julie MacDonald described him as dashing around her room, pushing off the walls as he rounded the corners and said that he reminded her of nothing less than a motorcycle racing in a steeply banked arena. She found him a very affectionate animal, more so than a domestic cat, and with a type of intelligence more like that of a dog. She had no trouble in training him with a leash and quickly taught him to use a box or newspapers for his toilet. He liked to retrieve and to have things thrown to him, which he would catch in midair. He was normally nocturnal, but she was able to switch him into a daily routine.

I have mentioned that the presentation of the rear end is a token of submission, but in Abu it was also used in some cases as a general greeting. Abu used this procedure if she wanted an object she was not

Gelada baboon family.

allowed to have. She would distract her owners by turning up her bottom and then quickly grab the forbidden button, cookie, sip of wine, ice cube, or magazine when her owner's attention was diverted with her bottom. She could steal, when she wished to, with a very great speed and dexterity. Apparently, female baboons have also been known to use this technique to distract the male baboon so that they can eat food which he normally would have the right to eat first. This in a sense might be termed a type of prostitution.

Julie MacDonald collected a considerable amount of detail about baboon language. She refers particularly to the bark which is used in the wild as a greeting whenever one of the troupe rejoins the troupe after an absence. Abu would bark at the sound of her owner's car, and she also used a bark to greet the day. Shortly after dawn, she would give a single bark, and in the daytime when the sun came out from behind a cloud, she would bark. On occasions, her owner would bark to her from a neighbor's house and would get a volley of five or six barks in return from Abu. There are about fourteen other vocalizations recorded by various authors for the baboon, and there are probably many more. The fourteen, however, are the principal vocalizations.

Abu also showed a number of ritual movements which were apparently threats. The first threat was to be a raising of the eyebrows, sometimes with the head thrusting forward; then there was the flicking of the hand, primarily a movement of the wrist, and simultaneously she could partly open her mouth. These threat gestures are well known to baboon students. If the animal being threatened is of a lower rank, it is possible that the eyes would be averted from that animal. A more severe form of threat is slapping the ground with both hands and using the feet as well; the mouth open on top of that adds, of course, to the intensity of the threat. Then there is a yawn in which the animal's teeth are completely exposed. It appears that the supreme threat gesture is the grinding of the teeth. Abu was only seen to do this once, and it was when the owner's Doberman Pinscher approached the pygmy marmoset that Abu had wanted to adopt. With the grinding of the teeth, there was a raising of eyebrows, flicking of hands, and eventually Abu screamed at the dog. The dog got the message and retreated without any hesitation.

Abu showed many interesting reactions; for example, with her first Christmas, the presents for her were a toy telephone and a new monkey doll which was larger than she was. Both these articles had been gift-wrapped and had been placed under the Christmas tree with all the other packages. The owner's children unwrapped their presents first, and Abu showed great interest in the contents of each package as it was undone. Finally, she was given her presents and insisted that she do her own opening, pushing away any hands that stretched out to help her. Now it is interesting that when she removed the phone, she wrapped the cord around about her body and dialed the phone, as though she recognized it as being similar to the phone in the house, which she must have observed the humans dialing. The monkey doll, however, since it was bigger than she, scared her, and she would not make contact with it until the rest of the family had patted and stroked it. She sniffed first at its bottom and then touched it, and it was not long before she had it in her embrace and refused to be parted from it. She was very fond of balls and liked to play catch with them and would give a pleasurable grunt when she made a successful catch. She also invented her own ball game.

Other people have studied the mental and manipulative abilities of captive female baboons. They are very observant and learn quickly, even compared with humans. But their abilities are definitely limited. The manipulative skills which the animals use in captivity and under tests are all those that are based on innate patterns of action, such as food gathering, biting, fighting, walking, and climbing, so that she had grasping and lashing movements, but otherwise the manipulative scope of their hands was limited. For example, they can twist the forearm, but could not twist it as much as humans, and they can move their hands less sideways than humans; however, the hand can be bent ventrally a great deal farther than it can in humans, but they cannot bend the hand backward. The fingers, however, can be bent backward passively, to a point at right angles to the dorsal surfaces of the hand; that, of course, is not possible in most humans. Baboons cannot spread the fingers as humans do, and they cannot work their fingers independently. However, they have been known to use sticks as tools and rake in objects out of reach.

One of the most interesting types of baboon is the gelada, known scientifically as *Theropithecus gelada*. Geladas are very large and rather heavily built animals. They look like baboons, but they can be distinguished from both the ordinary baboons and the mandrills by the shape of the face, especially the arrangement of the nostrils. Both sexes have a naked area in the pectoral region (on the chest) which is colored bright red. Desmond Morris suggests that this red patch is a sex signal. He points out that the female gelada indulges in a self-copying device:

> Around her genitals there is a bright red skin patch, bordered with white papillae. The appearance of the vulva in the center of this area is a deeper, richer red. This visual pattern is repeated on her chest region, where again there is a patch of naked red skin surrounded by the same kind of white papillae. In the center of this chest patch, the deep red nipples have come to lie so close together that they are strongly reminiscent of the lips of the vulva. (They are indeed so close to one another that the infant sucks from both teats at the same time.) Like the true genital patch, the chest patch varies in intensity of color during the different stages of the monthly sexual cycle.

Morris suggests that since the wild gelada baboons spend a good deal of their time sitting upright, more so than most monkeys, the sexual signals which would normally be transmitted by colors and swellings on their bottoms are hidden so they are reproduced on the chest and used to transmit sex signals to the other members of the group. As he points out, a number of species of Primates have brightly colored genitals, but very few of them have these bright colors transferred to

Male gelada baboon at the Yerkes Primate Center.

the head and chest. His suggestion about the sexual significance of this red spot on the chest of the gelada is of interest, but it is difficult to understand why the male has the same red patch.

The gelada was first discovered in 1835, and even by 1840 it had been classified among the baboons, but it was elevated, in 1943, to a separate genus. It did not reach Europe until 1877, when a group was put on exhibition at the famous Alexandra Palace in London. Later on they became an exhibit in the London Zoo.

The male gelada has a beautiful mane on his back and shoulders. The various investigators who have studied this animal have pointed out that the male geladas will examine the chests of female geladas who are in estrus prior to copulating with them. However, experts say that there is not a complete correlation between the estrus changes which take place on the skin of the chest and those which take place in the perineum. In fact, the reddest perineum is said to occur in nonestrous females and in those which are lactating. Copulation, in any case, occurs only after the male and female have had a very long session of grooming each other.

Geladas seem to be completely quadrupedal and they live entirely on the ground. Their habitat is mostly grassy land high in the mountains, especially in the central and northern parts of Ethiopia. They scarcely ever climb trees, even when they are in captivity and are supplied with artificial treelike structures. At the Yerkes Center, we provided the gelada colony with a series of wooden structures for them to climb, but all they did was gnaw around the supports until all the structures had collapsed. Dr. Osman Hill reports that there is a colony of these animals at the zoo in San Antonio contained in a compound with a single tree, which the animals do climb and use as a lookout post.

189

Geladas like living together, and their troupes may contain as many as 400 animals. These troupes, I suppose, would probably best be described as herds, but they are not especially organized as such. They simply consist of units composed of one male with two or three females which represent his harem. There are also all male groups, infant play groups, and juvenile groups. It is interesting that males that do not have a harem of their own will often play with and look after the young animals.

The heads of geladas are rounded, and their muzzles stick out like dogs'. The muzzle is not pointed; it is rather globular, and the nostrils are upturned and behind and above the bulbous lips. Geladas have white eyelids and, as we have mentioned, red chests. The canines are three inches in length and extremely sharp. But though they are so terrifying to look at, geladas are not really any more aggressive, and possibly less so, than other baboons. The group of geladas at the field station are quite friendly to humans. They will often come over to the wire which separates them from the rest of the world to greet and will vocalize for quite a long period of time in a sort of grumbling low undertone. I have searched the literature to see if anyone who has ever made a pet of a gelada has published a detailed written record but have not been successful. Ivan Sanderson mentions that they show a great deal of affection for each other when they are in captivity and that they also are known to adopt other animals, particularly small ones, as their own personal pets. Sanderson says, "A friend of mine who has observed them in Abyssinia offered the suggestion that they, like the lioness, are overwhelmingly powerful in their own sphere; that they seem, again like that animal, to have developed a certain aloofness to all lesser breeds and are consequently rather indifferent to their activities unless they interrupt the even tenor of their own lordly lives."

The other types of baboon are divided up into dominant males, subordinate males, females, weaned juveniles, younger juveniles, and infants. When a baboon group moves across the savanna, the subordinate males act as the protectors at the front and rear of the group, and the infants and the females remain in the center of the group, with the dominant males close to them. Even when resting, the group more or less retains this arrangement. However, when danger threatens, the dominant males will move through the troupe to take the leading positions as scouts and leave the subordinate males to do battle with the enemy. It is difficult to surprise a group of baboons because every few seconds they look up from whatever they are doing and scan the country. Since there can be as many as 200 baboons in a group, there are 400 eyes keeping watch on what is going on. It, therefore, becomes very difficult for a predator, certainly a human, to get very close to such a group. When a group of baboons is on the move, sometimes a mother baboon with a baby finds that she has to use one arm to hold the infant and so has to walk on three legs, causing her to move more slowly than the group. When she has to drop back, an adult male usually drops back and stays with her until the three of them regain the main group. The role of the adult male as a defender is absolutely vital to the survival of the troupe as a whole, especially to the survival of the more or less helpless animals. These would include females with their young babies, small juvenile animals, and sick or injured animals. The males weigh more than twice as much as the females, and this is an obvious advantage for defenders. A group of males will attack even a leopard if they feel the animal is menacing the troupe.

The baby baboons at play prepare for their role as adults. For example, the older juvenile females do not take part in the semi-serious, but actually mock, fighting which is characteristic of the play that the older juvenile males indulge in. It is in this type of play that the males are taught how to fight. By the time they have grown up and developed the powerful muscles that move the lower jaw and by the time their canine teeth have come through and grown to a respectable size they have already had many years of practice at fighting. The whole setup of the group is directed toward the juveniles' being prepared for their adult roles.

The troupes of baboons have a number of subgroups based on the age, sex, dominance factors, and personal preferences of individual animals. When the group is resting or when it is quietly feeding without danger, most of the adult members of the group will gather in small groups and will just sit or perhaps groom each other. Similarly, juveniles of the same age will gather into small groups, which have been called play groups by Sherwood Washburn and Irven DeVore. These play groups of juveniles will spend the day eating, playing, and even resting together.

During periods when very little happens, other members of the groups may come to the area where the mothers and the adult males are sitting and perhaps sit alongside them or groom them. There is no herding necessary for the male baboons because few of the members of the troupe will stray very far from where the big males are. The more peripheral members of the troupe include adult males who are subdominant, some of the older juveniles, and females that are pregnant or are in heat. These peripheral adults may, in fact, leave the troupe for various, usually short periods, and sometimes the peripheral groups will, if the troupe is moving slowly, move on faster than the main group to find a new area for feeding. This means that they could be separated from the main troupe by as much as a quarter of a mile and may remain away from it for as long as an hour. It is rare for any other members of the troupe to leave the main group unless they are escorted by the big males.

A troupe among the trees does not seem to be particularly organized, but when it moves out into the open plains, a pretty clear order quickly becomes obvious. A troupe approaching the trees from open plains uses very special caution, particularly since this is an area where leopards may be found. They like to sleep in tall trees, which are found only where there is water near the surface, usually where there is a pond or a river. At the base of the trees in these situations there is fairly deep undergrowth, and this is where the predators such as the leopards often lurk. If a predator attacked a troupe head on or even around the periphery, he would run head first into the adult males and then into a second group of adult males before he could actually reach the defenseless members of the troupe in the center. When the peripheral adult males run into a predator, they give a series of alarm calls. In an emergency, all the adult males will be actively on the defense. Washburn described how on one occasion he saw two dogs run up behind a troupe of baboons, barking at them. At this, the females and the juveniles moved on ahead hurriedly, but the males continued walking slowly so that in a very short time there was a group of something like twenty adult males interposed between the dogs and the more defenseless members of the troupe. There was need for only one male to turn on the dogs, and the latter decided that there were more interesting things in the opposite direction. Once Washburn also saw three cheetahs approach a group of baboons. It also took only one adult male baboon to turn around toward the cheetahs and give a loud defiant bark and display his canine teeth, and the cheetahs also found other urgent business in the opposite direction and trotted away without further ado.

If, when baboons are moving toward the trees in which they sleep, they come upon an animal that is dangerous to them, the troupe will stop and wait, while some of the males move ahead and perhaps look for an alternative route which will not bring them into direct confrontation with the predator. While they are doing this, the young juveniles and the mothers will stay back in the safekeeping of the peripheral adult males. When the dominant males finally get back, the original order is reestablished, and they move off along the route that has been reconnoitered by the dominant animals. Baboons are probably safest when they are in the trees, although leopards have been known to attack them even in this situation. They do not seem to be too disturbed by cheetahs, dogs, hyenas, or jackals, but the sight of a lion is likely to lead to flight for the entire troupe. They make no effort to resist the lions when they are on the ground, but once they are safely in the trees, they will bark at the lions and threaten them. However, if they are close to trees, they will feed on the ground even if they are only a hundred feet from lions.

Drills and Mandrills

The drill and the mandrill are very similar to the baboons in many ways, but there are some fundamental differences. The drill was recognized by the famous French naturalist and anatomist Baron Cuvier in the early part of the nineteenth century, when he was superintendent of the Paris Menagerie.

Drills and mandrills are known popularly as forest baboons or stump-tail baboons. The drill, which has a black face, is called scientifically *Mandrillus leucophaeus,* and the mandrill, which has colored stripes on the sides of its face, is called *Mandrillus sphinx.* They have a relatively limited distribution in Africa. Drills are found mainly in the forests of the northern part of the Cameroons. Mandrills are found throughout Gabon and as far south as the Congo; they also extend as far east as the Ubangi River. The drills and mandrills are immensely strong animals with tremendous muscles. They are very heavy and very compactly built. The shoulders are set higher than the pelvis so that they slope backward when they stand on all fours. They both have tails, but the tails are small and usually erect. The buttocks are hairless, red to pink in color, and in the female this skin swells immensely and reaches an apex during the period of tumescence in the middle of the sexual cycle.

The hair on the body is thick and straight, but the individual hairs are rather fine. The coat varies in color according to the species of the animal. In the case of the drill, the hair is a sort of olive brown, and in the mandrill it is either charcoal or blackish gray. The color of the face of the mandrill is described by Ivan Sanderson, who says:

> The skin is bright blue, that on the swollen involuted areas of the face being a vivid sky blue. The nose region and nostrils are lacquer red, and this color may extend to the lips. The whiskers are pure white and the beard and parts of the ruff behind the angles of the jaws are orange. Fur on the top of the head and of the main body are

193

Portrait of a drill at the Yerkes Center.

Close-up of male mandrill at Yerkes Center showing details of markings on face.

brownish, but the undermane, chest and limbs are jet-black; the naked and monstrous buttocks are purplish-blue surrounded by brilliant scarlet.

The young of the drills and the mandrills are very difficult to distinguish from each other, and very often the baby mandrills can only be differentiated from the young drills by the fact that there are creases in the black skin on the side of the face. Both species are completely omnivorous. The mothers are very careful of their offspring. The males are constantly on watch for danger. They are likely to attack humans who approach the group, sometimes throwing stones and sticks. They have also been known to scrape material from the ground with their feet and propel it by the scraping motion at an animal or a human being that has frightened them.

Drill at the Yerkes Center giving a yawn threat—note the saberlike size of the upper canines.

Both the drills and the mandrills look extremely ferocious, and yet, in my experience with these animals in captivity, this is not so. At the Yerkes Primate Center, we at one time had a very large, immensely powerful drill called Ajax, with canine teeth three or four inches long. Dr. Irwin Bernstein was able to maintain a friendly relationship with this animal. Unfortunately, Ajax developed a sudden fever which turned out to be an attack of infectious arthritis. It left the poor animal completely crippled, and from being a big, muscular, powerful creature, he dwindled to a mass of skin and bone with his knees "set" in a bent position so that they could not be moved. He hobbled around for quite a time until we thought it would be a kindness to put him out of his misery.

I first came in contact with drills in the laboratory of Lord Zuckerman, at that time Professor Solly Zuckerman, in the Department of Anatomy at the University of Oxford. There were a number of young and older drills in this laboratory, and one of them, a very big male, would rush, with his mouth open, at your hand if you put it in the cage. But if you had the nerve to leave your hand there, he

A male mandrill at the Yerkes Center poses for camera.

would put his mouth around it, close his teeth on it, and just gently pinch the fingers or else simply mouth the hand with his lips. His appearance was very misleading, and of course, this was a fine trick to be able to demonstrate to visitors. There were not many of us, however, who had the nerve to undergo this remarkable test of the animal's good intentions. Some of the technicians, unbeknownst to the powers-that-be in the Anatomy Department, used to put this big drill on a lead and take him walking through the University Parks, where I am sure he caused the greatest of amazement to the elderly dons and their families during their Sunday afternoon stroll through the parks.

There seems to be little evidence that the drills or the mandrills were known to the people of ancient Egypt or Greece, although there is some evidence that Aristotle described a mandrill, possibly from reports of the Phoenician sailors who had ventured outside the Strait of Gibraltar and gone down the west coast of Africa. It has been suggested that the Chinese of the Ming Dynasty (1368–1644) knew about the mandrill, but this is not certain.

Conrad Gesner, a Swiss scientist, published a book called *Quadrupeds* in 1551, and in it he has drawn an animal which is undoubtedly a mandrill. This is probably the first certain reference in history to either the drill or mandrill. Some authors, however, think that Gesner was confusing the mandrill with the hyena. During the seventeenth century, the Royal Menagerie at Copenhagen had a male mandrill on view, and when he died, the famous anatomist Caspar Thomèson Bartholin dissected the carcass and represented the animal in one of his publications.

An anatomist named John Ray described the first mandrill to be exhibited in England, or at least the first of which we have any information. His description, which follows, is taken from Osman

197

Drill at the Yerkes Primate Center.

Hill's work on the Primates, volume 8: "a sort of dog-faced monkey or rather Baboon with tusks or long Dentes canini, sho'd about London, called a man-teegeran. 1702, blue about the cheek; red about jaws, belly bare, his testes of a vermilion color, so his naves."

In addition to the man-teegeran or the man-tiger, the female mandrill was described as a woman-tiger by the famous anatomist John Hunter during the eighteenth century. Other mandrills were exhibited around London during this period, and there was one at Vienna at the same time. Also, the famous naturalist Buffon, in the middle of the century, had both male and female mandrills in his Paris menagerie. He, too, found that although they were bigger than other baboons, they were much gentler animals and easier to handle. Buffon, in fact, used the name "mandrills" for these animals for the first time.

In the drills, the lower lip has a pale patch between the back and the beginning of the lip, which becomes red in the adult animal. The drills are forest animals and usually travel in family parties. They are usually not very aggressive toward man. They vocalize a lot and use a great variety of sounds in communicating with each other. They also have a bark not too different from that of a big dog, and they give a kind of roar on occasion or sometimes a growl.

The mandrills also live primarily in the forest, but they will leave it to climb about the hills and rocky areas and, like the drills, travel around in small groups, although troupes of up to fifty animals are known. The males often are at some distance from the group, probably acting as scouts for any aggressor. When they see one, they will growl and open the mouth in a type of yawn which shows their

198

lethal equipage of teeth. They would be a match even for a leopard. When they are in a defensive mood or enraged, they are said to be pretty much as powerful as a gorilla and can do their damage very much faster than a gorilla can.

It is probable that the colored faces of the mandrills help them distinguish themselves from the black-faced drills. The red coloring is a result of the distribution and the richness of blood vessels. The more excited the animal gets, the brighter the red. Mandrills look for ants, small animals, and insects by turning over rocks and debris. In the forest at the western edge of the Congo basin, which is their habitat, they raid the farms of the native tribes in the dry season because of the shortage of fruit. Males may weigh as much as 100 pounds and females about 50 or 60 pounds. The males are also much brighter in coloring than the females. The gestation period varies between 220 and 270 days, and the females are very dedicated to their infants and take good care of them.

Their life-span is probably between forty and fifty years, probably not much over forty. The male has enormous canine teeth and will stand his ground against a leopard or the best-trained hunting dog. They do not seem to be fazed by guns fired near or at them, and when they are kept in captivity and have continuous contact with humans, they, like the drills, form very good relationships.

Sabater Pi of the Barcelona Zoo has said that although dogs have been attacked by mandrills in the wild, he has never heard of a man being attacked by a mandrill, either in the wild or in captivity. The mandrills, however, get a bad time in West Africa because in the area where they live, their meat is preferred over any other type of meat by the natives. Dried and smoked mandrill meat is found in quite a number of the West African markets. Because of the encroachment by the tribes into the forest home of the animals, they have a very difficult time ahead and in a few years will probably become an endangered species.

VIII. The Old World Monkeys—
Guenons, Snub-Nosed Monkeys,
and Leaf-Eaters

IN AFRICA AND ASIA there are a very large number of different sorts of monkeys in addition to those we have already described. The macaques, mangabeys, and baboons are known as Cynopithecoids, and the Cercopithecoid monkeys and the Coloboid monkeys constitute the rest of the Old World monkeys. The guenon monkeys, the talapoin monkeys, and the swamp monkeys are Cercopithecoid monkeys.

Talapoin monkey (dwarf guenon) and baby.

Guenon monkeys were very well known to the Mediterranean peoples before the birth of Christ; in fact, they were probably the first monkeys to be known to the people of that part of the world. Two or three thousand years ago the Primates of Africa were distributed much more extensively to the north than they are now. They were almost certainly in Libya, possibly in Mauritania, and certainly in the northern part of Egypt. The green monkey is probably one of those which was very well known in the Mediterranean; it is known scientifically as *Cercopithecus sabaeus*.

The guenons represent one of the largest groups of monkeys. There are at least 100 sorts of guenons and possibly some that have not yet even been discovered.

The Egyptian monuments and many Egyptian writings depict or refer to the number of different monkeys, even as far back as 1500 B.C. Monkeys, as well as baboons, were used as pets; they were sometimes the object of adoration and sometimes used in barter. They were also used as presents from one Egyptian potentate to a potentate in a distant country.

The grivet monkey (*Cercopithecus aethiops*) is very difficult to distin-

Allen's Swamp Monkey.
Courtesy of the San Diego Zoo

guish from the green monkey. The grivet figures in a number of the illustrations on the various tombs, particularly at Giza near Cairo. Some of them even show a giraffe with a monkey on its neck, apparently riding it. Whether the knowledge of monkeys came first into Egypt and then spread to other parts of the Middle East, it is difficult to say. There are no monkeys, for instance, in Mesopotamia, although it seems pretty certain that the Pharaohs in the first and second millennia before the birth of Christ probably sent them as presents to the kings of Babylonia. Ashurbanipal II, King of Assyria, was known to have a collection of large monkeys. The frescoes on the island of Crete, which were originally discovered in what was the town of Knossos, date back to the Minoan period, and the monkey depicted on them looks very much like the African green monkey. The Pharaoh Amenhotep II, who lived about 1400 B.C., is shown in some paintings with a monkey sitting on his shoulder.

Monkeys were undoubtedly known to Aristotle, since he made the classification of manlike animals. He almost certainly knew the Barbary ape from North Africa, and some of his descriptions seem to fit the type of guenon known as the mona monkey. The Roman author Pliny has described an animal which seems to have been a colobus monkey; this is not a guenon but belongs to the next group of monkeys to be described. It has already been mentioned that most of Galen's anatomical work was based on the studies of the Barbary ape, but there is no doubt that Galen was familiar with a number of the African tailed monkeys as well.

Although the famous Arabic author Avicenna, who lived toward the end of the first 1,000 years after the birth of Christ, described tailed monkeys, we are not certain what type of monkey he was talking about. But in 1550 a traveler known as Leo Africanus published an account of his travels in Africa, and he describes an animal which was most likely the mona monkey. He even went so far as to say that the internal organs of the monkey he described were very similar to those of the human.

After 1550, the guenon monkeys began to appear more and more in the writings of European naturalists, and then during the seventeenth century there were records of the apes having been discovered and brought to Europe. In the latter half of the eighteenth century a number of monkeys were exhibited at the Jardin des Plantes of the Royal Menagerie in Paris. It was a very poorly organized place until around 1730, when Buffon was appointed supervisor, and from then on, it developed and became a fine menagerie. A number of the monkeys that died in the Royal Menagerie in the century before were dissected by the anatomist Claude Perrault (1613–1688). He is responsible for some of the earliest anatomical studies of monkeys.

In 1791, the English naturalist Thomas Pennant classified and listed a number of monkeys. Of the guenons, he included the Diana monkey, the green monkey, the mustache monkey, and the white-nosed monkey. After that, guenon monkeys became progressively better known; they are animals remarkable for their beauty and variety, and their pelage, or coat, is in many cases very attractive and often has a pepper and salt effect because of a series of alternating zones or bands of color along the length of contrasting hair shafts.

The weight of the brain compared with the body weight in different guenons is about one-thirtieth of the body weight, but the *Myopithecus talapoin* (the talapoin monkey), which is in effect a pygmy guenon, has a brain weight/body weight ratio of one-nineteenth, which suggests that it is a highly intelligent animal. In fact, mental and behavioral studies by Dr. Duane Rumbaugh on this animal have shown that it scores relatively high, in some cases equal almost to that of a senescent chimpanzee. It still has, however, a moderately large olfactory or smelling area in its brain, but the vision and hearing area and

Talapoin father makes a threat.

association areas are well developed. The brain in these animals shows the adaptation to life among the trees and the beginnings of a substantial intelligence. The talapoin is the smallest of the African monkeys—only about 10 inches in length—and it has an olive-green fur.

The guenon monkeys live in most of Africa south of the Sahara, thriving particularly in the tropical forest areas. Usually they have long tails. Their muzzles are greatly reduced, compared with animals such as the baboons and the rhesus monkeys, so they have flattish faces. They usually have ischial callosities on their bottoms, but these are not developed to the extent that they are in the baboons. Their hind legs are long, the opposite of baboons, and they are usually built in a slender fashion and do not grow to be very big monkeys. Their average size, in fact, is about that of a domestic cat. Most of them dwell in the trees and come down onto the ground mainly to drink from streams or pools of water or to steal from the farmers' crops on the outskirts of the forests. They have well-developed cheek pouches, like the rhesus monkey, and they can raid a cultivated plot, stuff their cheek pouches full of food, and then run for the safety of the tops of the trees to enjoy their meal.

The arrangement and length of the hair and the distribution of colors are very characteristic in the different species. They often have white bands across the brow and sometimes white eyelids, and at times the brow bands are colored; some may also have colored nose patches; some show different kinds of whiskers or beards, and some even have a fluffing of the hair around the neck to form a kind of ruff.

Guenons are usually very noisy; a group of them will make alarm calls at any threat of danger or anything unusual and they can move along the treetops at very high speed. They do not actually swing from

204

Mustache guenon at Yerkes Primate Center.

branch to branch, but they will jump from one tree onto the fairly thick patch of smaller branches in the neighboring tree and, having grabbed the small branches, will run along the branch, as it bends, to a more solid branch farther along and then make another jump from that. They can move with surprising speed through the trees by this system. They are usually very well camouflaged, and if they keep still, it is very difficult to see them.

Probably the biggest groups of guenons are the vervet or green monkeys. They usually inhabit secondary forests and the forests of acacia which are found in different parts of Africa, including South Africa. The green monkey has a mixture of black and yellow hairs in its coat, and the facial skin is very often black. The hairs around the face are white and frame the face in a very attractive way. Green monkeys are probably the most commonly imported of all guenons. They are quite hearty, and sudden changes in temperature do not seem to worry them too much, since they are able to acclimatize themselves to the stiffer conditions of the northern latitudes much better than most other species of monkey. The green monkey is an aggressive and bold animal, and it is very expert in stealing from cultivations. It harmonizes extremely well with its surroundings, and it is very hard to see and follow in the jungle.

The mona monkey is probably as good-looking as any of the guenons. It is usually found in West Africa and extends right across Africa to the western part of Uganda. It has a kind of olive color on the back and on the under surface of the body, the inner surfaces of the limbs are white, and there is quite a sharp division between the two sets of coloring. In addition, the upper lip and chin are yellow, and the skin of the face is black. The mona monkey seems to be fairly aggressive and to get angry easily, but it is quite playful. When it gets

Greater spotnose guenon.

Courtesy of the Cheyenne Mountain Park Zoo

Courtesy of the Fort Worth Zoo

Spotnose guenons.

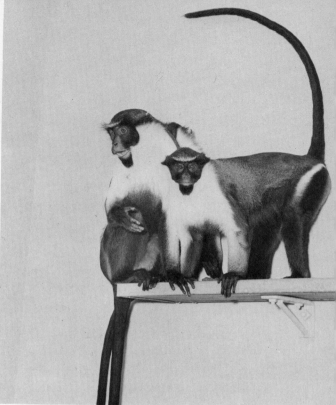

Diana monkeys.

upset, it is liable to strike with its fists and then try to bite. Mona monkeys were released on the islands of St. Kitts and Grenada in the Caribbean a good many years ago, and their descendants are still found on these islands. The monas are great insect eaters, but they also enjoy fruits, nuts and leaves as do other guenons. They live mainly in the tops of trees; the noises they make are more reminiscent of birds of prey than monkeys, and they often travel in large groups and make a lot of noise. They sometimes misjudge when moving from tree to tree and on occasion have been known to fall into pools of water after missing a branch. The Spanish word for monkey is *mona*, a word that probably came into the language from the Moors.

The Diana monkey is probably the best-looking and definitely the largest of the guenons, sometimes reaching two feet in height. It is more highly colored than the other monkeys of West Africa and has a white crescent on the forehead. There are two types: One is black, with white speckling on its coat; the inner surfaces of the thigh are a kind of chestnut color, and the animal has a black beard with a white

207

Hamlyn's owl guenon.

tip and looks rather like a medieval Spaniard. The other type has a slight orange color or white on the hair on the inner part of the thighs. This variety has a pure white beard. In keeping with their fine appearance, Diana monkeys indulge in a good deal of personal grooming, and they usually look immaculate. As young animals the Dianas, like a number of guenons, are very gentle and appear to be quite intelligent. However, after they have become sexually mature, it is very difficult to achieve any rapport with them.

Other guenons include the mustached monkeys, which often have red mustachelike hairs on the upper part of the face. Then there are the white-nosed monkeys, which have a little white nose and often have a red tail.

All the guenons are very social animals, and they live in either large troupes or in smaller groups and even in family parties. The troupe can sometimes be formed of groups of different species. Others, however, are very territory-minded and will not make any cooperative arrangements with other species. Guenons make very pleasant pets when they are young because they are very gentle, but as they grow older, they become much more unpredictable and erratic and can inflict dangerous bites.

Guenons have a wide range of foodstuffs; they eat wild fruits and leaves, shoots, and various nuts. Like all Primates, they eat a wide variety of insects and when they get the chance, wild honey. Again like other Primates, birds' eggs and young birds are popular. They are curious animals and will investigate anything that seems to be unusual.

One of the monkeys that is classified either with or close to the guenons is Allen's swamp monkey. This is known by the scientific name of *Allenopithecus nigroviridis*. It is small and very agile and lives in the forests that grow in association with the swamps found in parts of the Congo River and in one or two other places. Though they live in these watery areas, they do not voluntarily go into the water and some authorities think they cannot swim, but I doubt that. There has been a special genus established for this animal because it seems to be different enough from all the other guenons, most of whom have the generic name *Cercopithecus*. Its coat is primarily black, speckled with a yellowish-green color. It has a black face, and the skin around the eyes is gray. Its chin, its chest, its throat, and the hair on the inner parts of its arms and legs are bright yellow. They usually live in small family groups and are not very easy to see. They have a unique voice, and the noises they make, which are very highly pitched, are similar to those of the guenons. Like most of the guenons, they will eat nuts, leaves, fruits, and insects.

We would need an encyclopedia to talk about all the different sorts of guenons, but there are one or two that might be mentioned further. One of these is the De Brazza monkey.

The scientific name of this animal is *Cercopithecus neglectus*, and it has a really gray fur with white markings on it. The white markings are distinctive and form symmetrical geometric patterns, according to the position that the animal takes. It has a white beard and a bright-orange patch on its brow. These animals live extensively in the basin of the Congo River, and they have also been found in the southern part of Ethiopia, the valley of the White Nile, and parts of Uganda and Kenya. They live in swamp forest and in bamboo thickets but can also live in dry mountain forests at an altitude of more than 6,000 feet.

A De Brazza will often feed on the ground, eat leaves and berries of a variety of plants, and will sometimes form the leaves, particularly in plants that have a latex kind of sap, into a plug before eating them. It will cross open country to get to food and will eat fruit and figs and also green cotton bolls, maize and roots, and the gum of some trees. It

De Brazza family.

will eat animal food and insects and is known regularly to eat lizards, especially geckos, and grasshoppers. When it eats geckos, the monkey pulls them apart with great care, removing and rejecting the viscera and the eyes. It then shreds some of the flesh away from the bone and throws most of the bone away.

The fully developed male De Brazza monkey, with full coloring, is said to be very confident and very aggressive and likely to attack other monkeys if he is kept captive with them in the same cage. In the wild he displays very boldly and will be very protective of other members of his troupe. Animals that do not have the fully developed color are much more timid animals, both in the wild and in captivity. The female is much smaller than the male, but there is very little difference in coloring between the two except that the female perineum (the skin between the anus and the genital region) is colored brownish red. Like many other monkeys, they will, if they wish to submit to another animal, hide their faces, go onto their elbows, put their heads onto their arms and raise their hindquarters and tails toward the animal or the person they wish to be submissive to. This has been seen in

De Brazza monkey.

Courtesy of the Cheyenne Mountain Park Zoo

captivity when an undeveloped juvenile male has been put in with a fully developed adult male. This submissive gesture is said, by some authors, to be more stereotyped in the De Brazza's monkeys than it is in other monkeys.

Dr. Jonathan Kingdon, in his book *East African Mammals*, makes the following comment about the De Brazza monkey's submission:

> I have seen a very similar ritual in an adult captive male, although no trace of brown colouring remained on the rump and the animal appeared to display this behaviour as an appeasement ceremony towards me. He initially presented with raised rump and feet widely parted after a bout of threats and snatching; the rump was presented several times after the threatening and on each occasion the animal looked over its shoulder. Later, when I confined the animal in the large aviary which I entered, he was uneasy about my closeness and after he had been followed about a few times, he initiated the following display; he suddenly backed towards me, meanwhile looking between his legs and, having wrapped his tail around my arm, reached out with his hind legs to take the weight of his hind quarters on my hand, he was then standing on his head. This performance, which was repeated several times, sometimes led to the monkey losing his balance and falling over, when he was willing to be groomed. The gesture was obviously more elaborate than a simple presenting and it is possible that the animal was attempting to force an appeasing gesture, as had been reported by Simmons in 1965 with macaques.

It is of interest that when this animal was presented with cloths of different colors, he was particularly responsive to the orange-brown color which he touched eagerly with his fingers and then smelled; probably the reason this color elicited so much interest was that it was similar to the perineal brownish color of the females. The bright-blue

211

Green monkey (grivet guenon).

color produced an aggressive reaction, which took the form of violent chattering and a general aggressive response; perhaps he identified this color with that of the male scrotum. There are two types of threat. In the first the animal will rise up on his hind legs and bob his hindquarters and beard up and down. In a second type of threat, which is usually regarded as a milder type, the animal spreads the legs widely to show the white insides of the legs, and this, together with the white beard, makes the threat easily interpretable and sometimes the blue scrotum and red penis are also well displayed. Jerky erections may also be present as part of this threat. The animal often responds to sudden alarm by repeatedly rising and spreading his thighs so that the white may be easily seen.

A legend told by the Pygmies of the area where the De Brazza monkey lives is that when a male displays in this fashion, often an excited troupe of De Brazza monkeys gather around him, and while they are observing, another animal will circumcise the animal that is giving the aggressive display. It is hard to believe this, but it is an interesting legend and leads one to speculate what sort of behavior by these animals would give such an impression to the Pygmies, who are pretty keen observers.

The De Brazza monkey groups contain up to about thirty-five individuals and rarely fewer than fifteen. On rare occasions, the groups are really family groups, but solitary males also occur. Some troupes contain only one fully developed male, and in this case, a young male growing up in the group will stay with the troupe but will behave and look rather like a female in the earlier stage of his development; but the moment that he develops the adult coloring, he will most certainly be driven out of the group. If the group is large, he may take a nucleus of juveniles and females with him and form a new troupe. This is probably the built-in mechanism which stops the troupe from getting too large, and it provides for the proliferation of troupes.

The male De Brazzas are fast runners and good swimmers. When they are moving leisurely on the ground, they adopt a sort of swagger. Because of their swimming ability, they do not find rivers, even large rivers like the Congo, much of an obstacle.

The adult animals of both sexes have more to say for themselves than the other members of their community. If they are startled, they emit a sort of chattering croak of alarm which has been compared with the "wooden rattle of a football [rugby] fan," and they follow this up with a series of separate barks. They also make a hooting noise which seems to be associated with the sudden onset of relaxation after tension, *e.g.*, after a predator turns away. They also give a grunt when they see something that looks like food or when a caretaker appears with food if they are in captivity.

De Brazza monkeys breed easily in captivity and are well known in zoos. In East Africa they are often used as pets. When young, they are very tameable and gentle animals.

There are a number of monkeys related to the green monkeys; they are often called savanna monkeys. There are probably about twenty subspecies of savanna monkeys, and there are only very slight differences between any of them. They extend all over Africa, from the savanna belt south to the Cape and west into Angola. They are called by different names in different parts of Africa. For instance, in South Africa they are known as vervets, in the northern parts of Africa they are known as grivets, and in West Africa they are called green monkeys or African greens. They all have a similar habitat, primarily in the trees; but although they are arboreal, they can run around very actively on the ground, and they are also active in rocky areas such as the boulder-strewn side of a hill. They are very fast runners and are good jumpers. They have short, sleek, shiny hair of a greenish-yellow color, their tails are long, and their hands are small. The skin on their faces is black.

Vervet monkeys move in troupes, but if a troupe has the temerity to penetrate the territory of another troupe, it is treated as an act of war. A large-scale fight develops which can go either way.

If a troupe is occupying territory which becomes uncomfortable to live in by an increase in predators or a decrease in food supplies, scouts move out to spy over the surrounding country. If a good piece of territory without the above drawbacks is found by the scouts, they mass into it. If the scouts report that the desirable piece of real estate is already occupied, the troupe will still move into it if the occupiers are a small group. The latter will be attacked and put to flight.

Like most monkey groups, there is a dominant male, but he seems to be subject to frequent challenge by other males. If the dominant male loses such a challenge but does not lose his life, he is banned from the group and spends the rest of his life in solitude. Such animals live very lonely lives without female company and become very morose but also become very daring, especially in robbing farmers. They become so skilled that it is rare for them to be caught.

The farmer also suffers from raids by whole troupes of vervets which not only eat a lot of food but destroy and waste even more than they eat. One example of this waste is when they move into a mealie (corn) field. They prefer mealies when they are still in the milky stage of development, and after a raid, hundreds of mealie cobs can be found scattered on the ground and many more broken and hanging from the plants. Not one cob is completely eaten; in many cases only one bite is taken out of the cob before it is thrown away.

A farmer would sometimes bring in a group of helpers to kill off the monkeys which were destroying his fields and would usually wait until there was a group of vervets actually in the field actively eating; then they would fire at them with shotguns from several different regions. Those that were wounded were dispatched quickly by the beaters who accompanied the riflemen, and those which were not injured were chased and slain if they could get near them. The author (F. W. Fitzsimmons) of the *Natural History of South Africa* describes, very poignantly, the behavior of a mother monkey which was wounded in one of these onslaughts against the vervets:

One of the wounded was a mother monkey; she had been struck in the lower part of the back, and her hind quarters were paralyzed. She clasped the baby in her arms, and was begging piteously for life—for the life of her child. Casting both arms around it, she made the most frantic efforts to shield it with her body. Looking over her shoulder, she glared at her enemies, the muscles of her face assuming a variety of forms, and her teeth glittered in the early morning light. Those who have seen the grimacing of a mother monkey defending her child are never likely to forget the sight.

Presently the menacing mien subsided; she ceased to chatter in a menacing way. Her face relaxed, softened, and she gazed at her enemies with a piteous expression in her eyes, meanwhile giving vent to a succession of low mutterings which sounded sad and plaintive. It was evident she was appealing to the higher emotions of her enemies.

We took the wizened faced, terror sticken babe from her dead body, and wended our way back to the homestead feeling guilty and mean. There was no rejoicing over our victory at the breakfast table. Each of us seemed to see the face of that dying mother monkey, and hear her plaintive voice begging for the life of her child.

A charming account of a vervet monkey in captivity on an African farm has been written by Charlotte Truepenny. She and her sister acquired their vervet monkey in a similar fashion to that described above. A man arrived at their house one day with a tiny bluish baby. The vervets do have a rather bluish skin, and since the baby had very little hair on it, the blue tinge in the skin was very obvious. The man who brought the monkey pointed out that this was a young animal whose mother had been killed during a drive against the vervets. After a little negotiation over the price, the Truepennys bought the animal and it became a part of the family. Charlotte Truepenny's description of the baby as handed to her is a very touching one. It goes as follows:

> His mother killed by dogs. A waif. In need of care and protection. His eyes staring at me were not those of an animal. They were human and they disturbed me. He went on crying, then curled in my hand, exhausted. I thought he was better dead in a forest. I did not believe he could survive without his mother. It would not be possible, I thought, to rear such a morsel as this. But he clenched his hands and feet, crying violently out to live. It was the scream of one born to fight.

She found that she could not put the monkey down. He clung to her; obviously she was his substitute mother, and unfortunately they had great difficulty getting him to drink anything. He ignored milk and would pay no attention to any other sort of food, but eventually they were able, with a hypodermic syringe, to squirt some glucose and water into his mouth. Then, once they were able to get milk in a bottle with a teat on it, they were able to get the animal to nurse very well.

Once the animal was strong enough to climb out of the shoe box in which they kept him while he slept, between feeds, they found it absolutely essential to carry him about. He, not surprisingly, wanted contact with these foster mothers. For the next two months either Charlotte Truepenny or her sister, Ann, had to carry him every hour of the day and night.

It is of considerable interest that when he began to get a little older, although while still a baby, he reacted in his sleep as if he were having dreams. He also talked in his sleep. He was very restless and grasping with his hands and would sometimes beat the bed with his tail, which by this time was 14 inches long, and the noise could be heard through the house. It emerged later, however, that this restlessness at night was simply due to the fact that he was teething and his canines were coming through first.

As the monkey grew a little older, he was given the name Zephyr and was introduced to the two cats in the household. One of these was a black tom, and the other was a rather aristocratic Persian cat called Smokey. The owners were rather frightened that the cats, particularly the tom, would give him a bad time, but this was not so. When the tom first saw the monkey, he went up to him and butted him with his head and then circled the chair and did it again. When the monkey reached out for the cat's hair and grasped his fur, the tom gave no evidence of being concerned. This was taking place while there was a violent storm; the rain was falling heavily on the tin roof, and the monkey was obviously frightened of it and the tom gave the impression that he was trying to comfort the little orphan. Later on the monkey met Smokey; when she came near the monkey for the first time, he reached into her fur, and she did not like it and walked away. However, she came back and looked him over and sniffed him again and later on withdrew. On another occasion, when she was at rest and rather sleepy, they put the monkey alongside her, and Zephyr dived immediately into her fur and got in between her paws and her bushy tail, where it was nice and warm and comfortable. The only reaction of the cat was to lean forward and begin to lick the monkey's head. Subsequently she accepted the monkey completely. She let him climb over her, pull her fur, and one day he even grabbed her by the nose and looked inside it. When the two cats were together, curling against each other, Zephyr really enjoyed himself, getting between the two of them and luxuriating in the warmth and the fur of the two animals. As the author says, "A combined purring of the cats hummed across him like a miniature roll of African drums."

The cats, especially the Persian, spent a good deal of time licking the baby. During a session of licking, the authors noted that there had been 209 successive licks. When the baby monkey emerged from the fur of the cat after this treatment, it came out rather wet. Zephyr knew what to do to get this treatment. If he pushed his body under the cat's chin and waited, it would not be very long before his head began to be licked. To get the same attention for a new area, he would simply roll himself over and would lie with his eyes half closed in a kind of dreamy position and become very drowsy. However, if the cat stopped licking, he would wake up immediately. After the cat had stopped licking, an arm held close was enough for the cat's tongue to start work again.

The very young baby vervet holds closely to the breast of its mother. It has a rather frail-looking body with a disproportionately large head and thin spidery-looking limbs. It has a pink wrinkled face, which is described as resembling that of an aged Hottentot, and its expression is that of fretful anxiety. When the baby suckles, it sucks both teats at the same time; they are close enough together to permit it to do this without effort.

At three months, the Truepenny monkey was old enough to be taken almost anywhere; he was frequently taken out in the car when his owners were shopping and visiting. On one occasion when some friends visited them, the monkey was put in the room belonging to one of the sisters, and all movable objects and objects likely to be broken were put away carefully in the drawers. However, as the friends were departing, they saw the monkey through the window of this room and called out to the Truepenny sisters, "Did you know you have a red monkey?" When they went in, they found Zephyr with a crimson face owing to the fact that he had one of the sisters' lipsticks in one hand; his chest, arms, and stomach, as well as his face, were liberally smudged with red. He had put lipstick on the soles of his feet and all over his hands. It was obvious that the discovery of this lipstick must have been a great occasion in his life. He had, of course, to be cleaned up with soap and flannel. While this was being done, he was allowed to keep the screw-on top of the lipstick, and with this compensation, he submitted to all the indignities of being washed without any complaint. He had also eaten a lot of the lipstick and some hours later was seized with an attack of vomiting and diarrhea, and everything he produced was, of course, bright red.

Zephyr. A green monkey.

From *Dear Monkey* by Charlotte Truepenny (London, Victor Gollancz, 1963)

Zephyr was intrigued by his own image in a mirror. He had been ten days old when his mother was killed and he was brought into the household of these people, so he had very little idea of what monkeys looked like. It was interesting to speculate on what he thought he saw in the mirror and if he had any way of recognizing or identifying himself. He put his hand behind the mirror on several occasions to try to catch what he saw there. This is the type of behavior many Primates have shown when given a mirror.

On one occasion, the sisters were brave enough to take Zephyr into town for a portrait. First they had him on the knee of one of his foster mothers, and he sat there for a while with his fingers in his mouth. In between times, he was pulling her hair or trying to climb up the inside of her sleeve. He would not sit still at all. The photographer crouched on the floor trying to get shots with his camera and giving instructions on how the monkey was to be positioned. The photographer said, "I'll take a reel of pictures." Then the animal's lead slipped, and he started off, ranging around the room with the photographer after him. He then went up a curtain, and everyone wondered whether the next step would be the disintegration of the curtain as he made a flying leap at the camera. He was offered toys in an attempt to bring him back; he simply took them up somewhere higher. He took a flower and some tin oddments and littered the carpet with petals as he broke up the flower. The photographer was in an awkward position, trying to shoot him from below, and the monkey then threw a chrysanthemum at him and dropped a trail of tin lids, whistles, hair curlers and screws. He suddenly jumped back on his foster mother's knee. To calm the animal, she gave him a plastic cup with milk in it; the animal took the cup and threw the milk away and put the cup on his head and then flung it on the floor. They moved to pick up the cup, and he was gone again. This time they found him with his tail twisted around the tripod of the camera, his face peering between the legs. There was complete chaos while he was chased again and finally caught and brought back to his former position on the knee of his owner. By the time the reel of photographs was taken everyone was exhausted, and they started home, all looking like complete wrecks. It was agreed that this one expedition was positively to be the last appearance of Zephyr in town.

At six months old, Zephyr was 12 inches in length, including head and body. His tail was 17 inches long and had a black tip. He had a complete set of teeth with two impressive canines. Although he had been brought up with the two sisters, he had now become very possessive of the one, Ann, who had done the most nursing of him when he was very young. If he was carried by her, he found the close presence of Charlotte an intrusion. On one occasion, when she was close by while he was being carried by Ann, he leaned over and bit her. So they got the message that Charlotte was not to touch Ann while she was carrying the monkey. He now could not be passed between Ann and anybody else. In fact, no one could take him from Ann—his attitude hardened with time. The only way they could transfer him from one to the other was for Ann to tie up his leash and retire from view. Then Charlotte could come and pick him up, and he would be quite happy with her until Ann returned. He would dismiss Charlotte with a small nip and rush chattering to his favorite, Ann. The nips that he gave Charlotte became progressively more severe.

They also found that he got very angry if strange cars drove up to the house. He would give alarm cries if he sighted a strange car a half mile or so away. They learned that when he gave the cry, the best thing to do was put him in Ann's room with Smokey for company. Otherwise, he would greet the visitor with a considerable amount of hostility. He got very upset if he saw Ann shaking hands with someone he did not know.

Although the animal ate a number of leaves and plants around the house, instinct prevented him from ever eating those which were poisonous. He was equally discriminating with insects, some of which were dangerous. In the wild, the vervet monkey will drop onto the back of a duiker or bushbuck and hunt through its fur for vermin which they eat while the antelope stands and relaxes.

Zephyr had up to this time been thought by the sisters to be a boy but was now found to be a girl. She was a very curious animal and showed great interest in boxes and would work at them until she could open them. She would swipe books and look through them or grab a toy and examine it. All these things she would take up into some high place, preferably up in the trees, where she could examine them in great detail and leisure. She reacted to close-up pictures of faces of either human beings or animals. On occasions, when she was shown a propped-up photograph of a dog, she would approach it quietly and suddenly grab it with both hands and press her nose against it.

Charlotte Truepenny gives an excellent description of Zephyr's appearance, a description which fits the vervets very well: "Trim, soignée, neat waisted, she sometimes looked as clear-cut, exaggerated and tapering as a fashion plate drawing. With the subtle moving color of her gray-blue-green coat, her slender, black gloves and the white bar across her brow, she looked, at times, sophisticated."

Jealous reactions are known in captive vervets, but these reactions also occur in the natural state, according to Drs. J. S. Gartlan and C. K. Brain. This jealous reaction would be demonstrated, for instance, when the mother started to groom another monkey or be groomed by one. When the infant saw this, it would stop whatever it was doing and run over to its mother and attempt to suckle from it. If the mother permitted it to do this, the grooming session was obviously terminated. On one occasion, the authors saw a mother indulging in copulation with a male, and the infant, upon seeing this, ran over to the copulating pair and jumped at the male, who was at that time mounted on his mother and thrusting. Under the attack by the infant, the male dismounted and took off without showing any signs of enmity toward the infant that had broken off his activities.

Captive vervets clean food that has fallen to the bottom of a cage and become covered in sand, by rubbing the piece of food with their hands or by rubbing it against the inside portion of the forearm. An exceptional animal, however, has been seen to pick up pieces of food (*e.g.*, beetroot), carry them to the water trough, and wash them in the water. Over a three-month period, one animal established this procedure as a definite pattern and took all its food, whether it fell into the sand or not, over to the water trough to wash it before eating it. Over the same period, none of the other animals showed signs of imitating the vervet although one of them had been over to the water trough while the vervet was washing its food and had taken some of the washed food out of the water and eaten it itself. It may be that this washing behavior will be passed on to the offspring, who will pass it on to other young animals. If a new idea is passed only among the younger members, then a generation gap exists in monkeys just as it does in humans.

We have been emphasizing that smell plays a relatively less

important part in the life of Primates than it does in pre-Primates, but it is still significant at many levels of Primate evolution. In the green or vervet monkeys, scent and smelling still seem to play a role in determining the boundaries of territory. For instance, captive animals will rub their cheeks and the inner part of the lower jaw against many parts of their cage, and wild animals will do the same with rocks and tree branches in order, presumably, to mark them with their scent. Apart from this role, the smelling of strange objects is always a precursor to handling by these animals. Even the great apes tend to sniff strange objects when they are first given them. When vervet babies come back to the mother, the mother will sniff them presumably as one method of identifying them. Mouth-to-mouth sniffing may occur between two individuals. In this case they gingerly move close to each other, stick their necks out, and purse up their faces; then they gently touch each other on the mouth to the accompaniment of lip smacking. During this process they sniff each other's mouths, and it seems that from this they get information about what the other animal has been eating. An example of this is the account of an animal which had been eating the flowers of a saba plant; its mouth was sniffed by another animal, which then went directly to a saba plant and ate its flowers.

The threat face of the green monkey consists of what is really an attack face—the mouth is tightly closed, the eyes are wide open, and the eyebrows are drawn together in the midline to form a frown. This attack face is usually used immediately prior to actual physical combat. If the animal is scared, he may "crouch close to the ground with lips retracted into the grin face and eyes almost closed. In exceptional circumstances the animal assumes the fetal position of complete submission with eyes tightly closed."

Drs. Gartlan and Brain, referring to an expedition one of them made, say:

> When danger threatens or when they desire to cross to the opposite bank of a stream, the vervet monkeys have no hesitation in plunging into the water and swimming across. I startled a large troupe of these apes one day in the bush on the bank of the Umgeni River in Natal. With loud cries of alarm they swung themselves up into a tall tree, one of the topmost branches of which spread out over the stream. Running along to the end of this branch, the apes, one at a time, in rapid succession, sprang with hands and legs outspread and tails straight out behind the distance of twenty feet or more into the top of the smaller tree on the opposite bank. It was a most interesting sight to see them leaping gracefully, and with such apparent ease.

Thirty-six distinct vocal signals have been recorded from vervet monkeys. One of the problems of forest life is to communicate efficiently in conditions where the visibility is very poor. The solution that has been adopted by many Primates has been the development of high-pitched calls—some of them quite elaborate. Some of the apes, for instance the orangutan and the gibbon, developed air sacs connected with the trachea (the tube carrying air to the lungs); the animals can fill these with air and use them as resonators to carry their voices a considerable distance through the forest. The smaller Primates, however, mostly seem to depend on the high-pitched type of penetrating call.

When vervet monkeys sleep, they usually sleep in small groups and put their arms around each other to keep warm. In a caged group the high-ranking monkeys constitute one group, and the subordinate monkeys constitute another. The colder the night, the more closely the groups cohere.

The infant in a vervet monkey group—and probably in a lot of other monkey groups as well—is an important factor in integrating the group. The mother-infant bond is very strong and enduring. The other females are also very attracted by infants. Many of them will keep close to a particular infant, looking for any opportunity to pick it up and have the pleasure of handling it. If any danger threatens, any female in the neighborhood will come to the aid of the infant.

One way that females without infants of their own get close to an infant is by approaching the mother and grooming her. Females who have not had babies themselves are the ones most interested in the infants, and they try the hardest to get them away from their mothers. These females are often very clumsy in the way they handle babies—holding them upside down or not holding them up high enough on the chest, so that they can only walk by supporting the baby with one arm. As a general rule, mothers do not like to let their babies go until they are at least four weeks old, but there are some unsophisticated mothers who will let babies that are only two or three days old be held by other females in the group. Once an infant is held by another female, it is never handed about among the remaining females but is always passed back to the mother. The subadult male monkeys of the vervet groups never show any interest in the babies. I have not seen any record of the kidnapping of babies and the keeping of babies away from their mothers for a long period of time, such as we have seen in the rhesus monkey groups at the Yerkes Field Station.

Sex behavior is also said to help to hold a group of monkeys together; this hypothesis was originally propounded by Lord Zuckerman in *The Social Life of Monkeys and Apes*. He was, however, talking primarily about baboons. The vervet does not copulate as often as the baboons do, perhaps because the external signs of the animal being in heat do not exist in vervets as they do baboons. In the latter there is an enormously swollen and reddened area of sexual skin in the perineal region which gives a cue to the males that the animals are in estrus. In the case of the baboons and even of chimpanzees, when a female is in estrus, there is often multiple copulation with her. This apparently does not occur in the vervet monkeys, at least in captivity, and in the wild copulation of the vervet is difficult to see. The animals are very shy, and if any animal produces an alarm bark, copulation is immediately stopped.

Cercopithecus mitis lives in the forest canopy, but *Cercopithecus aethiops* lives primarily in the woodland savanna and are prey, therefore, to many different sorts of predators. The crowned eagle attacks them, and so do various other birds of prey, so the animals live a very hazardous life. In addition, there is the problem of a regular supply of food and water. The *Cercopithecus mitis* probably never needs to drink since it gets a great deal of moisture from the food it eats in the forest. The vervets, however, eat much drier food and have to drink a good deal to counteract this dryness.

As in the baboons, the adaptation to life on the savannas leads to a much tighter group organization and much more of a visual communication than a vocal communication, except when an alarm has to be given. There are two typical alarm calls. One is a low-intensity noise which is produced when a predator is seen in the distance or if there is some unusual object; according to Gartlan and Brain, this call is "a quiet, whispered call with lips drawn back. Its effect in alerting other monkeys is enormous, while its quietness is clearly important because it does not readily draw the attention of the predator to the position of the monkeys." There is another and much louder call when immediate danger threatens.

The green monkey in South Africa is an example of the success of an animal that is flexible enough in its habits to be able to adapt itself to a wide variety of habitats and to change its pattern of living as the environment changes. In this way it was able to make a success by the exploitation of the complex life of the savanna. The same kind of problem attended man as he developed through his Australopithecine stage.

Another interesting story about one of the guenons is that of the white-throated guenon, *Cercopithecus mitis albogularis*, also known as Sykes' monkey. In 1943, a little female Sykes' monkey fell into the hands of a woman, Dr. Hope Trent, at Mombasa in Africa, who named her Audrey and lived with her for many years. The animal had black fur with russet strands on the body, but the collar and the front were cream colored and tufts of cream fur grew from her ears. Her face, ribs, and tail were all black, and she weighed about 13 pounds. They tried to mate her on a number of occasions with a handsome male Sykes' monkey, but she did not accept him. When she was young, she played with young male vervet monkeys. She had been used as a surrogate mother for a number of baby monkeys that had to be brought up without their real mothers; some of these were vervets, and one was a young baboon. Audrey would carry these young monkeys, hanging to her belly, for a year or so until their backs were dragging on the ground. She gave a good deal of love and grooming to them. She would not, however, share her food with them and would push them away if they tried to get to it, but if she heard any kind of distress call from the animals, she would rush to defend them. In this state she would bite whoever she thought was responsible for the call. When she was not bringing up small monkeys, she would be satisfied with a substitute baby, such as a small toy rabbit.

Dr. Trent described her medical experiences with Audrey in *Medical World News*, July 23, 1965. Among the things Audrey acquired was a variety of wounds from fighting with other monkeys or from attacks by dogs; she had smallpox; she was bitten by a dog with rabies; she had a disease called "metropathia haemorrhagica"; she had diabetes; she had hemoglobinuria (this is to say, blood in the urine) after having been stung by bees; and she had pyorrhoea.

As far as the wounds are concerned, she had the skin badly torn in several places, and on one occasion it took an hour to suture all the wounds together. Monkey skin is pretty thin and is relatively easy to tear.

In 1956, Dr. Trent was working in a mission hospital in Mwenzo in the middle of Rhodesia. There was an outbreak of smallpox at that time. Every day Audrey was taken by her mistress to the hospital, where she was tethered to a tree. Her owner did not know whether monkeys could contract smallpox or not, so she had never been vaccinated. One day the cook told the doctor that the monkey had smallpox. She went out to see Audrey, who was sitting in a dejected fashion on the corner of the veranda; on her face were little pockmarks. She would eat nothing for several days but would drink lots of water. Finally, she accepted blackberries which were growing in

Sykes' monkey at the Yerkes Primate Center.

the garden, and these were the only food that she would take for a time. Fortunately, it was not a serious attack, and she picked up. In ten or twelve days she seemed to come back to normal and showed no unpleasant aftereffects.

Later on there was an outbreak of rabies in the district, and all dogs either had to be on a leash or shot on sight. One morning, when Audrey was walking along with her mistress, a savage black dog rushed out of the grass, bit the monkey on the nose, and then rushed into the house of an African, where it bit four puppies before it was killed. There seemed to be little doubt that this animal was rabid, so it was shot; Audrey's wound was thoroughly cleansed, and she started on a course of twenty-one antirabies injections.

When Audrey was about fourteen, she showed signs of excessive amounts of blood during the time of her menstrual periods. Sykes' monkeys' menstruation usually starts at about five years; it goes on for about three to five days and occurs about every twenty-eight days. With Audrey, the interval between the periods gradually got less and less until she began to lose blood every day. She apparently suffered a good deal of pain because she was very miserable and remained hunched up in a corner most of the day. She also developed a very serious anemia. She was treated with the hormone progesterone which is normally produced in various stages of the human female's sexual cycle and in substantial amounts by the human female during pregnancy. At the time that Audrey was suffering from this problem, she and her owner were living in a town on Lake Victoria in Tanganyika. At a mission hospital at Sumve, 25 miles away, there was a woman gynecologist who was asked to perform a hysterectomy, which she agreed to do. At the operation the uterus was found to contain a number of fibroid growths just like some human uteruses. Since the animal had had a good dose of barbiturates for the operation, it was very delayed in coming out of the anesthetic; twenty-four hours passed, and she was still asleep. A number of drugs were used to rouse her, but she remained asleep until the following morning when she recognized her owner's voice, opened her eyes, gave a grunt, and was finally fed with a little water from a teaspoon, and from then on she began to recover. That night she slept in a little bed near her owner's bed. In the middle of the night her mistress heard a thud and found that Audrey had tried to get out of her bed to come to her mistress and had fallen onto the floor. The doctor took her into her bed and tucked her in. She was there safe and sound in the morning and still in good condition. The wound healed after that without any complications and Audrey was soon full of energy, so much so that she had to be restrained by a harness. In fact, when her benefactor, the gynecologist, Dr. Schroeder, came to visit her, she attempted to bite her. This sign of rambunctious good health was a good sign because Audrey had years of excellent health after this.

In 1962, by which time she was middle-aged, she began to become very thirsty for water in the morning and also tended to produce rather a lot of urine. Her mistress had been very busy and it was a while before the significance of these symptoms dawned on her. When they did, she tested the animal's urine and found there was sugar present in it which suggested that Audrey had diabetes. Her owner was not anxious to give the animal insulin, and she tried cutting down on the sugars and starches in Audrey's diet, but this did not help the situation. She would not take oral tablets to control the disease and finally her owner had to put her on insulin. With insulin she became quite well and happy. She was given only one injection daily. They had thought that perhaps more than one injection would be desirable, but on one occasion the extra insulin caused the animal to go into a coma and have a convulsion, so they decided that although one injection a day was barely enough, this would be adequate to keep her in reasonable condition.

On one occasion she was tied to a tree by a leash and attacked by a swarm of bees. She managed to break loose from her leash and rush over to her mistress' car with the bees chasing her. The car was quickly driven away, but with a number of angry bees in it. By the time they got to the house it was obvious that Audrey had received a number of stings, and she sat in a corner moping and would not take any food. That evening she passed red urine. She continued to pass red urine (which, of course, was due to blood pigment getting into the urine) for a couple of days and then recovered. Subsequently Audrey got pyorrhoea. Although she was getting plenty of vegetables and fruit which provided her with plenty of vitamin C, she was given injections of vitamin C, which did not do the pyorrhoea much good (sometimes pyorrhoea relates to an inadequate intake of vitamin C). Most of her incisor teeth had already fallen out by that time so that it was apparently an infection of the gums unrelated to the lack of the vitamin.

This account of Audrey's vicissitudes is interesting because it is the only account of the life of a pet monkey which accurately records its medical history.

Military Monkeys or Red Monkeys

There is a group of monkeys known as military monkeys, red monkeys, Hussar monkeys, ground monkeys, or patas monkeys. They are called Hussars because their "red" coats are somewhat similar, to those humans with a vivid imagination, to the red coats worn in the nineteenth century by the regiments of Hussars. They are also called military monkeys because they move about together in groups in what seems to be a military fashion. To me, the face of these monkeys conjures up the face of a retired English colonel with his close-cropped mustache. There are two groups of these monkeys: the patas monkeys that live in the west part of Africa and the nisnas monkeys that live in the east. Both types of monkey have the generic name *Erythrocebus.*

The patas monkey is exceptionally well adapted to living on the ground; it ranges over the western and eastern areas of Africa especially in the grassland and the woodland savanna areas, and it has a "desert coloration." The top surface of the coat is colored a kind of darkish red with rather off-color white underneath.

Patas monkeys were studied extensively by the late Dr. K. R. L. Hall. He found that in any particular group there was only one adult

Patas monkey.

From J. A. Allen, "Primates of the Congo," *Bulletin of the American Museum of Natural History.* Vol. 47 (1925).

male, even in a group as large as thirty-one animals. The other members of the group included one or two subadult males, females of various ages, juveniles, and infants. The adult male played watchdog. Whenever the group was disturbed or when it was coming near a new area, he would proceed to some elevated area, maybe some hundred meters away from his group, and look the situation over. If a human observer disturbed him, he would give a noisy display, jumping on bushes and trees, and then would run away from the group, presumably with the intention of leading the observer away from his fellow monkeys, so he functioned as a watchdog or sentry and also as a decoy. During the day, the adult male scampers about on the periphery of the group, returning only during the resting, feeding, and grooming periods and, of course, during mating. However, there is a strong matriarchal control in these monkey groups, for it is the adult females that decide when and where the group moves at any particular time.

Individual groups of patas monkeys are usually pretty well spaced so that they do not encounter each other very often. Usually the larger group will chase a smaller group away. A whole patas group will run away at high speed from a group of baboons without even a threat or a bark from the baboons. The patas themselves are not very vocal animals and have very little to say; such vocalization as they make can rarely be heard more than about 100 meters away, and they make a noise only two or three times a day. When they are disturbed, they produce a kind of a chirrup and if a patas in captivity hears this noise, it immediately becomes alert and watchful. Sometimes a patas will utter a contralto type of bark, but it is produced only by the adult male and seems to be made only when the group is making contact with patas monkeys outside their group.

Patas monkeys are very fast. They are called the Primate cheetahs, and Dr. Hall says, "They are adapted to fast locomotion, to silence, concealment and dispersal. This contrasts very clearly with those of the much noisier, larger, more aggressive baboons in whose group there are several adult males."

During the day patas monkeys rest, especially in the shade of tamarind trees, from about one to three in the afternoon and sometimes for longer than that. During that time there is very little feeding. The juveniles and infants carry out a good deal of social play during the first three hours of the day; but then they get tired, and when the group is resting in shade trees during the hot part of the day, play is very uncommon. Similarly, social grooming is most active in the first three hours of the day; some occurs during the resting period, but during the last three hours of the day there is practically no grooming at all.

During the rest period, the animals may climb into the nearby trees; sometimes if there is only one tree around, the whole group may climb up into it and sit and recline among the branches. At night the group disperses and the individuals distribute themselves throughout a whole series of trees.

They do not seem to be preyed upon by any of the large birds. There is some interaction with birds, such as the gray kestrel and the gray hornbill, but the older patas simply behave as if these animals do not exist; young patas monkeys sometimes strike out fiercely at the birds with their hands if they come too close to the group, and they also chase after them once they have settled on the ground.

The patas eat a wide variety of foods—grasses, berries, fruits, beans, and seeds—and will also eat mushrooms, lick up ants, catch grasshoppers and small lizards, and eat birds' eggs and small birds. They have also been seen occasionally eating chunks of red mud. Mushrooms seem to be highly prized foods. If, for example, they are eating mushrooms and an adult female finds herself without one, she will run at a youngster that has one and make him drop it. Sometimes the youngster will run away in erect posture, holding onto the mushroom with both hands. It is very unusual for animals other than the adult males and the senior adult females, especially those with infants, even to get hold of a big mushroom. An animal may take as much as twenty minutes to eat an entire mushroom, including the stem. Sometimes infants sitting near the females will pick up and eat small pieces that are dropped. Sometimes they will grab larger pieces but are never allowed to keep them.

Wild patas monkeys drink surprisingly little water and probably get most of their water from the dew or the rain on the things they use for food or from the water that is actually present in the food that they eat. Nevertheless, patas monkeys in captivity drink in a regular fashion.

There is no doubt that the patas monkey has good speed on the ground. Dr. Neil Tappen, in 1964, reported that he was able to drive his station wagon alongside a patas monkey, and the speedometer on his vehicle registered about 45 miles per hour. The animals can keep up a fast pace across the savanna grass for quite considerable distances. Even adult females carrying babies can move with great speed. Patas monkeys have adapted to savanna life by the development of great speed, and in this they are greatly superior to the baboons. The baboons have adapted to the savanna life and the threat of predators by forming very highly organized groups with a number of associated big males and by developing a bigger, stronger, and more vicious behavior. The reaction of the patas, on the other hand, is to run away from trouble, not to meet it. When they are running away from some object that has startled them, some of the animals will leap into the vertical position as they are running and look back over their shoulders at what they are running away from; they then drop back into the four-legged gallop without a pause. When they are on the ground feeding, they can drop suddenly into an erect position and maintain this for half a minute or more. They commonly stand up on their legs when they are feeding in order to bring down the grass heads so that they can eat the seeds or pull berries off bushes.

The sexual behavior of the patas monkey does not differ very much from that of most other monkeys. They will mount each other on the branches of trees or on the ground at most hours of the day, though the most common time is when the group is resting or when it is feeding around trees. Grooming, which the adult male does especially to the adult female, seems to occur mostly as a preliminary to sexual behavior. When the female wishes to solicit copulation by the male, she has a special behavior which has been described very well by Dr. Hall:

> The female's cringing half-run with short quick steps towards the adult male was usually seen in the wild as she approached him along a tree branch during the day resting period. On nearing him, she would sit with her back to him and arched forward, her cheek pouches out, and the jaw thrust forward, as she turned her head to look at him. This puffing out of the cheeks is accompanied, in the captive females, by a wheezing, chortling sound going with the

Patas monkey.

expiration of breath—the sound never heard in the wild because it is inaudible at about six meters.

In the wild group, a female has repeated this soliciting behavior several times to the adult male without him necessarily responding. It has never been seen to occur except to the one adult male of the group.

Copulation results in a series of mounts with pelvic thrusts repeated about every two minutes until ejaculation takes place. It seems that the male will do this several times a day for two or three days with a particular female, and then he will cease to have any interest in her and relax until the next female comes into heat.

The life of the adult male during sex time in the group is not all a bed of roses, as the following description taken from Dr. Hall indicates.

An adult female approached the adult male several times with soliciting behavior sequence; he mounted her, and was at once

231

threatened by two other adult females and by some juveniles; he dismounted, and made a threat charge at one female who ran away; he then leapt against the small bush, in the side bouncing display . . . and rebounded from it; he went up to another adult female and put his nose up to her anal-genital area; two other females in turn did the soliciting sequence on him. On another occasion a female he had mounted herself turned on him and was joined in threatening him by two other females. It looks as though this animal just could not do anything right.

The other group of pataslike monkeys, the nisnas monkeys, have faces with black skin, and the hair across the browline is darker than the rest of the head hair, giving a distinct line. Nisnas monkeys are very social animals; they usually run in quite large parties and are said, when they are resting or eating, to post sentries at high points so that they may alert the group to any approaching predators. They are very cooperative as pets and can be trained to walk on a leash or even to heel like a domestic dog. They do not have the same destructive tendencies as other monkeys and can be toilet-trained. Apparently they will challenge the biggest dogs if the dogs trespass on what they regard as their territory, which, if it is your pet, will be your garden. They seem to be very even-tempered animals and quite reliable.

The Leaf-Eaters

There are three groups of monkeys that we still have not talked about. These are leaf-eating monkeys—those that live predominantly on leaves. There are three sorts. The first comprises the guereza or colobus monkeys. Then there are the beautiful langurs, of which there are five types. Finally there are two groups of monkeys with unusual noses—one is the snub-nosed monkey, and the other is the proboscis monkey of Borneo. Some of these leaf monkeys are found in Africa and some in Asia, from western Pakistan to Tibet and China, and they are also found in Ceylon, Java, and Borneo.

The colobus or guereza monkeys were well known in antiquity and probably were known to the Romans. They are black and white; the black is jet black set off with white bushy tails and long capes. When the animals are in the trees, these capes look like hanging mosses and lichens—such good camouflage that it is very difficult to see the colobus monkeys in the wild. Some colobus monkeys come from Africa and others from Ethiopia.

They are big monkeys, 30 to 40 inches in length, with the tail varying from 2 to 5 feet. Their fur is thick and makes them look much

Courtesy of Thüringer Zoopark, Erfurt, German Democratic Republic

Colobus (guereza) family (*Colobus abyssinicus*).

Colobus (guereza) monkey and baby (*Colobus abyssinicus*).

Courtesy of Thüringer Zoopark, Erfurt, German Democratic Republic

Colobus (guereza) monkey pair.

bigger than they are. In the process of evolution they seem to have lost their thumbs; their buttocks are bare (they have ischial callosities), and their muzzles are foreshortened. They are very agile and can get about in the tops of trees with great speed. They are also fine jumpers.

Colobus monkeys have a very unusual stomach. In other monkeys the stomach is simple and rounded, rather like the human stomach, but in the colobus monkeys there are a successive series of stomachs, the connections between them being quite narrow. In some respects this multiple stomach resembles that of the cow that chews the cud. However, there is a fundamental difference between the way the two types of stomach operate. Most of the coloboids eat leaves almost exclusively, but despite this vegetarian tendency, they also eat insects. The multilayered stomach is directly related to digestion in the colobus monkey. It eats the leaves of trees and passes them into the first stomach, where they are attacked by bacteria, fermented, and broken down into simpler products. With this fine supply of food, the bacteria thrive and grow and multiply until they pass down into the subsequent stomachs, where they are digested by the monkey's digestive juices and the animal absorbs the products of the digestion of the bacteria as food.

It is a little different in the cow. In this animal the bacteria which live in the rumen actually break down the cellulose of the grass and other material that the cow eats, and the products of the digestion then pass down further into the intestinal tract, where they are absorbed. In other words, the cow uses bacteria to digest the material which it does not have the digestive enzymes to do itself. In the

234

Colobus (guereza) monkey and baby.

Courtesy of the San Diego Zoo

colobus monkeys, the bacteria, after digesting the leaves, are themselves then digested and used as food by the monkeys.

Black and white colobus skins have been greatly prized in the past and have supported a flourishing fur trade. The skins were first brought to the furriers of Venice and Genoa from the Orient. The furriers thought that the Oriental craftsmen had developed a technique of inserting long white hairs into what were naturally black skins. It was not until the Europeans got into East Africa and explored it that they found that these skins were completely natural. Apparently they had been exported from Africa by the Arabs to Zanzibar and Mombasa, and from there they had gone to India. Then they had been shipped to Central Asia, where they were greatly prized by the various khans and the wealthy merchants. When they came back to Genoa and Venice from Central Asia, they were believed by the furriers to be fakes. By the middle of the nineteenth century Europe and America had discovered these skins, and they became very popular. The demand for them by the fur trade became so great that many of the African tribes were persuaded to send their hunters out to kill these animals for their furs. Some of the native tribes actually used the skins as headdresses for the tribal chieftains and for a variety of other decorative purposes. For this latter purpose, however, they had been hunted only with the bow and arrow, and not many were taken, but the advent of firearms led to a great slaughter of these animals; in 1892, 175,000 colobus monkey skins reached the European market alone. Since only perfect skins were accepted for the market, tens of thousands more must have been slaughtered but discarded as unacceptable.

After about ten years, the wholesale slaughter was stopped, and some kind of legal protection was secured for the monkeys. However,

235

Nilgiri langur (*Presbytis johni*) and baby.

Courtesy of Thüringer Zoopark, Erfurt, German Democratic Republic

236

Front view of spectacled langur shows white rims round the eyes which give animal its name.

Spectacled langur and baby.

Hanuman langur.

by this time more than 2,000,000 of the skins had been sold on the fur market, and the population of the colobus monkeys had been severely decimated.

Langurs

India is the home not only of the rhesus monkey, but also of a variety of coloboid monkey known as the langur. These long-limbed animals with long tails are graceful in appearance and in movement. In most of India they are regarded as sacred. They can be found all over the place, but particularly in the sacred city of Benares. They are said to be dedicated to the god Hanuman and are called Hanuman langurs. Because they are dedicated to the god, they enjoy absolute protection from human beings. As a result, they are not wild animals, but neither are they tame. They just wander around the country in enormous crowds at their own pleasure and obviously very much aware that they are immune to humans. Sometimes they live in the jungle for a time and then come into a town and rob shops and private

Hanuman langur.

homes, and nobody dares to interfere with them. They even rob the temples, but despite this, the priests encourage them to reside there and feed them regularly with plenty of grain even though Indian babies and children may be starving. Foreigners have been killed in India by crowds of Indian people because they have killed or interfered with a langur.

The scientific name of the common Indian or Hanuman langur is *Presbytis entellus*, and they are also found in Ceylon. They live in a variety of habitats, from dry scrub to low trees and thick damp forest, from almost sea level to 11,000 feet; in fact, they are known in the foothills of the Himalayan Mountains and have even been seen in fields of snow.

Among the predators of langurs are tigers, panthers, and wild dogs. In some areas rhesus monkeys live in the same place as the langurs, but they usually ignore each other, even when they are actually feeding on the same trees. Two rhesus macaques have been seen living

239

Capped langur.

with a troupe of langurs in one place, and in another place juvenile langurs have been seen playing with juvenile bonnet macaques.

The troupes vary in size from about 5 to 120, and some adult males live outside the troupes. Troupes can move as much as two miles in the day. In the morning the animals are most active, and they resume activity in the evening; but at midday they rest. Infants are passed about between the mothers and other females in the troupe but have been known to be kidnapped by a female from another troupe. They are weaned between about the eleventh and the fifteenth month after birth. By the time they are two years old the ties are completely severed with the mother, and this is also about the time the mother gives birth to a second infant. In some troupes, however, two-year-old juveniles can sometimes be seen suckling side by side with an infant that has just been born. Adult males are not interested in newborn animals and rarely are seen near any of them.

If a newborn infant has been kidnapped by a female from another troupe, it is very difficult for the mother to get the baby back. For instance, if she approaches the female that is holding her infant, the

Family of douc langurs.

alpha male and the other females of that troupe will chase her. Sometimes the alpha male of the mother's troupe will go with her and threaten or attack the troupe to which the kidnapper belongs.

An infant has to be ten months old before it can make contact with the adult males of the group. At this time it follows a prescribed procedure. It will run up to an adult male who is standing or walking and will mount it. As it dismounts, the male will sit down, and the infant will then run around to face the adult, and the infant and the adult male will embrace. One adult male may be mounted and embraced by a number of infants and juveniles in rapid succession.

Dominance among the langurs is not as clear-cut as it is among some other monkey groups. For instance, a dominant male in the langur group can copulate with a female in estrus, taking precedence over others; but this sexual dominance is not firmly established, and one male may have a continuous series of fights with the others to secure this right. Following the fighting some of the defeated males may actually leave the area. Among the females, a very dominant female seems to be able to take food when she wants to, and when she

241

Silver langurs.

Courtesy of the Fort Worth Zoo

is irritated or uptight, she may be avoided by females who are less dominant. The dominance is relatively unstable and poorly defined, and an adult female dominant in one situation is not necessarily dominant in another; temporary alliances of two or more females in these dominance interactions have been seen.

When two troupes of langurs meet, the dominant males leading each troupe will threaten each other by grinding the teeth, grimacing, and biting the ear. The tension will increase until one of these leader males will rush into the other group and make what is called a display jump, often followed by chasing a member of that troupe. The leader of the troupe which has been invaded will then chase the intruder, and the intruder will soon return to his own troupe. The two troupes will stand for a little longer, continuing to threaten each other and then will usually move on. Although it is mostly the leader males that do this interacting, sometimes adult and subadult females will fight the leader male or even the females of the other troupe. Juveniles have an

242

Family of douc langurs.

Courtesy of the San Diego Zoo

interesting reaction when the leader males threaten each other. Sometimes the male and female juveniles of one troupe will squeal and run forward, still squealing, and embrace the juveniles of the other group. It is only very rarely that these encounters between troupes develop into serious fights.

When there is a change of leader, the new leader sometimes comes from outside the troupe. The new leader may attack and injure or even kill all the infants in the group. The significance of this activity is difficult to understand, but the presence of a new leader is often followed by a marked increase in sexual activity in the group. There are not only mixed or bisexual * groups, but also groups exclusively composed of males. Sometimes a group of males that do not belong to the bisexual group will expel the male leader of the group. Often they fight together until only one of them remains, and he then becomes the leader of the troupe and, as mentioned earlier, is likely to kill all the infants. If the male leader is expelled in this way, sometimes the young males of the group will go with the expelled leader.

Another group of langurs are known as the purple-faced monkeys. They are black with a brown crown and whiskers. One variety, found on the island of Ceylon, has a coat that varies from silver-gray to pure black. It has long white whiskers and a brown crown on its head. All these purple-faced langurs have bright purple hairless faces. They are about two feet in height and have three-foot tails.

The other types of langurs are found outside India and occur in southern China, Burma, Thailand, Malaysia, and Indonesia. There

* The term "bisexual" used for monkey groups does not mean the same as it does in human behavior. In the case of monkeys, it means both sexes are present in the group.

Pileated langur.

are nineteen species of langurs in this area, including the capped langurs, the banded leaf monkeys, and the douc langurs. The capped langurs live mostly on the mainland, the leaf monkeys live mainly on the islands, and the douc monkeys live mainly in Indochina. The douc monkey has beautiful coloring; there are naked areas of the face which are yellow in one species and black in the other. They have oblique eyes, and they stand about two feet high, with tails that are less than two feet long. Their fur is glossy and very thick with a variety of colors and patterns. A collar of light yellow rings the neck, and behind this a belt of gray, brown, or black. The main part of the body and the arms are grizzled gray; the sides, the feet and the palms of the hands are black, while the forearms are pure white and the lower legs are reddish chestnut. Finally, there is a white triangular saddle across the rump, and the whole tail is white. The colors are well defined and very vivid.

We cannot possibly give a detailed discussion of every type of langur, so we will have to move on from these fascinating animals and talk about the two remaining monkeys.

Snub-Nosed Monkeys

One of these is the snub-nosed monkey, of which there are several species, and the other is the proboscis monkey, of which there is only one species. The first of the snub-nosed monkeys is called the Mentawai Islands langur or the Mentawai leaf monkey (*Simias cocolor*). The use of the word "langur" is inaccurate for this animal because it is not a langur at all and comes not only from the Mentawai Islands, which are near Sumatra, but also from some other Sumatran islands. They are about 24 inches long, with very small naked tails only 6 or 7 inches long. In many respects these animals look more like

244

Douc langur and baby (*Pygathris*).

Courtesy of the Thüringer Zoopark, Erfurt, German Democratic Republic

Douc langur and baby (*Pygathris*).

Courtesy of the Thüringer Zoopark, Erfurt, German Democratic Republic

Snub-nosed langur.

From Ernest P. Walker, *The Monkey Book* (New York, Macmillan Company, 1954)

macaques than any other type of monkey, but they differ from the macaques in having black, naked faces with brownish grizzled fur. They have a general doglike appearance. They are primarily arboreal, and very little is known about them.

Other types of snub-nosed monkeys known as *Rhinopithecus* (various species) vary in color; they usually have a brown or golden coat and sometimes have a white cape over the shoulders. They are found in Central China, in Kachin in upper Burma, and in the Hunan and the Szechuan provinces of China; some of them are also found in the eastern part of Tibet. They may reach three feet from head to bottom, not including the fairly long tail. Their fur is very long and thick. Most of them live in rather high mountains, and since they are resistant to cold, many live up near the snow line and migrate regularly from one range of mountains to another. Although they are relatively unknown, they have been hunted for their pelts in the past. The long, silky golden fur was a special prize and was apparently plucked and used for weaving into the cloth used for the upper classes in Chinese society.

Courtesy of the San Diego Zoo

Proboscis monkey family.

Proboscis monkey at Yerkes Primate Center.

Male and female proboscis monkeys.

Proboscis Monkeys

The proboscis monkey is found only on the island of Borneo. I have had the pleasure of observing these animals in the wild near the mouth of the Kuching River. Very few have ever been in captivity, so they are rather rare Primates even in the zoo. We had one animal at the Yerkes Center for some time which had some skeletal deformity, probably caused by improper nutrition (*e.g.*, vitamin D deficiency) when it was young. It lasted a few years at the center and then died. Proboscis monkeys spend a great deal of time lying on their backs or sunning themselves on the treetops. When they are relaxed in this way, they are apparently resentful of any interference either by any other animals or humans. The animals I saw were scampering about in the treetops and appeared to be very active, although when they saw me, they moved off to a distance from which I found it difficult to observe them. The males are big monkeys and may be 2½ feet from head to tail with a tail just as long, and their weight may go up as high as 45 to 50 pounds. The females are much smaller and lighter than the males. They have a sandy red color, which is deeper on the back, and pinkish faces. Their eyes are very small, their noses are large, and the tips are directed upward. The nose looks like an enormous, bulbous version of the human nose of the type characteristic of an alcoholic. In fact, the native name for the animal is the *orang blanda* which means a "white person" and obviously refers to the Dutch colonial white men whose large red noses, accentuated presumably by their consumption of alcohol, once made them outstanding figures in that part of the world.

IX. The Apes Take the Stage

SIXTEEN MILES south of Cairo there is a depression which is known as the Fayum depression. In Oligocene times, 45,000,000 years ago, the Fayum was a swamp which lay on the border of the Mediterranean Sea and was covered in jungle. A number of Primate fossils have been discovered here, including an animal called *Oligopithecus* which had thirty-two teeth like a monkey and is accepted by most scientists as being the first and most primitive of the monkeys. Not only did it have thirty-two teeth, but it had molar teeth with four cusps—a characteristic of monkey molars. The animal was tiny, only six or seven inches high. At this time, the New World and Old World were joined together, and the New World monkeys separated from the Old World monkeys around the same time. The progenitors of the New World monkeys went to America and were eventually forced by the Ice Age down into South and Central America. The land bridge between the Old and New World was lost. The two sets of monkey ancestors developed into the characteristic forms found in the two continents today.

Another discovery in the Egyptian Fayum was a fossil jaw. Eventually it was decided that this jaw belonged to an animal which was called *Parapithecus*. This animal seemed to be about the size of the modern-day squirrel monkey. Although its teeth were primitive in shape, it had the simple dental formula which we find today in Old World monkeys and in anthropoid apes. What made this jaw interesting was that the pattern of the cusps on its molar teeth was a simplified form of the pattern we find in the molar teeth of anthropoid apes and man today. This suggests that *Parapithecus* was either the ancestor or closely related to the ancestor of the animal which preceded the modern-day apes and man. Although *Parapithecus* was about the size of modern-day squirrel monkeys, the appearance of the jaw suggests it looked rather like *Tarsius,* which is found today only in Borneo and the Philippines.

The distinguished anthropologist and anatomist Professor Frederick Wood Jones, with whom I had the privilege of working during the 1930's, claimed about that time (1929) and even earlier (1916) that humans and apes were directly descended through animals related to *Tarsius* (the tarsier). Professor Sir Wilfrid E. Le Gros Clark of the University of Oxford also pointed out the similarities between *Parapithecus* and the tarsiers. It is evident that sometime between 45,000,000 and 35,000,000 years ago the monkey ancestor and the ape-man ancestor were already in existence and separated from each other. The monkey's ancestor went on to father the whole series of animals which are now the modern monkeys, and the ape and man ancestor went on to father the various anthropoid apes and man. In other words, apes and man did not have monkeys on their family tree. They developed separately from the monkey. Although the monkeys show many similarities to us, they are not our ancestors nor were their ancestors our ancestors. The distinguished Russian anthropologist Dr. M. Nesturkh does not favor the development of man through a *Tarsius*-like ancestor.

About 100 years ago, another small skull was found in the Egyptian Fayum. This belonged to an animal subsequently called *Propliopithecus*. This animal was about the size of a modern-day small gibbon and also had teeth like the modern gibbon. It was obviously a stage following

Parapithecus in the development of apelike characters and was a little later in the time scale.

By the beginning of Miocene times, about 35,000,000 years ago, there is some evidence that a great variety of different Primate-type creatures existed. Fossils of apes from the early Miocene epoch have been found in Kenya and other parts of central Africa. They are of many different sizes, ranging from that of a small gibbon up to that of a large ape with many similarities to modern-day chimpanzees and gorillas. Limb bones of the Miocene apes that have been found indicate that they were of a much larger build than the big apes today, and there is evidence that they were more active and probably much more agile than present-day apes. The bones of the limbs also suggest from their length and shape that the animals were probably much better at running along the ground than they were at swinging in the branches of the trees. The elongated arms associated with tree-swinging habits which we find in the modern orangutan, gibbon, or chimpanzee had not yet developed and are thus a more modern specialized development.

On an island in Lake Victoria in Kenya, the late Dr. Louis S. B. Leakey, the internationally famous anthropologist, discovered a fossil skull known as *Proconsul*. A number of varieties of *Proconsul* have since been found, and there have been suggestions that they may have given rise to the modern gorilla.

During the Miocene times an ape called *Limnopithecus* also flourished. *Limno* means "lake," so its name means the "lake monkey." The *Limnopithecus* skeletal remains show many similarities to the modern gibbon, and it may well have been the progenitor of the gibbon. Around the same geological time or a little earlier there was another creature which left fossil remains and subsequently became called *Dryopithecus*. Many specimens of *Dryopithecus* have been described, and they vary considerably in the anatomy of their teeth. In some of them the teeth appear more like a chimpanzee and in others more like an orang and in others more like a gorilla. It is possible that *Dryopithecus* may have been the originator of the three varieties of modern great ape. However, its skeleton shows it to have been a slightly built animal, and probably because of this, it had considerable agility; its arms were much shorter than those of the modern-day apes. Fourteen or fifteen million years ago, near the end of the Miocene and the beginning of the Pleistocene epoch a creature lived in India whose fossil remains have been christened *Ramapithecus* and which is now accepted by most anthropologists as being the earliest true ancestor of man.

251

2.5-MILLION-YEAR-OLD SKULL

Richard Leakey displays the skull that he believes is at least 2,500,000 years old—the largest complete skull of early man ever found. Leakey, holding the fossil's upper jaw in his left hand, unearthed the skull in fragments near Lake Rudolf in Kenya. Preliminary study indicates that the skull came from a creature of man's genus, *Homo*. Mr. Leakey's work is supported by the National Geographic Society.

The time scale used for dating these fossils is obtained by dating the rock strata in which the fossils are found. This dating is done now by radiodating—in other words, according to the radioactivity in the rocks. Recently, some University of California scientists, Drs. Allen C. Wilson and Vincent M. Sarich, have been studying the hemoglobin (the red pigment of the blood cells) in the gorilla, chimpanzee, and man. Hemoglobin is primarily a protein, and protein is made up of a number of units called amino acids. There are a number of different sorts of amino acids, and the sequence in which they are placed in any particular protein decides the nature of the protein and the animal it belongs to. The sequence of the amino acids of the same protein (*e.g.*, hemoglobin) is similar in related animals. For example, the hemoglobin of all the different types of men would have an identical sequence of amino acids—that is, they would follow each other in a specific order. One would expect that the amino acids in particular proteins, such as hemoglobin, would have an identical or very similar sequence in man and closely related animals like the apes. The two scientists found that in the 300 amino acids of the hemoglobin of human and gorilla the sequences varied in only two positions. Man and monkey, on the other hand, differed in twelve positions in the 300 amino acids.

This led Drs. Wilson and Sarich to suggest that the great apes and man may have split away from each other a very much shorter time ago than the 25,000,000 to 30,000,000 years suggested by paleontologists. They suggested possibly 5,000,000 years may be more likely. Subsequently, Drs. M. C. Clark and A. C. Wilson, studying the electric charge on the serum proteins in man and chimpanzee, concluded that the two species diverged from each other 5,000,000 to 11,000,000 years ago. However, the difficulties of reconciling all these facts can be shown by the fact that Dr. Richard E. Leakey has recently found a skull, as yet unnamed, which is very much more human than any of the ancient skulls that have been described so far and which is 2,500,000 years old. Although this skull has a fairly sloping forehead and is rather prognathous (in other words, the upper jaw sticks out a bit like a snout), it lacks the heavy eyebrow ridges which are characteristic of all the early fossil skulls such as *Homo erectus,* which has been found in Indonesia, China, and Australia and dated only 1,000,000 years ago. The cranial capacity of this skull was about 800 cubic centimeters—twice the cubic capacity of the brains of the great apes or the early fossil man, *Australopithecus.* This skull is said certainly to belong to the genus *Homo,* and in view of its age, its owner must have been living at the same time as the ape-man. The current view of the origin of man is that he developed through the line of *Australopithecus* to *Homo erectus* to modern man, but this skull shows us that an almost modern man was living 2,500,000 years ago, with *Australopithecus,* and lived with him for about 1,000,000 years.

Richard Leakey was the son of Dr. Louis Leakey, who made fantastic contributions to our knowledge of fossil man. Louis Leakey died on October 1, 1972, in a London hospital in England at the age of sixty-nine. He had produced a number of theories of human evolution. He promoted the theory that man's earliest origins could be traced back to Africa, although the idea did not originate with him. It was expressed by Charles Darwin in *The Descent of Man* in 1871. Charles Darwin said, "It is . . . probable that Africa was formerly inhabited by extinct apes closely allied to the gorilla and chimpanzee; and as these two species are now man's nearest allies, it is somewhat more probable that our early progenitors lived on the African continent than elsewhere."

Dr. Leakey was born on August 7, 1903, near the village of Kabete, not far from Nairobi in Kenya, where his parents were missionaries. His early childhood was spent with the youths of the Kikuyu tribe. So intimate did he become with them that at thirteen and a half he was given the manhood initiation by their tribal rite and was named Wakaruigi, which means "the son of the sparrow." A reporter once asked him if he thought that his childhood among the Kikuyu had helped him in his subsequent profession of the hunter of fossils. Dr. Leakey said it taught him "Two things. Patience—especially patience—and observation. In Africa, survival depends upon your reaction to irregularities in your surroundings—a torn leaf, a paw print, a bush that rustles, and patience."

At the age of sixteen, Dr. Leakey was sent to an English public school in Dorset and then went on to Cambridge University. In the 1920's and early 1930's, he led a number of archaeological expeditions into East Africa. In 1937, he began his excavation at the Olduvai Gorge, which is located in the Serengeti Plain in Tanzania. From that time forward he made a number of spectacular finds. He worked in Olduvai Gorge for twenty-eight years, finding many fascinating animal fossils before his wife, Mary, found the first human fossil, on July 17, 1959. On that day Dr. Leakey had a fever and was resting at the camp. His wife had gone off to search in the gorge. She suddenly turned up in her Land Rover, shouting out to Louis, "I've found him! Found our man!" Dr. Leakey hurried with his wife to the cliff in the Olduvai Gorge where she had been looking, to find some pieces of human teeth and skull there. When they put the pieces together, it was obviously a skull similar to those of *Australopithecus* (originally

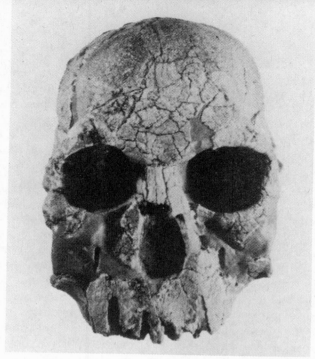

New fossil humanoid skull found by Richard Leakey in Africa.

discovered many years before in south Africa by Dr. Raymond Dart), and he was named by Dr. Leakey *Zinjanthropus,* also known as the "nutcracker man." He was dated 1,700,000 years ago, which means he was at the latter end of the Australopithecine series. Among *Zinjanthropus* material Dr. Leakey found other skeletal material which he said was sufficiently different from *Zinjanthropus* and sufficiently similar to the human skull for the creature to be called *Homo;* it thus became *Homo habilis.* The discovery caused a fair amount of controversy. There were some experts who thought that the new skull did not differ enough from the Australopithecine skulls to be put in a different genus. Australopithecines have now been found in the Middle East, in Israel, and in the Sahara, and recently, there is some evidence that they also existed in Java and China. So they were a fairly successful race of prehumans. *Homo habilis* apparently walked erect, and a number of stone tools have been found near his remains. Whether he was an intermediate between the Australopithecines and the next stage in man's development, *Homo erectus,* remains to be seen.

In 1962, Leakey discovered the upper jaw of an apelike creature in some Miocene deposits which are dated approximately 14,000,000 years ago. He gave this specimen the name of *Kenyapithecus*. He suggested also that this animal was a tool user since its skull was found in association with a number of bones fractured in a particular way. It seems likely now that *Kenyapithecus* was a variety of *Ramapithecus* and that this type of creature was fairly widespread over Africa and Asia during Miocene times. It seems that where both *Ramapithecus* and *Kenyapithecus* lived, there was a forest, and it may be that these creatures were not entirely terrestrial but spent some time up in the trees. There is anatomical evidence that early hominids in general lived in the trees. But apparently *Ramapithecus* and *Kenyapithecus* were making a greater use of the legs and using their arms less for support and probably not at all for locomotion along the ground. So when they came down out of the trees, they came down as true bipeds and not as quadrupeds like the gorilla, the chimpanzee, and the orang, which use all four limbs for support and for movement on the ground.

Elaine Morgan, a Welsh housewife who wrote *The Descent of Woman*, has pursued the theme originally suggested by Sir Alister Hardy, which we will talk about in more detail later, that man had a marine origin. Her hypothesis is that man came out of the water some 10,000,000 years ago to become an Australopithecine, a suggestion for which there is no evidence. There are, however, one or two things of interest about man which are a little difficult to explain. One of these is the fact that he has salt tears, and this is a means of excreting salt. Among mammals, this is found only among those in a marine habitat.

From all these evolutionary activities which took place millions of years ago, there emerged on the present stage the human being, the gorilla, the orangutan, the chimpanzee, and the gibbon, with its close relative the siamang, which is, in some respects, intermediate between the gibbons and the other apes. These animals are classified into three groups: the lesser apes, which include the gibbon and the siamang; the great apes or anthropoid apes, which include the chimpanzee, the gorilla, and the orang; and man himself.

X. The Lesser Apes—
The Gibbon and the Siamang

GIBBONS ARE slender animals with beautifully soft fur of many different colors. They have very long arms and no tails and have the ability to swing in a beautiful and rhythmical way, hand over hand, through the trees. They are found in only limited areas of the world now—in Indonesia, Indochina, Thailand, and Burma. Some of them, the concolor gibbons and the siamangs, have a large vocal sac in the throat, connected to the trachea and inflatable at will, that is used for calling in the wild. Some of the gibbons differ from each other in

257

"What do you want?"—
Siamang expresses doubt.

Yerkes Primate Center

the presence or absence of fur on the rump. For instance, the hoolock gibbon has fur on its rump, but the lar gibbons and the agile gibbons have a partly naked rump and a fringe of light hair around the face. The wau-wau gibbons have a naked rump but do not have the fringe around the face.

The siamang is jet black. Its fur is rather shaggy, and it has a naked face except for a few hairs. Underneath its chin its red vocal sac is also devoid of hair. There is only one species of siamang, and it is found in the island of Sumatra and in the Malay Peninsula. It is completely arboreal and moves with great speed through the trees. At dawn and at sunset, the forests resound with siamang calls. Of all the gibbon group, the siamang is closest to the great apes; it is, in fact, intermediate between the true gibbons and the apes. Siamangs are surly animals and get especially vicious as they get older. They never really settle down in captivity even after ten years. People who live in Indonesia and who keep gibbons as pets rarely keep siamangs. They are hard to control and have to be kept chained, nor are they pleasant or safe with children.

Gibbons and siamangs can often leap as far as 45 feet from one tree branch to another. When they are moving rapidly through the forest, they are in the air more than they are hanging onto anything. They tend to live in pairs with the young and establish what might be called feeding territory, and they warn their neighbors of its existence by their penetrating calls. When the sun rises and sets, they make a tremendous noise with a series of barking hoots.

A family of gibbons usually consists of a mother and a father and as many as three offspring, usually spaced about two years apart. Only one infant is born at a time. When a gibbon reaches the age of about six, it is chased away from its family and has to pick up a mate from

Portrait of a siamang.

Siamang showing partly inflated vocal sac.

among other gibbons who have been chased out and begin the nucleus of a new family. Their period of maturity seems to start at about seven years of age. Their exact life-span is not known, but it is said to be about twenty-five years.

The gibbon was first described in the eighteenth century by Buffon, the French zoologist, whose animals came from Indochina. He said, "It always holds itself erect even when walking on all fours, because its arms are as long as its body and legs." However, the Chinese have known the gibbons for a very long time. Jack Fooden, in an article in the *Bulletin of the Field Museum of Natural History*, quotes R. H. van Gulick, a Dutch scholar who spent a good deal of time in the Far East in the diplomatic service of the Netherlands. He wrote:

> From the first centuries of our era on, Chinese writers have celebrated the gibbon in prose and poetry, dwelling in loving detail on his habits, both in the wild and in captivity. Great Chinese painters have drawn the gibbon in all shapes and attitudes; until about the fourteenth century from living models, and when thereafter the increasing deforestation had reduced the gibbons' habitat to southwest China, basing their pictures on the work of former painters and on hearsay. So important was the gibbon in Chinese art and literature that he migrated to Japan and Korea together with the other Chinese literary and artistic motifs, although neither Japan nor Korea ever belonged to the gibbons' habitat.

The gibbon thus occupies a unique place in Far Eastern culture, it being possible to trace the extent of its habitat, appearance, and mannerisms for more than 2,000 years. It has been thought by the Chinese, from a very ancient period, to be the aristocrat among the apes and the monkeys. They regarded the macaque, or the monkey, on the other hand, as "the symbol of human astute trickery but also of human credulity and general foolishness." The Chinese philosopher Huai-nan-tzu, who died a little more than 100 years before the birth of Christ (120 B.C.), said of the gibbon, "If you put a gibbon inside a cage, you might as well keep a pig. It is not because the gibbon is then not clever or swift anymore, but because he has no opportunity to display his abilities."

Jack Fooden says that he first became interested in the colors of the coat of the gibbon, which varies enormously from silver gray through blond and brown to dark brunet or even black, in 1967 when he conducted an expedition to western Thailand for the Field Museum of Natural History. The idea of the expedition had been to make a study of the monkeys present in the forest of this area, but a few of his

Lar gibbon: buff phase on left, black phase on right.

colleagues also collected a number of specimens of gibbons. The color of the coat is not related to the sex since blond and brunet males and blond and brunet females exist at all stages of development from infancy up to old age. Blond animals are called by the Thais *cha-nee khao,* which means "white gibbon," and the darker animals, the brunets, were called *cha-nee dan,* which means "black gibbon."

Dr. Robert van Gulick, whom I mentioned earlier, searched Chinese literature and art for a reference to the gibbon from 1500 B.C. to the end of the Ming Dynasty (A.D. 1644). It appears that 1,000 years ago the gibbons ranged to the northeast in China as far as the yellow River. Their present northern limit of distribution is 800 miles south of that.

The disappearance of gibbons from China during historic times is probably the result of deforestation, which is of course correlated with the development of agriculture. The gibbons depicted by paintings on the Chinese scrolls are very lifelike and are portrayed with a good deal of detail, and they appear to be anatomically accurate. The species of

261

the gibbon that occupied China 1,000 years ago was probably *Hylobates agilis* which now does not extend beyond the island of Sumatra and the northwest of Malaya.

Louis le Comte gave a description of a type of ape, which probably was the gibbon, seen in the Strait of Malacca (part of Malaya) in the year 1697. Le Comte commented on how nimble the animal was and what great distances it could leap.

The gibbon is not particularly manlike to look at, and as a result, it did not get as much attention from the early scientists as did the gorillas and the chimpanzees. During the eighteenth century, particularly in the latter part, a number of authors wrote about it; and even its behavior and the sounds it made have been recorded. The name "wau-wau" was given to this gibbon, as it was said, because of a fanciful similarity to an old woman. The Javanese word *wau* means "an old woman."

The Siamese missionary Tabraca, quoted by Richard Turpin in 1771, has this to say about the gibbon:

> One takes pleasure in raising this species of monkey (the onke, siamese gibbon) because they are gentler than the others. . . . Economical and dextrous, they never break or smash anything. Friends of peace, and full of compassion, they wish to embrace those who weep, and their pity redoubles in proportion as they hear the groans of the unhappy; they never abandon those whose tears have not dried.

H. O. Forbes, writing in 1896, gives the following account of the young siamang found clinging tightly to its dead mother and then adopted as a pet:

> Its expression of countenance is most intelligent and often very human; but in captivity it generally wears a sad and dejected aspect, which quite disappears in its excited moods. With what elegance and gentleness it takes with its delicate tapered fingers whatever is offered to it! Except for the hairiness, its hands, and, in its use at all events, its head, seems to me more human than those of any other apes. . . . The gentle and caressing way in which it clasped me around the neck with its long arms, laying its head on my chest, and watching my face with its dark brown eyes, uttering a satisfied crooning sound, is most engaging. Although it often inflates its laryngeal sac, it rarely gives utterance to more than a yawn-like noise or a suppressed bark; but this dilatation has no reference apparently to its good or bad temper, although when very eager and impatient for anything, a low pumping bark is uttered. Every evening it makes with me a tour

Pileated gibbons.

around the village square, with one of its hands on my arm. It is a very curious and ludicrous sight to see it in the erect attitude on its somewhat bandy legs, hurrying along in a most frantic haste, as if to keep its head from outrunning its feet, with its long free arms seesawing in a most odd way over its head to balance itself, and now and again touching the ground with its fingertips or its knuckles.

The gibbon will be jealous of attention given by its master or mistress to another animal. In this respect Sir Stanford Raffles says, "It is the general belief of the people of the country that it will die of vexation if it sees the preference given to another; in corroboration of which I may add, as the one in my possession sickened in this situation, and did not recover until we moved from the cause of vexation by his rival the siamang when we moved to another apartment."

263

Gibbons have a very well-defined "fear" expression. A good photograph of a gibbon showing fear is to be found in Yerkes' book on the great apes, and Forbes also describes two occasions in which he interpreted grief and sadness in these animals. "The wau wau has a wonderfully human look in its eyes; and it is with great distress that the writer witnessed a death of the only one he ever shot. Falling on its back with a thud on the ground, it raised itself on its elbows, passed its long tapered fingers over the wound, gave a woeful look at them, at his slayer, and fell back at full length—dead." "Saperti orang" (just like a man), his Malay companion remarked. He kept a specimen in captivity for a short time, and:

> . . . it became one of the most gentle and engaging creatures possible; but when the calling of its free mates reached its prison house, it used, most pathetically, to place its ear close to the bars of the cage and listen with such intense and eager wistfulness that it was impossible to retain it in any durance any longer. It was accordingly set free on the margin of its old forest home. Strange to say, its former companions, perceiving perhaps the odor of captivity about it, seemed to distrust its respectability, and refused to allow it to mingle with them. Amid the free woods we may hope that this taint was soon lost and it recovered all its pristine happiness.

The gibbon has a voice which not only has remarkable range but is surprisingly musical and attractive. This was one of the things about it that attracted the attention of early naturalists. In 1841, a naturalist named Martin produced a musical notation of the gibbon call and gave the following description:

> The voice of this gibbon is extraordinary, not only for its power and volume, but for the succession of graduated tones in which its cry is uttered. In the room, it is overpowering and deafening: it consists of a repetition of the syllables oo-ah, oo-ah, at first distinctly repeated, and ascending in the scale but at last ending in a shake, consisting of a quick vibratory series of descending notes during which the whole animal's frame quivers to produce them; after this, she appears greatly excited and violently shakes the netting or the branch to which she may be clinging; which action being finished, she again traverses her cage, uttering the preliminary syllables oo-ah, oo-ah, till the shaking then concludes the series. It is principally in the morning that the animal most exerts this modulated cry, which is, probably, its natural call to its mate, and which, from its strength, is well calculated for resounding through the vast forest. . . .

The brain of the gibbon is very small, but then it is a very small animal. The size varies between about 90 and 100 cubic centimeters. The siamang brain, which is actually a little larger, is around 125 cubic centimeters. Because a big animal has a big brain, scientists usually compare the weight of the brain with the weight of the body in order to get an idea of the comparative size of the brain in different animals. Different Primates show a brain weight and body weight ratio: the marmoset, 3.82 percent of the body weight; the siamang, 1.16 percent of the body weight; the gibbon, 1.67 percent; the rhesus monkey, 1.43 percent; the orangutan, 0.54 percent; and an adult man, 2.81 percent. The fact that the brain weight to body weight is so large in the marmoset, which is actually the most primitive of all the monkeys, suggests that we can place a very limited significance on this ratio. In other words, the marmoset is too different for a comparison to be made. However, the orang and man are sufficiently similar in size that a comparison between their brains and body weight could be better related to their different levels of intelligence. There are some areas in the gibbon's brain, especially in the cerebellum and in the basal part of the brain, which suggest that the gibbon is perhaps more related to monkeys than to anthropoid apes.

Dr. Ray Carpenter was the scientist who first studied in detail and analyzed the method of locomotion in the gibbon. This type of locomotion is known as brachiation, which means "swinging by the arms." Dr. Carpenter analyzed this method of locomotion by direct observation and also by analyzing slow-motion movies. He found that the fundamental motion involved in brachiation is that of a swinging pendulum. When the animal finally came to the end of a brachiating run, it would oscillate backward and forward before coming to rest.

How does a gibbon start its swing? He stands up from the sitting position, extends an arm and hand, and falls smoothly forward and downward with his legs tucked up near his body. At the bottom of the glide, the hand that encircles a branch checks his downward movement and swings him forward. At the apex of the forward swing, he may let go and fly through the air till his outstretched hand catches another branch, or he may catch the branch at the moment of letting go with the first hand. The whole thing is done with incredible smoothness and rhythm. The gibbon can cover considerable distances this way at a pace at least as fast as a man can run. It is estimated that the animal will cover 20 feet in two and a half swings.

A gibbon jumping onto a broken branch has been seen to grab the stub of the broken branch which projected only about 6 inches; it swung under it and over and then, apparently with no loss of momentum, made a jump of 30 feet to the next tree. Where the jumps are very big, mothers with babies tend to make detours to avoid them.

Dr. Carpenter, on one occasion, saw a gibbon swing on a branch up to a bird's nest located on another branch, pick off the nest and place it on its feet, and then swing back to a branch on which it was able to sit in comfort. It then removed the nest from its feet with its hands, sat with the nest in its lap, and removed the eggs and ate them. Gibbons are very choosy about how they eat. They examine food with great care and look for small spots of decay, which they remove with their teeth and their tongues, and they also eliminate the skins and many of the seeds and pits of fruit. They have been known to place a hand in the path of an advancing column of ants and wait until the ants have crawled all over it and then pick up the hand and lick the ants off it. They also eat a variety of other insects, birds' eggs, and young birds. Among the fruits, the fig is probably the favorite, and they also eat mangoes, grapes, plums, and other fruits. They will drink water from the trees, and they will also lick their fur to obtain water. When they come upon a pool of water and wish to drink from it, rather than put their face in it, they clench their hands into fists and dip them into the water and then drip the water from the knuckles and the back of the hand into the mouth or even suck it from the hand. In captivity, however, they have been known to drink by sucking the water directly into their mouths.

In Burma Dr. Carpenter found there was a story about reincarnation connected with gibbons. Gibbons are believed to be disappointed lovers who have been reincarnated. It seems likely that the rather long and mournful call of the gibbon is the origin of this rather strange story. Some of the natives in these areas will not kill gibbons because of their beliefs that they are reincarnated humans.

When dawn breaks in the forest where the gibbons live, the first indication of their existence is the early-morning calls they make. The calls float back and forth between the various groups as the early-morning call of one animal will stimulate calls from all the other groups in the vicinity. They answer back and forth for up to an hour or more. Then the time comes to start feeding. The leader of the little group will lead the way by swinging along the trees for perhaps half a mile or so. They may stop along the way to fiddle around and perhaps pick up a thing or two to eat, and sometimes they encounter another group also out looking for food. They then give each other a series of greeting calls, which are much more drawn-out than the early-morning calls. If both groups are anxious to get to a particular food tree,

266

Gibbon in a tree, Atlanta, Georgia.

Yerkes Primate Center

they will both stand by and indulge in what Dr. Carpenter calls "a vocal battle." They will call at each other in this fashion until one group or the other gives up and leaves the tree to the one which presumably has vocalized the best. They may eat for two or more hours. As they become full, they slowly cease feeding, and the young will play with each other or groom each other. The mother will relax and groom her infant. The male will also rest, but at the same time he will keep a watch out for any predators or any gibbons strange to the group. Then they have a siesta period, which lasts for about three hours, until they start swinging off again in search of another food tree. Then, after feeding a second time, they move back the way they came to some familiar tree to sleep for that night.

I had an opportunity a few years ago in Malaya at least to hear the morning calls. I was awakened about 4 A.M. and dragged out into the pitch-darkness in a Land Rover and driven out into the jungle. There we left the Land Rover by the road and tramped through the jungle to a particular tree known only to my guide, and how he found that in the dead of the night, I will never know. Stuck in the tree were a series of spikes that I could use to climb up the tree, and my guide warned me that they went up for a considerable distance. Up in the tree was a platform reached by climbing over a large branch and dropping down onto it. I climbed the spikes, one after the other. It seemed absolutely never-ending, particularly in the dark where everything was by feel and I could not see anything around me. Eventually I got to the large branch and managed to slide over it and down onto the platform. Then completely exhausted, having got only three hours' sleep, I lay down on the platform and fell asleep. As the sun came, however, the calls of the gibbon aroused me, and I could hear the groups calling from one part of the jungle to another. However, the thing that startled me was the height above the ground. I had no conception how far I had climbed. The tree projected above the jungle for quite a distance, so the platform we were on actually overlooked the green sea formed by the top of the jungle that lay below. It was a perfect place, however, for listening to the gibbons calling, and we could spot these areas for a considerable distance around and got some idea of the delineation of their territory. Of course, when it came the time to descend, I was terrified at the thought of climbing down the spikes that I had climbed up so happily earlier in the morning without really knowing what I was doing. However, I made it down, and in the end all was well. I was richer by having heard the musical morning calls of gibbons.

Gibbons very rarely seem to get into fights with each other under natural circumstances. They will fight, however, if they are closely confined in a captive situation, and when they do, they are very savage fighters. In the wild it is possible that these animals have used the technique of vocalization to replace the physical contact of fighting; however, in the captive situation they can be very aggressive, and very often their attacks are of an enraged, explosive nature. At Orange Park where we had a number of gibbons, one of the scientists who had good relationships with them was once very badly bitten on the forefinger, so badly that the tendons connecting to the terminal joint of the finger were severed. This was a very unfortunate occasion for him since he played a violin in the Jacksonville symphony orchestra. Although the tendon was repaired, he never had quite the same plasticity or rapidity of movement in the joint as he had before.

Belle Benchley at the San Diego Zoo has described gibbons as very dangerous if they decide to attack because they move so fast that it is almost impossible to see them. They will grab a hat, a strand of hair, or a sleeve and pull it or tear it before you even realize that you are under attack. Mrs. Benchley describes how on one occasion she was feeding cherries through the wire to the female of what appeared to be a pair of very tame and certainly docile gibbons. Suddenly the male dropped from a high bough, and Mrs. Benchley felt a jolt and a vicious blow on the face. She then observed that there was blood on the floor of the cage and realized that her middle finger had been bitten open by the diving male while the female that she was feeding had at the same time slapped her face.

Emily Hahn has recently written an account of her own personal relationships with a wau-wau gibbon she bought in a shop in Shanghai in 1935. The animal was very young and, like all young gibbons, was extremely affectionate. In fact, it became so attached to Emily Hahn's housemate, Mary, that it could not be dislodged from her without a great deal of shrieking and screaming. After a time it rejected Mary and accepted Mrs. Hahn as a surrogate mother. The latter found that the Chinese term for gibbon was *shing-shing*. She used to take it everywhere with her and always kept a diaper on it. She found that it would eat fruit and almost anything else that she would eat, even hens' eggs. It was not very fond of meat although it would nibble at it a little bit. It was very happy with grasshoppers and really enjoyed mealworms. If it saw anyone eating, it always wanted to sample the food.

One of the principles of handling that Emily Hahn discovered very soon was that the hands were very important to the gibbon and it did not want to be picked up either by the arms or by the body, but had to be picked up by the hands. If they attempted to pick it up any other way, it would bite and swing away from them.

As it got older and its jaw muscles got stronger, she found that its play bites were becoming much more painful. Eventually she had to hit it every time it bit. It apparently did not get the message, but finally when it bit her, she *bit* it back. It was so surprised that it finally got the idea that its owners did not want it to bite.

On one occasion Emily Hahn's Siamese cat had a litter of kittens. The gibbon behaved as if the kittens were produced as special playthings. It would clutch one in its foot and then swing across the ceiling from curtain pole to the chandelier, from the chandelier to the doortop with the mother running along the floor crying in agonized fashion to have her kitten back. It never dropped any kittens, however. Eventually the cat took her kittens to what she thought was a safe hiding place. But the gibbon found the hiding place and started to steal them again.

Emily Hahn eventually acquired a number of other gibbons and lived with them in Hong Kong. She used to take them to the beaches across the island and would leave them in the sand while the family went in to swim. In the beginning the gibbons cried when she and her family went into the water. They would never go into the water themselves but would run up and down in great distress on the dry sand beyond where the waves were breaking. Eventually they learned that the family always came back out of the water, and then they simply entertained themselves by playing and wrestling in the sand and picking up bits and pieces from the beach and examining them.

Although one thinks of gibbons as being entirely tropical animals, they can be found in fairly high and rather cold elevations in some

Three stages in swinging from a rope: 1. Gibbon climbs up with rope.

Courtesy of Dr. Duane Rumbaugh

parts of the world. June Johns has referred to the gibbons at the Chester Zoo in England and drawn attention to the fact that when the moat surrounding their island freezes, the gibbons very quickly learn to use the ice for sliding. One, however, instead of going back to the island at the end of the sliding, came ashore. Eventually it was caught and kept in a cage until the thaw set in and could be returned to the island. However, the other animals by this time would not accept it. They greeted it with bared teeth and would probably have killed it if it had not been removed. While one would expect a gibbon group to behave like this to an animal from another gibbon tribe that penetrated into their territory, it is a little strange that after a few weeks they would not accept the return into their group of one of their own. This is the same type of behavior found in rhesus monkeys, and we might have expected the gibbon to be a stage above that.

Female gibbons have menstrual cycles, but they do not get the periodic swellings around the perianal region that are characteristic of the rhesus and other monkeys and the chimpanzee. The cycles are around thirty days, and it is perhaps significant that all primates seem

2. Gibbon puts rope over bar.

3. Gibbon holds onto rope with one hand and swings on it with the other.

to have menstrual periods of around thirty days. The age when menstruation begins is around ten years. Copulation can occur standing on branches or hanging from branches. Dr. Carpenter saw only two instances of copulation in his study in over a six-month period in studies in Thailand. It is known that gibbons will live at least thirty years in captivity, and so, presumably, a female will have at least fifteen years of reproductive life. Since she bears one infant about every two years, she could, during her reproductive life, produce seven or eight babies. It might be noted that our chimpanzees who live to forty-eight (the oldest so far recorded) still have their menstrual cycles; there is no menopause.

After giving birth, the mother eats through the umbilical cord and most of the placenta (afterbirth). The babies weigh about one and a half pounds and have very thin arms and legs that look almost transparent. The fingers and toes are very long and are able to grasp the hair of the mother very firmly almost from the moment of birth. The infant is carried by the mother astride the lower part of her abdomen. She brings her legs up so that the thighs are close to the body and serve as a kind of shelf to hold the baby. Since she moves through the trees by swinging by her arms, using her legs this way is a perfect way to carry the baby. The baby's arms are very long, so they go almost around the mother and so give the baby a very good grip. When it gets older, it accompanies the mother, swinging alongside her. One time a mother was moving through the trees in this way with her infant when she came to a spot where to get from one tree to the next one, it was necessary to jump at least 10 feet. The mother made the jump, but the baby was reluctant to do so and hesitated and cried at its mother. The mother, hearing this, turned back and gave every appearance of trying to coax it to make the jump. However, the baby would not accept it, so the mother recrossed, took the jump back, picked up the infant onto her belly, and made the leap with it. Immediately after the leap, the baby left her and began to brachiate on its own again.

For the first six months of the baby's life, it hardly ever leaves the mother, but it will move away from her to play with other young animals, at first only for short periods. However, the slightest disturbance will send it flying back to its mother at high speed. Eventually as it gets more anxious to get its own food, the youngster will leave its mother for longer periods, although it will still come back to her for warmth. Finally, its ability to travel in the trees becomes well enough developed so that it can live an independent life.

A mother gibbon training her baby to move on its own has been known to bridge a small gap between trees by holding onto one branch with her legs and another with her hands so that the baby could cross by using her body. Dr. Carpenter has described a female who upon hearing gunfire in the distance, rushed to her baby playing a few feet away from her and pulled it to her, whereupon it clasped her firmly around the abdomen, and the mother prepared to brachiate through the trees in fright. He also describes how a gibbon that had been wounded got up onto the top branch of a tree and forced her infant away from her so that it would stay there and then brachiated off herself into another tree. This was apparently an attempt to protect the baby by luring away the hunters. There have been a number of other reports of this type of protective behavior by the mother gibbon.

There are never more than four baby gibbons in a family at any one time. The family relationships are very intimate. The young gibbons play most of the day, indulging in wrestling, slapping, biting, and racing around in the branches in a kind of follow-the-leader game.

The father in the group has been known to take some part in bringing up the children. He has even been known on occasions to carry the babies through the trees. The babies, as they get a little older, attempt to play with the father, and they are usually very persistent about playing with him. Sometimes he gets jealous of the young, particularly if one of the male offspring is getting close to maturity and makes sexual advances to the mother. The father has a very definite defensive function, and if any of the babies gives an alarm call, he instantly comes to its side. In a group of gibbons in captivity, an attempt to catch one of the young will lead the father to make a vigorous attack on the captor, even interposing himself between the catcher and the baby.

When gibbons meet, they indulge in a period of examination of each other. They examine each other's genitalia, they touch each other, they look each other over, and they smell each other. This is a preliminary activity; following this, the animals are either friendly or antagonistic according to whether they accept each other or not.

Because of the precision with which the gibbons are able to brachiate through the trees, it is obvious that they must have first-class vision and first-class muscular control; otherwise, they would not be able to pick out a small branch some distance away and swing to it without any difficulty. The gibbon has been reported to be able to detect quite modest movement among people observing it from as far

Courtesy of Dr. Duane Rumbaugh

Gibbon on the ground—standing upright.

away as a quarter of a mile. There is no doubt that as far as muscle control and vision are concerned, these animals are extremely well developed.

There have not been many detailed studies of the intelligence of Gibbons. In captitivity they pay no attention to human conversation and make no attempt to imitate it. Nor do they show any interest in music. There are reports, however, that one captive siamang imitated some of the sounds produced by dogs and guinea pigs. If shown a mirror, it would try to grab the animal it thought was behind the mirror. It seemed to be surprised not to find it there. Then it would get bored, yawn, lie down, and go to sleep.

The intellectual level of gibbons has been compared with monkeys and other apes in a number of tests by Dr. Duane Rumbaugh. He has found that in most of the tests, the gibbons certainly score way below those of the great apes and in some cases very little better than some of the monkeys. Gibbons will, like most monkeys and like other apes, rake in food with a rake. But if the food is placed on the other side of the rake so that the rake has to be lifted over it before it can be raked in, neither the gibbons nor the monkeys get the idea. However, the great apes easily solve this kind of problem. Dr. Rumbaugh found that if he gave a gibbon a piece of rope, it would throw it over a beam or a perch in its cage and hang onto one end with one hand while it swung on the other end. This activity would seem to require a considerable level of intelligence.

The danger in interpreting the results of tests with primates to determine whether they are clever or stupid is well brought out by Allison Jolly in *The Evolution of Primate Behavior*. Here she points out that gibbons for a number of years were unsuccessful at pulling in food attached to strings and were called stupid because of that. In 1967 a tester elevated the strings so that the animals could seize them more easily with their long hands. They pulled them in just like a chimpanzee and thus obviously had this degree of intelligence, but the humans who had given them the previous type of test apparently had not been as intelligent as they should have been in devising it.

An interesting experiment was carried out with two cebus (organ-grinder) monkeys, some rhesus monkeys, and a gibbon. They were given a test known as the paired stimulus cards test. The cards had various sizes of black and white spots on them. The animal was presented with a card with the white spots and then had to decide if the other card presented to it was identical or not. If it gave the correct response, it was able to open a door and get a food reward. The cebus monkeys and the gibbon learned three degrees of complexity, but the rhesus monkeys only succeeded in two steps. In a later and more difficult phase of this test, the gibbon failed, but the cebus monkey succeeded. This adds to the generally accepted view that the cebus is a highly intelligent animal. When they were selecting the cards, the cebus monkeys would often turn their heads and look at one card and then the other card before making their choices. The gibbon did not do this. During some of these tests the gibbon proved difficult to get into the test cage, and on occasion the experimenter would threaten it with a small iron bar pretending that he was going to hit it. However, it was not intimidated, and when he put it down and turned his back to find some other device to encourage it to get into the cage, it grabbed the iron bar, carried it up to the top of the cage, and put it in a place where he could not get it. The gibbon reacted very strongly to the frustration it felt when it had carried out several unsuccessful test trials and failed to receive its food reward. On such ocasions it would cry out and flail around. However, this would last only for a few minutes, and it would very soon revert to its normal affectionate self.

Gibbons in general seem to be very nervous and excitable animals.

**Gibbon on the ground—
bipedal walking.**

Courtesy of Dr. Duane Rumbaugh

They will react very aggressively to any type of frustration. When they get angry, they hold the lips tight and open and close the mouth almost as if they were smacking the lips together. Young gibbons captured in the wild are very aggressive and will bite anybody who picks them up, and they may refuse to eat for several days after they have been caught. If they are caught young enough, they can become very rewarding and affectionate animals until they become sexually mature. If you capture an adult gibbon, however, it can never be really tamed; it remains very dangerous and very aggressive for many months and then becomes sullen and irritable. Gibbons are popular pets in Southeast Asia. They are often kept on leashes and sometimes enclosed in bamboo cages. They are fed the same food that the people eat, which is a staple of rice perhaps with a little dried fish and some eggs and, of course, fruit. The young ones are given milk, usually buffalo milk. It is believed that the natives of the interior sometimes use the gibbons for food.

The Karen and the Mia, Burmese tribes, will not shoot gibbons since they believe that these animals have a beneficial effect on crops. This is probably because they do not rob human gardens—they live so much in the trees that they never come down to the ground to attack grain crops. Gibbons are too quick in the trees to be really susceptible to predators. Pythons may drop on them occasionally, but as far as the big cats, such as the leopards, and other predators are concerned, the gibbons are so fast that it would be very difficult to catch them.

Dr. Ray Carpenter made a study of gibbons in the spring of 1937. In that year an expedition was launched jointly by Harvard and Columbia universities to study the gibbon in the wild. The gentlemen involved were Harold Coolidge, Jr., Adolf Schultz, Ray Carpenter, Augustus Griswold, and Sherwood Washburn. They spent about four

276

months in northern Thailand studying gibbons. The gibbon Dr. Carpenter studied (he was in charge of the studies on behavior) was *Hylobates lar*. He had had a good deal of experience in Primate field work prior to this expedition since he had studied Barro Colorado monkeys and the spider monkeys in South America. In an area in Thailand called Doi Dao he found a number of gibbons, well protected because they were thought sacred by the natives. The monsoon forest was scattered over the mountains, but fortunately there had been a number of trails cut into the forest by elephants engaged in logging operations.

Some twenty-five years later John Ellefson led a study of wild gibbons in Malaysia in the state of Johore, which is near the Strait of Johore separating the peninsula of Malaya from Singapore Island. Ellefson worked on the same species of gibbon as Carpenter, and he confirmed that they, like the Thai gibbons, live in small family groups—each with its own territory and each defending its territory from other groups. He calculated that the territory of a family group of gibbons was about 250 acres, but territories ranged from about 40 acres to more than 300.

XI. The Great Apes

The Chimpanzee

In 1948 Mrs. L. S. B. Leakey, the wife of the famous anthropologist, discovered a fossil skull which became called *Proconsul*. The name *Proconsul* was chosen because in England at that time a chimpanzee called Consul was making quite a name for himself in musical acts. So the name *Proconsul* was given to this fossil to indicate that it existed prior to Consul.

Further studies demonstrated that there were several sizes of Proconsul which was one of the Dryopithecine apes. One of these was as big as a large monkey and was called *Proconsul africanus*; another one about the size of a present-day chimpanzee was called *Proconsul*

278

nyanzae; a bigger one whose body size seemed to be that of a present-day gorilla was called *Proconsul major*. The skulls of these apes, which date back to the early part of the Miocene epoch, some 20,000,000 years ago, did not look like those of present-day apes. For one thing, modern-day ape skulls have very heavy eyebrow ridges, and the *Proconsul* skulls lack these ridges. There were some differences in dentition compared with present-day apes, but the size of the brain was very similar. The studies by Dr. John Napier, who is working in London, demonstrated that the animal's arms and legs were about the same length. Thus, it lacked the longer arms of the present-day apes which developed as they specialized in swinging or brachiating through the trees.

It seems pretty sure that *Proconsul nyanzae* was an ancestor of the present-day chimpanzee. This means that 20,000,000 years ago the chimpanzee and gorilla ancestors had already separated from each other. These early apes spread all over Europe and Asia, moving wherever there was tropical forest.

The orangs, of course, became progressively more specialized and remained in the trees, but the chimpanzee, as well as the gorilla, became much less dependent on climbing. As the animals got bigger, they needed more food. They had to travel greater distances to get it, and most of this travel, because of their weight, had to be carried out on the ground. The animal developed thick, fatty pads on its feet which supported it as it walked and fatty pads on the knuckles of its hands because it walked on its knuckles. The skull became modified so that its face stuck out more, and it became prognathic, a condition not common in its ancestors.

In addition to the chimpanzee with which the public is very familiar, there is also a pygmy chimpanzee. This pygmy chimpanzee is found south of the Congo River.

The ancients must have had some experience of the anthropoid apes, but they have not left much information about them in their records. Also, they probably did not distinguish between them and monkeys so that references to monkeys probably may have included anthropoid apes without specifying them as such.

The same could be said for the people of the Middle Ages. Up until the thirteenth century there is very little chance of any anthropoid apes ever straying into the Western civilization; if the medieval mind had come across such animals, it would have reacted to them by explaining them either as a result of an illicit union between humans and monkeys or would have regarded them as one of the series of partly "human monstra" which figured so much in the zoological and

279

geographical literature of that time. Even the Barbary ape was classified as a "monstrum" simply because it bore some resemblance to man. This is probably the reason why the first report of manlike apes attracted very little attention. The first report seems to have been from Marco Polo, who described the presence of "great apes of such build that they had the appearance of men." Marco Polo's travels carried him through India, China, and Japan. The animals he saw were probably orangutans, but it is possible that he had never actually seen the animals himself because if he had, he would have had a great deal more to say about them. At any rate Marco Polo's is the first extant record in Europe of the presence of the manlike ape. This was 250 years before Andrew Battell described the "gigantapes" of central Africa, which he called Pongo and which were probably chimpanzees.

H. W. Janson points out that the great zoologists of the Renaissance had no interest in the "great apes of Canary." Conrad Gesner in 1551 in his *Historia animalium* does not mention them, although he does cite an account by Girolomo Cardano of a large, rare "cercopithecus-like form with the form of a human." This presumably is a reference to some type of anthropoid ape. He also provided a picture of what was said to be an ape. However, the animal he produced had a tail, and from the protrusion at the back it obviously had sexual skin, so that it could possibly have been a chimpanzee although it is more likely to have been a baboon.

The English explorer Andrew Battell, was imprisoned by the Portuguese in the late sixteenth century. Eventually he was sent to a Portuguese settlement in Angola on the west African coast. There he obtained a good deal of information about gorillas and chimpanzees. His description was not published until 1625, and in the report he refers to the animals as the Pongo and the Engeco. This is what he says about them:

> The woods are so covered with Baboones, Monkies, Apes, and Parrots, that it will feare any man to trauaile in the land. Here are also two kinds of Monsters, which are common in these Woods, and very dangerous. The greatest of these two monsters is called, Pongo, in their Language: and the lesser is called, Engeco. This Pongo is in all proportion like a man, but that he is more like a Giant in stature, than a man. . . .

It is probable that the Pongo was actually a gorilla and the Engeco a chimpanzee.

The first anthropoid ape of which there is any record in Europe is

From *Tyson's Anatomy of Pygmy*
by Edward Tyson, Dawsons of
Pall Mall, 1699.

that described by Dr. Nicolas Tulp, a famous Dutch physician and
anatomist, who was the doctor in Rembrandt's "The Anatomy
Lesson." He published a medical work in 1641 and in the appendix
described an ape he had received from Angola. This ape had been put
in the menagerie of the Prince of Orange and was accepted by most
people as a chimpanzee, but there is a possibility that it may have
been a young orangutan; the use of this name by Tulp served to
confuse the literature for some time. Dr. Tulp would have liked to
dissect the animal, but he was unable to and instead described its
appearance and published a picture of it.

On June 1, 1698, Edward Tyson described a chimpanzee he had
obtained, again from Angola. He gave an official description of it at a
meeting of the Royal Society which was, in fact, the first scientific
demonstration of the close similarity between man and one of the
lower animals. It was, as Vernon Reynolds points out, "a theory that
was to wait almost 200 years, until Darwin's *The Descent of Man* was
published in 1871, before gaining acceptance."

Reynolds describes Tyson as follows.

281

Tyson was an English doctor who lived in London, and was a Fellow of the Royal Society. He was born in 1650 and died in 1708. He made studies of the porpoise, rattlesnake, tapeworm, and opossum, besides his routine medical work. In early 1698, a live infant chimpanzee was brought to London from Angola, the first recorded occasion that an anthropoid ape had arrived in England. Tyson realized its extreme importance for science, and when the little creature died, in April 1698 he took its body home to dissect it. An excellent anatomist and draftsman named William Cowper helped him with a detailed description of the muscles and did numerous accurate drawings of the ape and of the dissection.

Tyson published his findings in a book in 1699 with the title *Orang-outang, sive Homo Sylvestris: or, the Anatomy of a Pygmie Compared with that of a Monkey, an Ape, and a Man. To which is added A Philological Essay Concerning the Pygmies, the Cynocephali, the Satyrs, and Sphinges of the Ancients. Wherein it will appear that they are all either Apes or Monkeys, and not Men as formerly pretended.*

The distinguished anthropologist Dr. N. F. Ashley Montagu, a biographer of Edward Tyson, has pointed out that Tyson's publication is of such importance that its author should be placed on the same level as Vesalius, the reformer of anatomy, and Darwin. Janson has this to say about his impact:

If he failed to gain the popular fame of these men, it is only because his achievement was accepted without controversy. But the impact of the orang utan could be felt almost immediately; it not only attracted wide attention among natural scientists but stirred the imagination of poets and philosophers, arousing them to flights of speculation about the nature of this novel creature, which they viewed as more human than simian. Tyson himself, while emphasizing its extraordinary resemblance to man, had clearly treated the orang utan as an animal, but the very fact that he called it a Homo Sylvestris and a pygmie suggested to the non-technical reader that it was a human. Moreover, at this very time the west had just sustained a similar shock: travellers were returning from south Africa with the earliest reports of the Hottentots, a tribe said to be so much more primitive than any other savages encountered heretofore, that they appeared to be more bestial than human. Thus we need not be surprised that the eighteenth century found it difficult to distinguish the orang utan "the super ape" his physical and mental resemblance to man was believed to go far beyond the limits of a mere animal, and the lowest forms of man, such as the Hottentots, failed to live up to the accepted minimum standards for human beings.

If Tyson was the first primatologist, Linnaeus was the first person to use the word "Primates." In 1758 he published an edition of his *Systema naturae*, originally published in 1735, in which he used the word "Primates" for a group of animals that included man, monkeys and apes, lemurs, and bats.

Carl Linnaeus was a Swedish botanist and was the first person to make a serious attempt to classify different animals and plants. Even today his work forms the basis of the classification of both plants and animals. His use of the word "Primates" to include the four groups of animals is exactly the same as modern thinking except that the bats are no longer included. In his genus *Homo*, Linnaeus used *Homo sapiens*, which is man, and *Homo troglodytes*, a creature which was probably an orangutan and was originally described by Jacobus Bontius, a physician working in Indonesia. According to Reynolds, Bontius illustrated his description of the orangutan with a picture of a hairy woman. In his second genus Linnaeus included the monkeys and apes and called them *Simia*. This is where he placed the animal (the chimpanzee) which had been described by Dr. Tulp.

Another chimpanzee arrived in England in 1738, and the *London Magazine* gave this description:

> A most surprising creature is brought over in the "Speaker," just arrived from Carolina, that was taken in a Wood at Guinea; it is a female about four Foot high, shaped in every Part like a Woman excepting its head, which nearly resembles the ape: She walks upright naturally, sits down to her Food which is chiefly Greens, and feeds herself with her Hands as a human Creature. She is fond of the Boy on board, and is observed Always sorrowful at his Absence. She is cloathed with a thin Suit Vestment, and shews a great Discontent at the opening of her Gown to discover her Sex. She is the Female of the Creature which the Angolans call chimpanzee, or the mock man.

Reynolds points out that this was the first time that the term "chimpanzee" had actually been used for this species of animal. He thinks that it is derived from an African word and points out that N. de la Brosse in 1738 referred to the animal as a "quimpeze." Reynolds believes that these names have a common source in one of the central African languages.

We mentioned that Tyson's description of the pygmy stimulated a great deal of interest and the imagination of authors, philosophers, and poets. One of the publications which resulted from this stimulation was an *Essay of the Learned Martinus Scriblerus, Concerning the Origin of Sciences*. This was a collaborative work by members of the Scriblerus

Club including Alexander Pope and Jonathan Swift. It was a satire published in 1713, and it traced the origins of arts and sciences in the activities of apes and monkeys, an idea which actually is even more popular today than in those days.

Rousseau and Lord Monboddo took the liberty of putting man and the animal that Tyson had described in the same species. Lord Monboddo said that the speech was a natural character in both man and in *Homo sylvestris,* but only in man had it actually come to be used. Lord Monboddo thought the *Homo sylvestris* was more advanced than many of the natives of Africa. It always walked in the erect position, and it would use a stick for attack or for defending itself. He said it could also be taught the business of a sailor and how to play the flute. He himself produced an evolutionary theory—and this was 100 years or so before Darwin—in which he said man had actually begun his life or his career with a tail but without any form of speech or without any ability to reason. He lost his tail after a time, simply by sitting on it.

In 1817 Thomas Love Peacock published the novel *Melincourt,* in which he satirizes the concept of a *Homo sylvestris* type of animal which was unable to speak but became elected a Member of Parliament. Sitting in the benches of Parliament completely silent, according to Janson, "gives him the reputation of a powerful but cautious thinker."

Even Charlotte Brontë talked about the silence of simians as being of a voluntary nature and claims that "as monkeys are said to have the power of speech if they would but use it, and are reported to conceal this faculty in fear of its being turned to their detriment. . . ."

Probably one of the most famous of stories stimulated by Tyson's original paper was Jonathan Swift's *Gulliver's Travels.* Among the lands which Gulliver visited was the land of the Houyhnhnms. The Houyhnhnms were very noble creatures. They were able to speak like humans but had the bodies of horses. Among them, however, were the Yahoos—dirty, unpleasant creatures who looked exactly like human beings but behaved in a very horrible way. They were very hairy and had brown skin, and their behavior was more like animals than like humans. They were probably meant to be apes. One of the problems that Gulliver had in this country was he looked like a Yahoo, and the Houyhnhnms regarded him as one. It took him quite a time to demonstrate that he was clean, that he was able to learn, unlike the Yahoo, and that he was civil to them. Once he had demonstrated this, they began to accept him more kindly.

Apology Addressed to the Travellers' Club, or Anecdotes of Monkeys (1825) collected many anecdotes about Primates—monkeys in particular. They were told in a manner that had the double objective of making the ideas of the evolutionists look foolish and of satirizing human weaknesses. One of the chapters, "Republic of Monkeys," describes a group of monkeys that live in a village in huts and carry out agricultural activities.

In *Les emotions de Polydore Marasquin* by the French author Léon Gozlan, who lived from 1806 to 1866, the hero had a French father and a Chinese mother. After being shipwrecked, he found himself on an island on which the inhabitants were only apes and monkeys. The leader was a very tough and unpleasant baboon. The hero had a very unpleasant time in the beginnning but eventually was able to overcome the baboon and become king of the apes and monkeys. He was not able to do this entirely on his own, but he was able to gain the friendship of two chimpanzees who felt that they were more like him than they were like the lower crowd of monkeys, which they despised. Finally, after some time as king of the monkeys, he saw a ship passing near the island, attracted its attention, and managed to be taken back to civilization.

I mentioned the novel *Mélincourt* a little earlier. This deals with a young "orang utan" from Africa that was really a chimpanzee, but had a great success, according to the novel, in English society, becoming known as Sir Oran Haut-ton. Mr. Forrester was his tutor. Mr. Forrester, like Linnaeus, believed that the humans and the great apes are members of the same family. The fact that Sir Oran could not speak simply increased his reputation for wisdom. He had a nobility of feeling and was said also to have a great deal of gallantry. It was noted that he was able to rescue a maiden in distress but did not use this to take advantage of her. However, when he drank too much wine, he became rather gay and jumped out of the window and went dancing in the woods.

Another entertaining story is that "Ein Bericht für eine Akademie," by Franz Kafka, who describes how an ape, or a creature that was formerly an ape, addressed the members of a learned society about his experiences. According to the ape, his first memory was that of being in a cage below the decks of a steamer and then realizing with horror that there was no escape from this situation. Finally he decided that the only way that he could escape was to try not to be like an ape but to try to be like a human. These are some of the things that he has to say:

> "Gentlemen, you have done me the honor of inviting me to present a report on my early simian existence. I regret that I cannot adequately fulfill your request. It is now almost five years since I abandoned my apehood—a brief interval, perhaps, in terms of the

285

calendar but of infinite length for one who, like myself, has raced through it, sometimes accompanied by excellent people, good advice, applause, and band music, but fundamentally alone. . . . My accomplishment would have been impossible had I stubbornly clung to my origin, even the memories of my youth. On the contrary I, a free ape, had to repress every slightest personal impulse. Doing so, however, I also shut out all previous memories. At the start the gate through which I could have returned to my past (had I been permitted to do so) was as wide as the sky; because of my forced advance in the opposite direction, this opening gradually became narrower and narrower. I began to feel more and more comfortably enclosed in the world of man. The storm wind blowing from the forest was dying down. Nowadays it is only a slight breeze that cools my heels, and the distant opening from which it comes, from which I once came, has grown so tiny that I should have to scrape the pelt off my back in order to slip through it, if indeed I had the strength and willpower to retrace my steps over so vast a distance. Quite frankly, gentlemen, much as I like to find the proper image for this sort of thing: your own simian past, in so far as you have one, could not be farther removed than mine. After all, every one of us who walks on this earth feels the breeze around his heels, whether he be a little chimp or the great Achilles.

"So I learned, gentlemen. Yes, you learn if you have to; you learn if that is the only way out; you learn ruthlessly. You supervise yourself with the whip; you tear your own flesh if you sense the slightest resistance to learning. Thus my simian nature fled from me, helter skelter . . . until my trainer broke down and had to enter a sanitorium. Fortunately, he soon emerged again. But I used up many teachers, several of them simultaneously. As I began to feel surer of my faculties, as the public became aware of my progress, as my prospects brightened, I engaged my own instructors, placed them in five adjoining rooms, and learned from all of them at once, constantly jumping from one room to the next—Such progress: Such penetration of the raise of knowledge in all direction into my awakening brain: It made me happy, I admit. Still, I did not overestimate it, even then, and certainly I do not now. By an effort unequaled so far on the face of the globe, I have attained the level of education of the average European. That in itself would be no achievement at all, and yet it is something of the sort, since it released me from the cage and gave me this human way out."

Robert Hartman, writing about the anthropoid apes back in 1883, describes a little chimpanzee which Buffon possessed in 1740. This animal, about two years of age, apparently always walked upright

286

even when it carried heavy weights. This chimpanzee had a serious and melancholy expression, walked very slowly, and was said to be gentle and patient and very obedient. It would offer people its arm and walk with them in gentlemanly fashion. It would sit down at the table like a human to eat. It would open its napkin, put it on its lap, and wipe its lips with it. It used a spoon and a fork to eat. It poured out wine and clinked glasses with its neighbor. If it wanted tea, it would get a cup and saucer, put the sugar in the cup, pour the tea on it, and then let it get cold before it would drink. But, according to Hartman, it did not seem to be very happy with all this. It ate the food men ate, but it preferred fruit. It preferred milk and tea and sweet liqueurs to wine. It was a very friendly animal and took great pleasure in being caressed by humans. It took such a liking to one lady that when other people came near her, it would seize a stick and begin to wave it around. But Buffon intimated to the chimpanzee that he disliked this type of behavior, and the chimpanzee ceased to perform it.

Dr. H. Traill of Liverpool also had a female chimpanzee. He had bought it in Gabon, in West Africa. When this animal came on board the ship, it reached out its hand to hold the hands of the sailors and eventually got on good terms with the whole crew, particularly with the cabin boy. When the sailors were sitting down to eat, the chimpanzee would appear and beg for some food. When it was angry, it growled somewhat like a dog. It would also wail like a spoiled child if it was frustrated. It was very lively and cheerful while the vessel was in warm waters, but as it got more toward England, the chimpanzee became much less active and wrapped itself in a warm cloth. It was not happy standing upright, but it had very strong hands. After a while it acquired a taste for wine. It also was very fond of coffee and sweets and would drink from a glass. It was interested in articles of clothing, and if it found a hat, it would put it on.

There was a male chimpanzee in 1835 which was brought to the London Zoological Garden from Gambia and appeared to be a very faithful creature. They had a woman keeper for him, and the creature wore a little jacket most of the time and nestled in the lap of the keeper. He played with his toes just like a child and tried everything that he did not understand with his teeth, which we found that all our Yerkes apes will do. He was terrified when an anaconda, a South African snake, was brought into the room in a basket, and he would not take an apple from the top of the basket in which the snake lived.

In 1876 a male chimpanzee lived in the Berlin Aquarium. A very lively animal, he enjoyed playing with a young female orangutan, and they had all kinds of games together and many tender embraces. The chimpanzee got on very well with the son of the director. When the child came into the room, the chimpanzee would run to him, embrace him, kiss him, seize his hand, and lead him to the sofa so that they might play together. The child played very roughly with the chimpanzee, and the chimpanzee never lost his temper; but he was very much rougher with older boys. On one occasion when a number of schoolboys came to the office of the director, he ran at them, bit the leg of one, shook another, tore the jacket of a third, and gave the last a box on the ear. On one occasion when Dr. Hermes gave his nine-year-old son a light tap on the hand because he had made a mistake in his arithmetic, the chimpanzee, who was sitting at the table alongside the boy, followed suit and gave the boy a box on the ear as well.

The chimpanzee today occupies a very large area of Africa extending from Guinea on the west coast of Africa through the forests to western Uganda and Tanzania. It can live up as high as 8,000 or 9,000 feet and occupy many different varieties of forests, particularly in some areas where man is absent.

There are two species of chimpanzee. The true chimpanzee is called *Pan troglodytes*, and the pygmy chimpanzee is called *Pan paniscus*. The *Pan troglodytes* is divided into three subspecies: *Pan troglodytes verus, Pan troglodytes troglodytes,* and *Pan troglodytes schweinfurthii*. The *troglodytes troglodytes* is often called the tschego. The *troglodytes verus* is the common or masked chimpanzee. The *troglodyte schweinfurthii* is the eastern or long-haired chimpanzee. The *troglodytes verus* has black pigmentation which forms on a butterfly-shaped mask on the face, and the whole face darkens as the animal ages. *Pan troglodytes troglodytes* has a whitish face with freckles. As it gets older, it may become very much mottled and develop a muddy color. *Schweinfurthii* starts off by being white, and eventually the skin becomes a dark muddy color.

Adult male chimpanzees weigh about 110 pounds and stand about 5 feet in height. The females weigh about 90 pounds and are about 4 feet in height. They are enormously strong animals. Some tests carried out many years ago on some chimpanzees demonstrated that an animal was able to pull loads of 847 and 640 pounds with the right hand. With both hands the female was able to pull 1,260 pounds and on another occasion 905 pounds. In contrast, tests carried out in seven college football players, each of whom weighed, if anything, more than the weight of the male chimpanzee, showed that each of them was only able to pull 175 pounds with the right hand and about 368 and 377 pounds using both hands.

Chimpanzees have the largest ears among the anthropoid apes, ears certainly much larger than those of humans. The pygmy chimpanzee, however, has very small ears which are much more like the human's. The chimpanzee's eyes are recessed beneath very dominating eyebrow ridges, and the animal does not have a nose in the human sense. Its lips are very long, very large, and strong and can be elongated; for instance, it can bring its lower lip up over the front of the upper lip for quite a considerable distance. When a chimpanzee stands erect and hangs its arms down, its fingertips usually reach just below the knee. The knuckles of the chimpanzee are padded with thick pads of skin, and when these are studied under the microscope, they show the same kind of structure as the skin on the heel or sole of the human foot. This is because when chimpanzees walk, they use the feet and the knuckles of the hands, and the pads give the same protection to the knuckles as the thick pads on the soles and heels give to the feet.

In the chimpanzee the anus is right on the surface of the backside and is not recessed in the fold in the buttocks as it is in humans. It is only in humans that buttocks as we know them exist; none of the other apes have the muscles developed as we do to form buttocks. It is probable that in humans the buttocks are well developed because they play a very important part in the anatomical mechanism which holds us upright.

The male chimpanzee has a very thin, long penis, which, when erect, is a little thicker than a pencil. Since there is a good deal of long hair in the perineal region, it is difficult to see the scrotum in these animals. Female chimpanzees, when they are in heat, show a red, swollen genital area.

What do chimpanzees eat? In the wild they use more than ninety species of trees and plants for food and have been seen eating more than fifty types of fruit and more than thirty types of leaf and leaf bud; they also eat blossoms, seeds, and some of the bark and the pith of trees. Sometimes they lick resin from tree trunks or even chew on pieces of dead wood. They also eat a number of species of insects, including three species of ant, two species of termite, one species of caterpillar, and the grubs of beetles, wasps, and flies. When chimpanzees raid a wild bees' nest, they eat not only the honey, but all the bee larvae as well. They have been known to take and eat birds' eggs from nests in the trees, as well as young birds. They also eat a certain amount of meat. They will kill young bushbucks, young bush pigs, young baboons, and young or adult red colobus monkeys, and they have sometimes been known to catch a red-tail monkey or even a blue monkey. They have also been seen by Jane Goodall to eat small quantities of soil that contains common salt (sodium chloride).

The Japanese scientist Dr. A. Suzuki once observed an incident of chimpanzee cannibalism. He saw and photographed an adult male chimpanzee eating the legs of a living chimpanzee baby, which was screaming all the while. The mother chimpanzee was watching the performance very closely, but did not move or make a noise. Other dominant males came over, one after the other, and touched and smelled the body. Suzuki wonders whether this activity may have been connected with a period of excitement and tension which he says "coincides with the decline or end of a specific fruiting season and before a new consistent supply of food has been found."

There is circumstantial evidence from another scientist that a similar incident occurred. J. Kingdon describes how a half-grown colobus monkey was killed in the Buganda forest. The monkey was sitting in a tree surrounded by chimpanzees. It dropped to the ground, and a large male chimpanzee promptly killed and ate most of it. Females were allowed to touch and to take small portions. The juveniles were allowed to scavenge for pieces of skin and fur, which they sucked to get as much as they could and then spat out.

Jane Goodall was the first person to show that some chimpanzees use sticks, stalks of grass, or pieces of broken vines to insert in the mounds of termites, pulling them out when they are covered with termites and licking the termites off. Subsequently a number of scientists have found pieces of stick used for this purpose over widely scattered areas of Africa. However, there is some evidence that while quite a number of chimpanzee groups do this, not all have developed this habit.

Jane Goodall has also described how chimpanzees use crushed leaves as sponges to soak up drinking water that they cannot get from holes. They also use a variety of inanimate objects, such as stones and handfuls of vegetation, branches, and sticks, to express aggression or frustration. These are sometimes thrown at baboons or, as Adrian Kortland has reported, at stuffed leopards and at humans. There are some reports of chimpanzees that have used rocks as tools to crack open nuts, as in Liberia, where a rock was used to open a dried palm nut. Jane Goodall's chimpanzees were not observed to carry out this kind of activity although she saw them hold the food in one hand and try to open it up by banging it against a tree or rock.

Drs. T. T. Struhsaker and P. Hunkeler have gathered evidence that chimpanzees in the Ivory Coast area of Africa gather nuts of certain plants that have a hard shell and use other objects to open them. They carry the fruits to a tree with an exposed root. Sometimes there is a cuplike depression in the root. They place the nut in the cuplike depression and then smash it open with a stick or a stone. A number of blows are usually required before any single nut is opened. There is evidence, too, that the stones they use for this purpose are actually carried to the site where the nuts are smashed. The authors found a number of sites in the area where the remains of the sticks indicated that they were used extensively by the animals for crushing the nuts. This provides further interesting evidence that chimpanzees use inanimate objects around them as tools; this tool-using behavior once thought to be characteristic of humans is common among these animals.

An extension of this tool-using ability, described by Dr. Emil Menzel when he was at the Delta Primate Center in Covington, Louisiana, involved the use of poles as ladders for climbing. He had eight chimpanzees in an enclosure surrounded by a wire fence with an observation room to one side of it. He placed some poles in the compound so that the chimpanzees would have something to play with. They started off by manipulating the poles in an aimless way. Eventually one of them started to use the pole more as an instrument for climbing than for play. Then it began to brace a pole up a wall, use it as a ladder, and climb up it and peer in at the windows of the observation house to look at the experimenters who were supposed to be watching it. The animals would also erect these poles as ladders on the top of various other structures in the compounds, and as a result, they were able to get into trees they normally could not get to because a wire carrying electric current would give them a shock. The use of the ladder in this way was invented by one animal that was watched intently by the other seven members of the group; they all eventually copied the original inventor and began to collaborate with him. In the end all eight animals developed this ability to use the poles as ladders.

This communication of information about the use of poles from one chimpanzee to another is very interesting, but Menzel has also demonstrated that chimpanzees can convey much more complex information to other chimpanzees. For example, they can convey the presence, direction, quality, relative quantity, and preference value of an object hidden somewhere in the distance that the animal itself has not seen. How do they do this? What Dr. Menzel did was to take his eight chimpanzees and put them in a confined area where they could not see what he was doing. Then he would take something that they liked very much, such as some type of fruit, and bury it some distance away from where the animals were confined. He would take one of the leaders of the group over and show it the food, cover the food again, and put the leader back in the cage. The cage would be opened and all the animals allowed out. The leader would make straight for the food and would be followed by all the other animals. Perhaps this does not seem like anything very exciting, but it becomes more exciting when one realizes that even when, some food is put out in full view of the animals, along with the hidden food which the animal has been shown, the leader still tends to make for the hidden food. If it does, most or all of the other animals will follow even though they can actually see other food. If the animals are confined into the cage and then let out into the enclosure without there being any hidden food or the leader knowing anything about it if there is, then the animals do not follow the leader. So he must have been able to communicate to the other chimpanzees the relative quantity and presumably the quality of the hidden food.

Dr. Menzel set up an experiment which demonstrated this. He had two lots of food; one cache of food contained two pieces, and another four pieces. One leader was taken out and shown only the area where two pieces of food were hidden, while another leader was taken out and shown only the spot where four pieces of food were hidden. They were then put back with the other chimpanzees. Then the group was let out, and in the majority of cases, the animals appeared to pool information because they all went first to the larger of the two food piles and then all together to the smaller pile. So there must have been some sort of communication among them.

Then an experiment was done regarding quality. The animals prefer fruit to vegetables. If one cache was composed of fruit and another one of vegetables and each of the two leaders was shown only one of these, they would all combine and go first to the fruit and then to the vegetables. If, after confinement, they were let out without any animal's being shown any fruit, they would mill around without following any particular leader. So the leader had obviously communicated to them that he knew of something that was interesting and important.

·We know that honeybees have this kind of communication. For example, a bee which has found a store of nectar comes back into the hive and then does a special dance that indicates to the other bees the direction and the approximate distance of the supply of nectar which it has found. The bees do this by wiggling their abdomens. But although I have seen a belly dancer convey a message with her abdomen, I have never seen a chimpanzee use this method of communication.

When the leader, having been shown the food, is put back with the others, there is a tendency for it to wrestle with them. Dr. Menzel tried putting it in a separate cage alongside those so that it could not wrestle and it was still able to lead the group, which would follow it to the food as before. The method of communication presumably is by the firmness and the confidence of the behavior of the animal that knows where the food is, the subtler aspects of its movements, the way it moves, and the confidence with which it strides.

There were some occasions when the animals would not and did not follow. Presumably the leader on this occasion had not communicated by its bearing well enough. Sometimes if an animal which was not a top leader had been shown the food, it might have some difficulty in getting the animals to follow. In this case, it is intersting to see the methods that it used to try and get the animals to follow. They are best described by Dr. Menzel:

> If no one followed spontaneously in the food getting test, the leader glanced from one chimpanzee to the next, beckoned with the wave of the hand, or, more often, the head, or tapped one animal on the shoulder and presented "his back to solicit walking in tandem" (with one arm around the waist). He sometimes walked backward towards the goal while whimpering and orienting towards the group. If such a device failed, he tried to pull preferred companions to their feet, he screamed and bit them on the neck, he held their hand in his teeth and pulled them thus in the direction of the goal, or he physically dragged them along the ground by a leg. If the followers still did nothing, the leader sometimes fell down screaming and tearing his hair in a temper tantrum. At this, the followers ran to him, clung to him, and then groomed him—but by this time the leader was done in and no longer attempted to move out. After recovering from his upset he commenced to play.
>
> The only thing that might save the day after such a scene was for someone else to start walking out into the field in the correct direction. The leader would then get up too, and travel behind the followers as long as their course was accurate. If they started to change direction, he grimaced and glanced back and forth at them and toward the hidden goal, or he presented his back, got them in tandem, and steered them towards the goal. If the followers went along now, the move would start to pick up in earnest. Other followers started to come up once somebody got five to ten meters from the group distribution. The leader moved faster then, and as the followers began to too, he would break physical contact with them and run. This nearly always produced a pack run. As the pack came within a few meters of the food pile, the leader went into a "goal spurt"—and everyone within striking range followed suit.

293
.

In summing up this work Dr. Menzel says:

If chimpanzees are capable of fairly subtle hand, head, arm, and vocal signals, why do they not use these signals more efficiently and frequently—as humans use their verbal and sign languages? The answer is implicit in the previous discussion. Perhaps, among sophisticated adult chimpanzee leaders and perceptive chimpanzee followers and under routine conditions, this sort of language is no longer necessary.

One of a group of chimpanzees at the Delta Primate Research Center became a self-appointed dentist. It used fingers and thumb and twigs to clean its own teeth and the teeth of other chimpanzees. It was also seen to pull out nonpermanent teeth of other chimpanzees, using the forefinger and thumb as forceps. Some of the other animals resented these dental attentions; they would cover their faces and in some cases keep their mouths firmly closed. Some of the other animals have been seen to use a twig to lever up and remove loose teeth from their own gums. So the germ of dentistry seems to be present in the great apes.

In the latter part of the nineteenth century a zoologist called Robert L. Garner made the first study of wild chimpanzees. He took himself to Africa and built himself a large cage designed really to protect himself from the animals and not to catch them. He sat there for very long periods and watched animals of the wild go past. Needless to say, monkeys came and sat in the trees to observe him. From time to time a gorilla or chimpanzee walked past him, but he did not see them in any large numbers. He was very interested in the vocalizations of chimpanzees and counted something like twenty-five to thirty sounds which they used. He was able to learn ten of them, and he used them in such a way that he said the animals understood them. Most of the sounds that the chimpanzees make can, from the point of view of tone, pitch, and modulation, be made by the human voice, but some of them have too great a volume for humans.

Garner claimed that chimpanzees had dancing carnivals and had this to say about them:

One of the most remarkable of all the social habits of the chimpanzee is the kanjo, as it is called in the native tongue. The word does not mean dance in the sense of saltatory gyrations, but it implies more the idea of carnival. It is believed that more than one family takes part in these festivities. Here and there in the jungle is found a small spot of sonorous earth. It is irregular in shape and about two

From R. L. Garner, *Apes and Monkeys* (Boston, Athenaeum Press, 1900)

Chimpanzees observe Professor Garner in the cage he built for his own protection.

Professor Garner gets ready to look for apes.

From R. L. Garner, *Apes and Monkeys*
(Boston, Athenaeum Press, 1900)

From R. L. Garner, *Apes and Monkeys* (Boston, Athenaeum Press, 1900)

Kanjo Ntyigo—Chimpanzee Dance

feet across. The surface is of clay and is artificial. The clay is superimposed upon a kind of peat bed which, being porous, acts as a resonance cavity and intensifies the sound. This constitutes a kind of drum. It yields rather a dead sound but this is of considerable volume.

It is probably not the pounding of this claylike structure which makes the noise of chimpanzee carnivals but rather the pounding upon such structures as the buttresses of the ironwood trees:

> The chimpanzees assemble by night in great numbers and the carnival begins. One or two of them beat violently on this dry clay, while others jump up and down in a wild and grotesque manner. Some of them utter long, rolling sounds, as if trying to sing. When one tires of beating the drum, another relieves him, and in this fashion the festivities continue for hours. I know of nothing like this in the social system of any other animal, but what it signifies or what its origin was is quite beyond my knowledge. They do not indulge in this kanjo in all parts of their domain, nor does it occur at regular intervals.

Leonard Williams in *The Dancing Chimpanzee* is very critical of Garner's description of the dancing of the chimpanzees and the beating on the improvised drum. I do not agree with him, however, that the chimpanzees have no rhythmic sense. A number of the chimpanzees at our center appear to have a considerable rhythmic sense at least as far as beating on the steel doors which separate their inner sleeping quarters from their outside runs. I have had animals sit by these doors and beat them with such rhythm that I have difficulty in distinguishing their performance from that of the famous drummer Gene Krupa. To what extent Garner's description of the chimpanzees actually dancing while others drum is correct is difficult to say. We have one chimpanzee that starts an act it puts on for visitors by first doing rhythmical banging on the steel guillotine door and then doing a little dance, finishing up with a jump at the wire and sometimes with a clapping of its hands, all of which is done with a surprising degree of rhythm.

Wild chimpanzees do a "rain dance." Once during a wild storm with a heavy downpour Jane Goodall was watching some chimpanzees feeding in a fig tree. As the first drops of rain fell, the chimpanzees left the tree and plodded up the slope toward the top. These were mostly adult males with a few young chimpanzees. When they reached the ridge, they stopped briefly. Just at that moment the downpour began. It was accompanied by a clap of thunder. The chimpanzees acted as if this were a signal, and one of the adult males stood up and swayed in a rather rhythmic fashion and began a series of pant-hoots which could be heard above the noise of the rain. Then he charged down the slope toward the trees where he had just been eating figs. He ran about 30 yards and then swung around the trunk of a small tree, leaped into the lower branches, and sat without moving. Almost at the same time, two of the other males followed him. One broke off a low branch from a tree and waved it around in the air and then threw it ahead of him. The other animal, when he finished running, stood up and swayed and grabbed the branches of a tree and pulled them backward and forward before he finally broke off a large branch and dragged it farther down. The fourth adult male also charged and leaped into a tree and also tore off a branch and jumped down with it in his hand onto the ground and continued running down the slope. The last two males vocalized and also rushed down the slope, and as they did that, the first animal that had run down the slope began his slow plodding up the slope again. The others, who had also climbed into trees, when they had finished their run, climbed down from their trees and followed him up to the ridge again. Once they reached the ridge, they started rushing down the slope again, one after the other, just as they had before. The females and the young had climbed into the branches of the trees around the top of the rise as soon as the males had begun their activities, presumably a sort of grandstand seat, and watched the performance. The males continued this rushing down the slope and plodding back up for about twenty minutes. When they had finished, the females and the youngsters climbed out of the trees, and then they all went together over the crest of the ridge.

Chimpanzees move around in troupes of social units. There may be from twenty-five up to eighty or more of these animals. They usually occupy a fairly limited area of land varying from 14 to 16 square miles if the country is forest, but in the savanna containing patches of trees, they may occupy over 100 square miles. In the forest their range has a number of paths which are well beaten and which the chimpanzees follow when they move from one part of the range to another. In the savanna they will use preexisting game paths. The main social unit may actually be broken into smaller groups, but these smaller groups will get together to form a larger group when the food supply is localized. Subgroups are very unstable in the forest, and there is a great deal of individual independence so that the principal stable relationship is really between the female and her baby. There may be a number of different sorts of groups: all male groups, some nursery groups, mothers with young groups, groups of adults, mixed groups, and solitary animals. The size of these groups and the individuals which compose them are in a constant state of change. In the more open areas of country, the mixed groups predominate. The mixed groups contain at least one and usually more males, and they include both females and young animals. It is not known how stable that particular membership is, but they seem to have a greater social coherence in this wider range than in the forest.

Chimpanzees do not have specific mates and mate with any animal in the group. They threaten by grabbing at the vegetation, tearing it off, and throwing it around, by banging around, using the arms and feet, and by screaming. They also erect the hair along the neck and the back. The hierarchy is fairly loose. There usually are males which are more dominant than others, but it is not the tight hierarchy that occurs, for example, in a rhesus monkey group. Jane Goodall has described how a low-ranking male chimpanzee, rather timid in nature, found an empty kerosene can and rushed through the group, carrying the empty tin with one hand and banging it with the other. The resultant noise scared the other chimpanzees, and they took cover. With the aid of this kerosene can, this animal has progressed in the hierarchy of the group until he became one of the leaders. As Kingdon says, "A remarkable case of the self-made chimp." In a group of chimpanzees out at the Yerkes Field Station, the leader has an exclusive right to an empty oil drum, which it rolls around the compound and crashes into the fence. The other animals get out of its way at the last moment with very pained expressions on their faces.

Dr. Adrian Kortland has experimented with the reaction of wild chimpanzees to stuffed leopards. On one occasion he placed a stuffed leopard in the path which chimpanzees normally followed. When the

Yerkes chimpanzees gossip while keeping their babies under control.

animals saw it, they threatened it and then picked up sticks and swung them at the animal. Hunters have described how they were threatened and screamed at in this manner by members of a group of chimpanzees after they had killed one of the group. Personally, I do not blame the chimpanzees for doing so. Kingdon describes how a party of zoologists were once threatened when they came close to a group of chimpanzees that were on the ground. Apparently there had been no provocation which caused the animals to behave in this fashion, although if it was a group that had been subjected to the activities of a hunter, one can understand why they would react badly to other humans.

If an inferior animal approaches a dominant male, it may indulge in what is described as nervous panting. It may crouch or offer the palm of its hand as a form of appeasement or may turn its back and present. The dominant animal may hug the submissive animal or place its hand on top of its head, all of which seems to reassure the submissive animal.

In our experience, baby chimpanzees, if left with their mother, stay with her for five or six months and rarely go very far away. If there is anything to alarm them, they rush back to the mothers again immediately. Many other aspects of chimpanzee behavior and intelligence are to be found in my book *The Ape People.*

Studies of the development of chimpanzee infants at the Yerkes Center show that they start walking on all fours and then move onto the hind limbs. The mother often helps the baby learn to walk and climb. In the walking training the mother may actually take the baby by one hand and half drag it and partly lead it along. The mother has also been known to crouch down on the ground and encourage the baby to creep or try to walk to her. If the baby is not able to crawl properly, the mother has been known to place her hand under the little animal's body and lift it off the floor, thus encouraging it to start to crawl. Crawling apparently precedes walking. This is because the legs are usually not yet strong enough to support the whole weight of the body. The baby chimp, when it first crawls, has the palms of its hands and the soles of its feet flat on the ground, and it takes quite a number of weeks before it begins to use the knuckles of its fingers. The age at which the animal starts walking on four legs instead of crawling varies considerably with individuals. The chimpanzees at the Yerkes Center have shown that this occurs between the fifth and the twelfth month. It takes until about the second half of the first year for the animal to try to walk on its legs.

Chimpanzee mothers are very expert in anticipating the requirements of their babies. The infant makes a number of noises which have been described by various people as whines, screams, barks, shouts, and whimperings. These noises vary in the abruptness with which they are made, in their pitch, and in their volume. Each variation seems to mean something different to the mother and to call forth a different response. In the beginning she is very protective of the baby. Later on she takes a more relaxed role with it, but is still always ready to defend it if any danger looms on the horizon. The baby, on the other hand, is rather slow to learn the response to the mother's cues—vocal, manual, or facial.

The baby chimpanzee gets its milk teeth after about a year and three or four months. In the third year of life, it starts to get its second set of teeth, which is about two years earlier than man. By the time it has completed its tenth year the chimpanzee has all its permanent teeth. In man, of course, the wisdom teeth often do not appear until as late as twenty years of age. The chimpanzee, on the other hand, is a sexually mature adult by the time it is ten or eleven.

One of the most famous chimpanzees in the Yerkes colony was Alpha. She was the first chimpanzee to be born when the Yerkes

A recently born
chimpanzee poses
for its first picture.

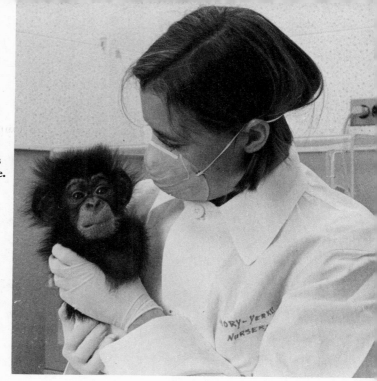

Dr. Josephine Brown studies suckling in a recently born chimpanzee.

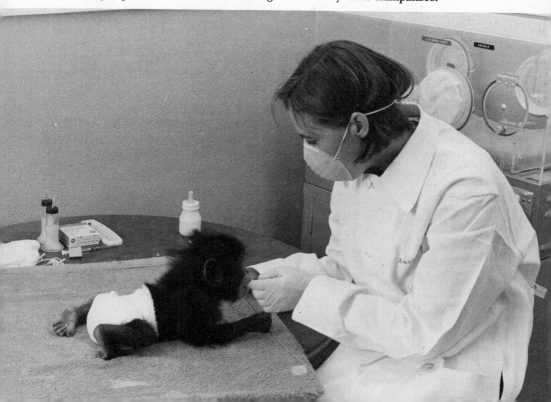

Recently born chimpanzee supports itself by one hand.

Center was established at Orange Park in 1930. Alpha was born in the same year, and her mother was Darwinia, shortened to Dwina, a name with an obvious derivation. Unfortunately, Dwina died from puerperal fever following the birth of the baby.

Alpha grew up to be a very highly intelligent animal. She was the first chimpanzee in the world used for study of drawing ability, an ability which she demonstrated in studies with Dr. Paul Schiller. She was also a very expert smoker. She smoked and enjoyed cigarettes. I often saw her smoke cigars, hold a cigarette to a match to light it, and stub out a cigarette and try to put the butt back into the packet from which she obtained it.

Alpha bore a number of babies in the center, and some of her offspring are there still. When she reached the age of thirty-four, she was pregnant again and seemed to be having a great deal of trouble in bearing the baby. The baby had to be removed by Caesarian section. After Alpha was sewn up, she recovered but seemed to deteriorate after this and eventually died. It was found in the autopsy that she had a tumor of the lower reproductive tract which had not been observable when the baby was delivered. This is what eventually caused her death. It was a very historic occasion at the center when she died.

Some interesting studies have been made of Alpha's first year of life. Since she was the first chimpanzee born there, the entire staff of the Yerkes Laboratories of Primate Biology at Orange Park, which eventually became the Yerkes Primate Center at Atlanta, tried to look after her. Her first year was recorded in very great detail. She spent her first eight weeks in a bassinet and was then put into a crib and then into an inside living room. The first nine months she met only human beings. She had problems with her teething, and she developed a liking for spinach. During all this time the accurate records of her physical development, including X rays of the development of her skeleton, were kept.

There are some interesting statistics on her development. The statistics are compared with those of human babies. On progress toward creeping: (1) lifting the head when on the stomach, the chin lifted free, three weeks in human babies, Alpha three weeks; (2) lifting the head when on the stomach until the chest is free, human being nine weeks, Apha five weeks; (3) pushing the knees or making swimming movements, twenty-five weeks in the human, seven weeks in Alpha; (4) rolling, twenty-nine weeks in the human, eight to ten weeks in Alpha; (5) rocking, pivoting, and worming along, thirty-seven weeks in humans, eleven weeks in Alpha. Progress toward assuming an upright posture: (1) lifting the head when on back, fifteen weeks for humans, five weeks for Alpha; (2) sitting alone momentarily, twenty-five weeks for human, twelve weeks for Alpha; (3) sitting alone, thirty-one weeks for humans, thirteen weeks for Alpha; (4) standing, holding to furniture, forty-two weeks for humans, fifteen weeks for Alpha; (5) pulling self to standing position by means of furniture, forty-seven weeks for humans, fifteen weeks for Alpha. Progress toward walking: (1) walking with help led by a person, forty-five weeks for human, seventeen weeks for Alpha; (2) standing alone, sixty-two weeks for human, twenty weeks for Alpha; (3) walking alone, sixty-four weeks for human, twenty-five to twenty-nine weeks for Alpha.

Chimpanzees are thus very precocious in their postural development compared with humans. Alpha developed a number of vocal and emotional responses, which included fear, anger, timidity, and even a mild excitement. During her first two months, Alpha did not indulge very much in play but spent a good deal of time sucking her thumb and her big toe. After two months she began a period of exploring, using her first finger for examination. She also used her finger for feeling and exploring the noses and the teeth of the humans attending her. From there she developed into a newspaper tearer, a puller-offer of eyeglasses, and a biter of furniture. She then began to play with grass, nuts, and vegetables. She picked up seeds, too, and continued this type of play for a year. She would tear pages from a magazine, lay them down, and pick them up and put them on her head. Although she used papers in this way as a kind of ornament, she did not use them in any way to build anything reminiscent of a nest. Her reaction to human beings was to approach them, then withdraw from them, then threaten them, and then do a swaggering walk. She would do a little tentative beating of her chest but not the strong beating that the gorilla would do. She would wrinkle her face and then eventually, when she came close enough to the person, would pat or gently strike the human's hands or face. She had a brief period when she was permitted to have contact with a thirteen-month-human infant, and she was interested in it and was moderately aggressive but quite gentle with it.

Later on Alpha was introduced to another chimpanzee a little younger than she was. This was Beulah, who is still at the Yerkes Center. Alpha completely intimidated Beulah by attacking her and biting her. They had to be separated. They were left together in adjoining cages for a while and eventually put in one cage again. Alpha was still the aggressor. She had Beulah confined to a corner and took her noonday food from her. Eventually she let Beulah eat her meals. Beulah reacted and resisted Alpha, and in due course the animals became very closely attached to each other, so much so that they objected to being separated.

It is of interest that during her first year Alpha was put through the Gesell Infant Mental Tests for the Preschool Human Child. Of course, in the tests designed to measure the control of her posture or her locomotion, she was not equal but superior to human children. In a test for prehension, she was about the same as a human child, but in tests that were more complex—for example, putting a cube in a cup or throwing an object on the floor—she was retarded by comparison with the human. In more complicated tests, such as throwing a ball into a box and spontaneously scribbling when given a pencil, she was a complete failure. Her eye-hand coordination tests showed her to be equal to a human child and in some respects better than one. She was completely useless in any kind of a test in which she needed to show "exploitive use of the materials." By the end of the year there were a number of tests in which compared with the human she was better,

Wendy (right), forty-eight years old (she died that year), and Soda (left), forty years old, confer.

but after a year she had reached the point where her progress was very slight indeed. Up to about a year, one could say that from a mentality point of view, there is a pretty close resemblance between chimpanzee babies and human babies, but after a year they certainly grow progressively farther and farther apart as the human infants spurt ahead.

As chimpanzees get older, they become very ill-tempered, introverted, and quite dangerous. My experience with our old chimpanzee, Wendy, and with Patti, a similar old animal, was that they were very unreliable and would attack you suddenly through the bars of the cage. You had to be constantly on the alert if you were near them to see that they did not do you harm. There has never been any real certainty as to the age to which chimpanzees live until the Yerkes Center began to keep them for long periods of time. Many years ago, Lord Zuckerman suggested eleven years as the average life of a chimpanzee in captivity and suggested that twenty-six years was the maximum. Whether those animals were fed an adequate diet or not, I do not know, but the two oldest chimpanzees which the Yerkes Center ever had were Wendy, who died recently at the age of forty-six, and Patti, who died a few years ago at the age of forty-four. We have a group of animals close to their forties right now.

Detailed studies were made at the Yerkes Laboratories in Orange Park of the ability of chimpanzees to distinguish color. Tested against four human beings, including two children and two adults, a chimpanzee has an ability to distinguish between hues of yellow and red which was approximately twice that of adult human beings, whereas its ability to distinguish in the range of the blue and green part of the spectrum is almost identical with that of humans. It is known that rhesus monkeys have almost exactly the same ability of color discrimination as chimpanzees. The cebus monkey, however, seems to have only a two-color vision.

Chimpanzees and other great apes are probably the only animals below man which are able to recognize themselves in a mirror. When you first present a mirror to a chimpanzee, it responds as if the reflection were another animal just the way other Primates do. For instance, the animal will bob up and down, or it will vocalize at the image and sometimes threaten it, often over an extended period of time. In other words, it responds socially to the image. However, if the mirror is left continuously in front of the animal, after a few days, this type of response to the mirror declines, and it tends to be replaced by a behavior which gives the impression that the animal recognizes the mirror image as himself. Some of the evidence for this is, for instance, that the animal grooms those parts of the body that it would be unable to see if it did not have the mirror. It picks bits of food from between its teeth while it is watching its image in the mirror. It also manipulates the anal-genital areas, being visually guided in this by the mirror. It picks material from the nose while looking at the image, and it makes faces at the mirror. It manipulates wads of food with its lips while it watches itself in the mirror. All this indicates very strongly that it is, in fact, an animal that is recognizing itself. How can you prove this? A scientist called Dr. Gordon Gallop did it this way.

What he did was anesthetize an animal that had developed the above behavior in front of the mirror. While it was unconscious, he marked the upper part of the eyebrow region and the top half of the opposite ear with a red dye. Once this dye dries, it has no smell and there is no way of giving any clue that it is there. The chimpanzee was put back in the cage, allowed to "come to," and was then exposed to a mirror. When it observed itself in the mirror, it began to inspect the areas where the dye was located with its fingers. It touched the

Auction of Yerkes ape art at Atlanta High Museum of Art. The mayor of Atlanta, Sam Massell (on podium), conducts the auction which raised $3,000 for the Zoological Society of Atlanta.

marked areas, and in one case, the animal not only touched the dye with its fingers but inspected the finger after it had touched the dye; smelled the finger as well. If, however, Dr. Gallop marked with dye an animal which had not previously been exposed to a mirror the animal did not notice the dye marks at all when it looked in the mirror and made no attempt to touch them. He then tried the same kind of test with the rhesus monkey, the stump-tailed macaque, and the Java monkey and found that they did not react to the presence of dye. The recognition of one's reflection in the mirror requires an advanced form of intellect, and anything below the chimpanzee does not have an intellect at that level. Some mentally retarded children do not seem to be able to recognize themselves in mirrors.

Chimpanzees that have been reared in isolation will spend two or three times as much time watching the mirror image as they will watching other chimpanzees. French psychiatrists have also shown that the onset of schizophrenia in human beings is often correlated with long periods of mirror gazing. In humans, the recognition of one's own reflection has to be learned. It has been shown that human infants, in fact, do not show signs of self-recognition in a mirror until they have reached about two years of age. Of course there will be exceptions to this as there are to almost any other type of activity.

Dr. Gallop speculated that since a number of animals eat more in the presence of a mirror in their cage than on their own, perhaps, the presence of mirrors in pubs, which are often placed behind the bar itself and directed towards the customers, serves the same function in encouraging them to drink more just as it encourages the isolated chicken to eat more. This may not indicate that bartenders are expert psychologists but perhaps the pubs with mirrors are given an opportunity of economic survival beyond those pubs which do not have mirrors.

There is a bar in Amsterdam called the Cortina Bar where they do not have any beautiful flaxen-haired Dutch girls dishing out the beer. Instead, they have a pair of chimpanzees called Otto and Jimmy. The owner of the bar, Gerard Cuyters, brought his two tame chimpanzees into the bar in the Kinkerstraat one evening as a treat for the customers. At the time Otto was seven, and Jimmy was six. They started sitting with the clients, helping them sip their beers, and sometimes accepting a pint themselves. The next time Cuyters and his wife decided to try the chimpanzees on the serving side of the bar. They soon became very good at stretching out their long arms for the correct bottles on the shelf and measuring out the appropriate tot of spirits. They learned to take the orders to the tables, and they also learned to light cigarettes with matches. They had come originally from Africa, where they had been exchanged for ten bottles of Dutch gin by Cuyters, who was then working as a ship's captain. They were only babies then, and he had to use a bottle to raise them and had to give them a lot of discipline. He said they are like children. "I have taught them to wash, sit at a table, eat, dress themselves—in fact, they can do almost anything humans can."

Otto was able to ride a bicycle. He could drive the car, and he was also permitted to take the controls of Cuyters' private plane. In the last few years, Otto and Jimmy also became Dutch television personalities. They once made a cancer collection in bars and cafés which totaled $31,000. There are a few things that they picked up which their owner was not very happy about. One of them was Jimmy's favorite trick of kissing the ladies and lifting their skirts in a naughty fashion. Cuyters says he has no idea where Jimmy learned that trick.

There is a fairly complex type of communication between Primates, particularly the higher Primates, using facial expressions. Lord Zuckerman, who has made many fundamental contributions to primatology, includes among his contributions a general account of facial expressions in a large number of animals. He noticed the faces of lemurs were "notoriously blank," although they do make flinching movements with the eyes and the upper eyelids if they are frightened,

Courtesy of the editor of the Holland *Herald*, June, 1972

Otto draws beer for a customer in an Amsterdam pub.

Jimmy lights customer's cigarette in Amsterdam pub.

Courtesy of the editor of the Holland *Herald*, June, 1972

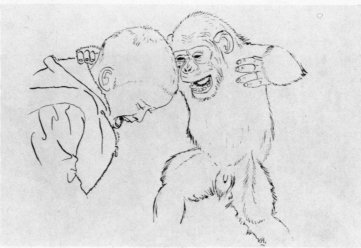

From "A Comparative Approach to the Phylogeny of Laughter and Smiling" by J. A. R. A. M. Van Hooff, in *Non-Verbal Communication*, ed. by R. A. Hinde (Cambridge, Cambridge University Press, 1971)

Interspecific social play. Chimpanzee, in active role, shows relaxed open-mouth smile.

and there may be slight movements of the forehead and the ears. There is very little other movement and very little other emotion expressed by their faces. Scent is a more important type of communication among the lemurs than facial expression.

A detailed study of facial expression in apes was given by Ladygin Kohts, who brought up a chimpanzee in her home and compared the facial expressions of the chimpanzee with those of her own child. Dr. Kellogg, who brought up a Yerkes chimpanzee in his home, was able to confirm her observations, and so was Dr. Yerkes. In general, it can be said that a number of emotional conditions are expressed by facial expressions in the Primates. This includes most of the monkeys as well as the apes. Certain facial movements are characteristic of certain emotions in all Primates. In happiness and joy there is a general lifting of the face. This is due to the important facial muscle called the zygomatic muscle which contracts and pulls the face up. This is the muscle which plays the important part in the process of smiling and in laughter. In the process of evolution, preparing the face for a bite has been changed into preparation of the face for smiling and laughter. A similar sort of process occurs when a bite is to be given and when an animal is happy. It will also indulge in play-biting, so that there is a very complex relationship among biting, smiling, and laughing.

When an animal is fearful or sad, the whole face tends to be

Girl produces "broad smile laughter" (no pun intended) with intense baring of teeth. This expression links up the silent bared teeth and relaxed open-mouth display of nonhuman Primates.

From "A Comparative Approach to the Phylogeny of Laughter and Smiling" by J. A. R. A. M. Van Hooff, in *Non-Verbal Communication*, ed. by R. A. Hinde (Cambridge, Cambridge University Press, 1971)

lowered. A muscle called the triangularis muscle pulls the face down. If the fear is not too great, then there will be a lifting of the eyebrows; this is due to the contraction of the frontal muscle. If the animal is really frightened to the point of feeling what we would describe as horror, most of the muscles in the face contract, and the whole face takes on a distorted appearance.

On the other hand, anger gives the impression of general tension in the face, and this is due particularly to the contraction of the muscles that surround the mouth. It is interesting too that since we often talk about hate and anger being related to love and affection that love and affection are also mainly related to the mouth since licking and kissing and love biting simply represent the modifications of the comfort which the baby gets from sucking.

The mouth is probably the most important organ of expression. The eyes are perhaps less important but do serve to modify the expression since they can be moved in a way which will add very strongly to the overall impression. For example, our pet ocelot, Cleo, sometimes fixes her eyes on me, and when she does, I know she is going to jump at my face and slap me very hard with both paws simultaneously on each cheek. It is a kind of intimidating attack. It does no harm, and she does it with lightning speed. Sometimes when I see this gaze, if I am quick with my hands, I can put them up fast enough to block her in her leap. But very often she moves with such speed that she has launched the attack and smacked me on the face and gone before I really have had time to move.

Dr. N. Bolwig, who has made some important contributions to the study of facial expressions and emotions, has made an interesting observation about a hand-raised baboon or mangabey. Such an animal often shows rather exaggerated movements or even mannerisms which are not seen in animals in the wild. He suggests two explanations: Either it attempts to copy the facial expressions of its human master, or perhaps the human is so unobservant or so slow in his interpretation in what the animal is trying to communicate to him that the animal, in order to get its message across, has to overdo its miming. It may be noted, he points out, that this often happens with adult humans when they are talking to children in order to get their point across or even to other adults who are deaf.

Charles Darwin lists the ten things that happen in laughter: (1) There is general relaxation of the body; (2) the mouth is open for vocalization; (3) the zygomaticus major muscles pull back the corners of the mouth; (4) the cheeks and upper lip are much raised; (5) the lower orbicularis muscles contract; (6) the nose is shortened, and the skin on the bridge is wrinkled in transverse lines; (7) the naso-labial fold is formed; (8) the upper front teeth are commonly exposed; (9) as the mouth is opened, there is an adjustment of the vocal bands for sound production and air is forced out by rhythmic contractions of the muscles of expiration; and (10) the eyes glow as tears are often shed in moderate and roaring laughter owing to pressure on the lachrymal duct by the orbicularis muscles and the levator palpebrum muscles.

The changes in the face in laughter are surprisingly close to those which we see in anger, rage, and sudden shock. It needs only a slight change in the contraction of some of the facial muscles to change from laughter to the various emotions listed above. This is often seen in children where the change from laughter to anger can be remarkably quick. In humans, laughter may occur because something funny appeals to the individual, but in Primates it seems to be used more in the nature of a friendly or a happy greeting.

At the International Congress of Primatology in Atlanta in 1968, a

The famous chimpanzee tea party at the London Zoo. Courtesy of the London Zoo

chimpanzee known as Judy was brought along to a cocktail party. Judy, a highly trained chimpanzee who came beautifully dressed in a cocktail frock for the party, sat with me for a time in a chair sipping rum and Coca-Cola. The woman who trained her and looked after her was with her, and she informed me that the animal could understand as many as forty separate spoken commands, including commands with such minor refinements of reaction as laughing and smiling. In other words, if her trainer said, "Judy, smile," Judy would smile by pulling back the corners of her mouth and exposing her teeth and giving what looked like a big grin. If she were told to laugh, she would add to these changes in her face the opening of the mouth and a certain amount of noise, together with the wrinkling of the skin around the eyes to give her much more the impression of actually laughing.

Smiling is generally regarded and spoken of as being a diminutive form of laughter. The smile that is due to humor developed from a social baring of the teeth is the kind of reaction that is often seen in lower Primates. This social baring of the teeth in monkeys is very

313

reminiscent of the close-mouthed smile that you give, for instance, at a cocktail party when you are introduced to somebody. Such a smile is not response to humor, but is purely a social acknowledgment.

The Dutch primatologist Dr. J. van Hooff says that laughter in man has developed through what he describes as the relaxed open-mouth display or is equivalent to the relaxed open-mouth display of the lower Primates—that is, Primates below the apes. It is related in these animals to mock aggression or play. Smiling, he feels, is the final stage in the development in what he describes as the silent bared-teeth display. This originally reflected the attitude of submission and can be seen in quite a number of monkeys. It gradually developed to represent nonhostility and then became used as an expression of social attachment or friendliness which is the very opposite of hostility. So the silent bared-teeth display which one gives at a cocktail party on being introduced to a new person is perhaps not always nonhostile.

Van Hooff points out that "we laugh when confronted with ludicrous things or situations, we may smile when confronted with something agreeable, tender or lovable."

The close similarity of human expressions to those of the apes has been shown in a startling fashion in Hans Hass' *The Human Animal*. In one of his illustrations he shows a chimpanzee with three very definite facial expressions and a girl using exactly the same types of expression.

To return to Dr. Menzel's communicating chimpanzees, the ability of the chimpanzee to convey information to the other chimpanzees by its manner and behavior is nothing new. The human has a whole complex of behavior which communicates information to those who take the trouble to read it, and many of us are able to read information contained in manner, behavior, and facial expression without even really thinking of how we are getting this information.

Human beings have a very considerable series of facial and body communications which are very meaningful and are much more widely used than people realize. Edward and Mildred Hall, in their article "The Sounds of Silence," in *Playboy* magazine, give four typical cases in which nonverbal communication was made and conveyed a good deal of information. One example is an individual who leaves his apartment and goes to a corner drugstore for breakfast. Before he asks for his breakfast, the man behind the counter says, "The usual?" and the man nods his head. While he is eating his danish pastry and drinking his coffee, a fat man sits on the stool next to him and crowds him to one side. The response is a scowl from the man eating the danish, and the fat man withdraws as much as he can. So two very different messages have been sent out by this one individual without his actually making any verbal or vocal communication at all.

The Halls cite another occasion where two men are to meet. The time of appointment is eleven, and one will arrive at eleven thirty. Although they are friendly and say nothing unpleasant, the one who has been kept waiting makes it obvious in his manner that there is still some hostility because of this fact. He has referred to it without actually telling the other person that he is hostile.

Another example is seen at a party where a man is talking to someone else's wife. Although what they are saying to each other is completely trivial, the husband glares at them with great suspicion. His suspicion is conveyed by his glare, and he can see that their closeness to each other and the way they move their eyes have revealed that they are very attracted to each other.

In another case the Halls describe a South American and an Englishman at a party trying to establish friendly relations because they are both in business and need to cooperate. Each tries to be very warm and very friendly with the other. However, their attempts with verbal speech are not satisfactory because the South American, as is common with Latins, moves in very close to speak. The Englishman does not like this closeness and tends to back away. The attempt by the South American to move closer is misinterpreted as pushiness by the Englishman. His reaction to that is interpreted by the South American as coldness. Although their words are cordial, their behavior toward each other, because of misinterpretation, is coded in quite a different way.

Most people are aware they are communicating only when they are talking. Much of the other communication is subliminal and is often scarcely realized. Similarly, when we are talking to somebody who gives the impression he is listening, we are able to tell whether, in fact, he is really listening and tuned in to us. My wife is particularly knowledgeable about when I have cut off what she is saying to me. She says I have a record I put on which enables me to carry on a conversation, at the same time taking no notice of what the other person is saying and being busy with my own thoughts. She can sense immediately when I have cut off in this fashion, even though I keep giving the correct responses to her verbal comments. Fortunately she is very patient with this eccentricity.

When a person engaged in a conversation nods his head as he is listening, this helps indicate to the other person that he is, in fact, taking in what is being said. He may give a very vigorous nod if he agrees, and if he has reservations, he may show his skepticism by lifting an eyebrow or by dropping the corners of his mouth. Similarly,

315

people who want to terminate a conversation can make movements that should indicate this fact to anybody with them. Such an individual shifts his body position, stretches his legs, bobs his foot, crosses or uncrosses his legs, or looks away from the speaker. If the speaker is at all alert, he ought to be aware by this time that his audience is completely lost. If he is still not getting the message, the person who wants to get away can give a final intimation to him by starting to look at his watch.

Experts in mime, such as Marcel Marceaux, can convey volumes with their movements without speaking a word. Some of you may have seen the film *Monsieur Hulot's Holiday* in which Jacques Tati stars; this picture is uproariously funny without a single word being spoken.

One of the cues which we may give, which is often not noticed, but is a very important signal is that the pupil of the eye widens when the person sees something which is pleasing to him. For example, if you are smart and look around quickly enough at the eyes of each of your opponents in a poker game as they pick up their cards, there is a very good chance that the player who picks up a very good hand will dilate his pupils subconsciously. It has also been found that the pupil of a normal man's eye doubles in size when he views the picture of a nude woman. So I guess somebody looking through *Playboy* or *Penthouse* has widely dilated pupils most of the time.

The new science of the body language is called kinesics. *Body Language*, by Julius Fast is a fascinating account of the messages that the body conveys. In the relationships between men and women, body language is particularly important. For example, a woman uses her hips and her shoulders to indicate her availability. If she sits with her legs apart, she is obviously soliciting attention, and she may supplement or amplify that by stroking her thighs as she talks or, if she is walking, may roll her hips. If, on the other hand, she sits with her legs crossed or legs tightly together and her arms folded across her chest, she is most certainly not an available young lady. A woman who just crosses her arms on her chest may in fact be saying, "I am not available," or she could be saying, if she does not have the other signals to go with the crossing of the arms, "I am locked in and I want to be let out," that she can be approached and might be available. Mae West, Marilyn Monroe, and Raquel Welch are typical examples of ladies who have turned to commercial advantage an exaggerated version of female body language.

Other approaches by a woman which can even be of an aggressive nature are an adjustment of her skirt as she sits close, the uncrossing of her legs, the pushing forward of her breasts, or the pouting of her mouth. She may use odor in the form of perfume to snare her quarry.

A man or a woman, when either talking with each other or looking at each other, may use gestures which have been described as preening behavior. For instance, the woman may take out her makeup mirror and look at it and rearrange certain parts of her clothes, push her hair off her face, or perhaps stroke her hair. The man may button his coat, straighten his tie, pull up his socks, finger the crease in his trousers, all of which says to the other sex, "I am interested in you. Please take notice of me."

Dr. R. L. Birdwhistell, of Temple University, who is the founder of the science of kinesics, has pointed out that body language is best interpreted in association with the spoken word. In actual fact, the spoken word can be used to cloak an outragously sexual body invitation by the utterance of innocuous remarks coupled with body language signals of a sexual character. A person can be asking for a close association, yet be protected by innocuous conversation from being accused by the other person of offering this sort of relationship.

When one thinks of the complexity of the body language in humans despite the facility for communication which exists with their ability of speech and writing, it is not surprising to think that even very minor behavior and movements on the part of apes could be very meaningfully interpreted by them. Apes, in fact, are not able to speak, and attempts have been made to teach languages to them, not by using vocal effort but by using, in the first place, manual methods. The first people to use this type of communication with apes were the Gardners. They taught their chimpanzee Washoe the American sign language for the deaf totaling about 130 words. Washoe has now gone to the University of Oklahoma with the hope that she may actually be able to teach the American sign language to other chimpanzees there and so start up a generation of "speaking" chimpanzees. I have heard recently that she has had some success in teaching her language to the other chimpanzees.

Language is, in fact, very important. It represents the most recent evolutionary development that is responsible for separating human beings from other species of animals. The latter have means of communication, but they do not have a language in the sense or the terms that we have it.

How is it that monkeys and apes cannot talk? Can a newborn baby talk? These are the questions that Dr. Philip H. Lieberman and his colleagues have tried to solve.

Human infants make a noise from the moment they are born. This noise is called crying, and they differ in this respect from infant apes, which do not cry. The development of speech in the human infant starts with a cry, which develops gradually to a babble and finally to the acquisition of words. In the neonatal period, the human infant is very limited in the number of sounds it can make. Various types of cry have been listed by Dr. Lieberman and his colleagues: birth cries, fussing cries, angry cries, gurgles, hunger cries, shrieks, and inspiratory whistles. Most of these cries can be made spontaneously.

The limitation on the types of cry that the neonate can make is due to the structure of its vocal tract, which makes it impossible for the infant to carry out vocal maneuvers characteristic of the human adult. The shape of the vocal tract in the human infant has a uniform cross section and is very similar to that of the adult great apes, more so, in fact, than it is to the human vocal tract.

Among the similarities are the following: The larynx in the human baby is quite high, being located in a position similar to that which has been described in the gorilla. Also, the infant's tongue is large by comparison with the oral cavity when it is newborn and virtually fills it; this makes the use of the tongue for vocal manipulation very limited in this early period. Later in development the lower jaw enlarges so that the tongue does not completely fill it and can then become a great deal more mobile. The mediation of speech is affected by a number of factors, and one of them is the ability for the root of the tongue to move and constrict the size of the vocal tract. Of course human infants, in distinction to monkeys and apes, eventually are able to produce a complete range of human speech. The monkeys and apes never get beyond the human neonatal condition.

Human speech involves the production of sound by the vibration of the vocal cords which produces turbulence in the air in the vocal tract; acoustic shaping of this sound is produced by the resonances which are intrinsic in the vocal tract. However, the shape of the upper part of the vocal tract has to change continually during speech in order to transform this sound into the various constituents of speech.

By studying the anatomy of the vocal tract system and of the skull and by knowing what is required in the skull and jaws of modern man who can produce speech, one can deduce from the skulls of other Primates and the fossil skulls of early man if they are, or were able to speak. The newborn human, the chimpanzee, and the skull of Neanderthal man have a number of similarities, and they all differ from that of modern adult man in a number of fundamental anatomical details. In addition the chimpanzee lower jaw has a shelf of bone across the inside of the front end which is called the simian shelf; this structure makes it impossible for the chimpanzee to produce some of the movements of the tongue essential for human speech.

The important vowels in human speech are *a*, *e*, *i*, *o* and *u*. It is probable that the vowel *a* could be produced by the chimpanzee if it were able to lower its jaw enough to produce it, but there are various factors that prevent it from doing this. The chimpanzee might be able to produce the vowel *i* by pulling the main part of the tongue forward and lowering the jaw a little bit. However, even with this it would not be able to produce a full *i* sound. The vowel *u* is virtually impossible for the chimpanzee to articulate. Since the anatomical factors of the upper part of the vocal tract of the human newborn do not differ very much from that of the chimpanzees, it is not surprising that the newborn human is also unable to articulate the three important vowels, *a*, *i*, and *u*.

Studies of the anatomy of the Neanderthal skeletons have indicated that the Neanderthal vocal tract could not produce the vowels *a*, *i*, and *u* either. Therefore, these vowels are absent from any kind of oral communication system of the chimpanzee, newborn human, and Neanderthal man. On the other hand, all three appear to have anatomical mechanisms which would allow them to produce what is described as labial and dental consonants. These are consonants like *b*, *p*, *t*, etc. However, various muscular or neural factors are absent which are necessary before the baby is able to do this. It is possible that these consonants are not produced in the baby because there is what might be described as a general inability to produce rapid articulatory maneuvers.

Chimpanzees do not produce dental consonants although they appear to have the anatomical machinery that would enable them to do so. We have had no evidence in the Yerkes chimps that the vocal communication patterns utilize differences between the labial (that is, produced by the lips) consonants and dental (that is, produced between the teeth) consonants. Kathy Hayes, however, was able to teach her chimpanzee, Vicki, to pronounce two consonants, *t* and *m*. I have also heard a chimpanzee belonging to Mr. Berosini of Las Vegas pronouncing the *m* consonant very well as it sounded the word "mama."

The fact that the chimpanzee does not produce some sounds that are anatomically possible for it may indicate that it has some perceptual problems that prevent it from doing so.

Discussing the problem of human speech, Lieberman and his colleagues point out that it is assumed that the communication factor in human speech is at the syntactic and semantic level. If this is so, almost any sequence of sounds would permit a transfer of the same information. They point out that a number of linguists have suggested that even the simple code such as the Morse code can be used to transmit complex linguistic information. If this is so, then Neanderthal man would need only two contrasting sounds to be able to produce a complex communication. However, as Lieberman and his colleagues point out, "Human speech is a special mode of communication that allows modern man to communicate at least ten times faster than any other known method. Sounds other than speech cannot be made to convey language well." They also point out that human speech is able to convey a good deal of information in a rapid period by what they call an encoding process, and the presence of vowels like *a*, *i*, and *u* appears to be one of the factors that make such an encoding process possible. They say, "In human speech a high rate of information transfer is achieved by encoding phonetic segments into syllable size units. These units can be strung into what is described as a 'surface structure.' This string of units can be easily transmitted by a speaker and can be perceived and stored in the memory of the listener." There is no other reason why adult humans do not speak in short sentences like "I saw the boy," "The boy is fat," "The boy fell down," instead of the encoded sentence "I saw the fat boy who fell down." The encoded sentence can be transmitted more repidly, and it transmits the unitary reference of a single boy within the single breath group.

Because Neanderthal man was not able to pronounce the vowels *a*, *i*, and *u*, he was not able to make use of syllabic encoding. And although he could communicate, he was not able to communicate in the sophisticated way that modern human speech permits. He was able to transmit information only at about one-tenth of the rate that can be transmitted in normal human speech. His language, therefore, can be described as being intermediate in the evolution of language— intermediate between the apes and the human, as one would expect—and his language communication was probably accompanied by extensive gestures as well. *Australopithecus*, as far as can be determined from its remains, had a vocal apparatus similar to present-day apes and thus almost certainly had not been able to acquire language although it certainly had some form of vocal communication.

Lieberman and his colleagues close their interesting article with the following:

Human linguistic ability thus must be viewed as a result of a long evolutionary process that involves changes in the anatomical struc-

320

ture through a process of mutation and natural selection which enhance speech communication. Modern man's linguistic ability is necessarily tied to his phonetic ability. Rapid information transfer, through the medium of human speech, must be viewed as an essential property of human linguistic abilities which makes human language and human thought possible.

Raymond Dart, who discovered *Australopithecus*, wrote a fascinating article on the evolution of language and articulate speech published in 1959. He draws attention to the fact that Sir Grafton Elliot Smith had, in 1916, pointed out that the most profound cultural change that occurred in human prehistory took place in the Aurignacian period some 30,000 years ago at the time that the Cro-Magnon people developed. It may be that this coincided with the development of modern speech. There must be some reason why the hand ax culture persisted for 300,000 years. Probably during this time early man's communication was restricted to babbling and pantomime; then, when the complex communication that became speech developed 30,000 years ago, man shot ahead.

Sir Richard Paget has suggested that before the development of true speech, man had a crude communicating system consisting of gestures, accompanied by cries and grunts, and he theorized that these could have lasted for hundreds of thousands of years. This is probably true and certainly Lieberman's work shows that the Neanderthal man was not capable of speech as we know it. It seems possible that speech may have had its beginnings in Neanderthal man, but there is no doubt that he was limited in the sounds he could make and that his "language" was very limited. Aurignacian times, 30,000 years ago, seem then to have been the golden period for the development of human language, although some form of speech must have existed many thousands of years before this.

In the absence of a sophisticated vocal communication, apes require a more extensive visual means of communication. Chimpanzees use many gestures. Drs. Beatrice and Allen Gardner, of the University of Nevada, working with the chimpanzee Washoe in an attempt to communicate with her, decided to use gesturing procedures because the early work of Dr. and Mrs. Hayes of the Yerkes Center with the chimpanzee Vicki showed that vocal communication was very limited. Generally, chimpanzees vocalize when they are excited; however, their hands are very prominent, and their use is obvious. Chimpanzees spontaneously use begging gestures when they are caged, and other gestures also spontaneously form a part of their behavior. Films taken at the Yerkes labs over the years demonstrate a variety of types of manual gesturing. Quite striking is the gesture used to gain the cooperation of another chimpanzee; this simply consists of taking the other animal by the hand and leading it in the desired direction. The soliciting chimpanzee may also put one arm around the other chimpanzee's neck and try to lead it by this method.

There are two methods of communication used in the American sign language. In one system there is an attempt to produce a manual alphabet; in other words, the position that the hands make with the fingers corresponds to the letters of the alphabet. The other system, however, consists of a number of manual configurations, coupled with gestures that are equivalent to specific words or specific concepts. Two typical examples of the American sign language are presented by the Gardners in a recent publication. One of these is the sign for "always," in which the hand is clenched with one finger extended and the arm is rotated at the elbow. In the sign for "flower" the fingers are all extended and touched first to one nostril and then to the other, as if a flower were being sniffed.

The babbling of human infants has been suggested as the origin of language, and it is shaped eventually into language by adults who rear babies. Similarly young chimpanzees, especially if raised in association with humans, undergo what might be described as manual babbling; the nature and use of this manual babbling and its transformation into the elements of language are very well described by Drs. Allen and Beatrice Gardner:

> We encouraged Washoe's manual babbling by being as responsive as possible. We would clap, smile, and repeat the gesture that she seemed to have made, much as people might repeat "goo-goo" in response to the vocal babbling of a human infant. When the babbled gesture resembled a particular sign of ASL [American sign language], we would try to initiate some appropriate activity. For example, during the period when manual babbling was common, Washoe was fond of touching her nose or her friend's nose with her index finger. This is very similar to the ASL sign "funny" in which the extended index and second fingers are brushed against the side of the nose. We could make a regular nose-touching game of this, sometimes she initiated the game and sometimes we did. Everybody laughed as though it were all very funny. At some point Washoe introduced a variation which consisted of snorting if the nose was touched. It was a simple step to initiate the nose-touching game whenever something happened that might seem funny to an infant chimpanzee. Gradually, Washoe came to make the funny sign in funny situations without any prompting.

In an article published in 1971, the Gardners listed eighty-five words that the animal had learned in American sign language. These included such things as—come, gimme, more, up, sweet, open, tickle, go out, hurry, listen, toothbrush, drink, hurt, sorry, funny, please, food, flower, cover, dog, you, bib, in, brush, hat, me, shoes, Roger, smell, good, Washoe, pants, clothes, cat, etc. Some examples of the type of gestures she has to make for some of these are as follows: For "listen" she closes her hand, extends the index finger, and touches the ear. For "toothbrush" the index finger is extended from a closed fist, and the side of the finger is then brushed back and forth across the upper teeth. For "drink" she extends the thumb from a closed fist and touches it to her lips. For "hurt" (pain), she takes the two fingers from two closed fists and touches the fingertips at the site of the injury. Washoe began to combine signs after ten months when she was between eighteen and twenty-four months old. This is quite close to the age when human infants begin to combine words into two-word combinations. The two combinations she communicated were "Gimme sweet" and "Come open." Washoe learned to use these signs for various actions and people in the correct association. The Gardners give a particular illustration. Four of the signs she used were one for Greg, one for Naomi, one for hug, one for tickle. "Hug" and "tickle" were signs that Washoe used for action and "Greg" and "Naomi" were two of the helpers who were obviously capable of hugging or tickling. Washoe was able to form the concept that Greg was Greg and was called Greg whether he was hugging or tickling or, as they say, coming or going. Similarly, the action of tickling remained tickle no matter who the person was who was doing it. Therefore, she was able to use the combination "Greg tickle" or "Naomi hug" but did not use meaningless combinations such as "Greg Naomi" or "Tickle hug."

The Gardners pointed out that in the training of young children, the child will often use one or two-word combinations, and the mother will then expand on it. They give an example of an actual conversation between a child and mother in which the child says, "More cookie," and the mother says, "Your cookie is right here on the table." The child says, "Cookie." The mother says, "No more cookies, Eve. Later we'll have a cookie." They described how they used this technique of expanding what the child says in the training of Washoe:

A good example was Washoe signing "out," when near the door of her house. Persons familiar with Washoe understood this as "I want you and me to go outside." Often we granted her request, but sometimes we denied it, signing a simple "No" or elaborating the

323

denial by explaining that it was "too cold," or "too dark," or "too close to suppertime." Washoe's response to the denial ranged from desisting, through persisting by signing again, to emotional outbursts such as whimpers or temper tantrums. If we tested the interpretation of "out" by partial fulfillments of the request such as thrusting her out by herself, or going out without her, we were certain to elicit strenuous resistance and the emotional outburst. We remained confident of this interpretation even though early combinations with "out" for this situation were of the form "please out" and "hurry out." Washoe began to use the two pronouns "you" and "me" in January, 1968 (month 19 of the project). By spring, 1968, she had signed "You me out" in the doorway situation and later produced many variants such as "You Roger Washoe out" or "You me go out" and "You me go out hurry."

When Washoe used one or more signs one after the other, they could be normally interpreted as a combination; but sometimes there were periods of various length between the individual signs, and it was almost impossible to be certain that they were meant to be a combination. It was, therefore, an essential that she have included in her vocabulary such joining words as "and," "for," "with," "to," when she was in her thirty-sixth month of training. The Gardners point out that this type of word is also not present in the early sentences of young children. It became necessary to distinguish when Washoe's communication had finished by noting the rest position to which Washoe returned her hands when she had finished making a communication.

Some of the combinations which Washoe developed were "Roger you tickle" or "Greg you peek-a-boo" or "Catch me" or "Tickle me" or "Drink red" if she wanted to drink from her red cup. "My baby" she used for her doll. She would say, "Kay open food," at the refrigerator, or "Open Kay clean" at the soap cupboard, or "Kay open please blanket" at the cupboard where the bedding was kept. She also referred to soda pop as "sweet drink" and would say "Please sweet drink," "More sweet drink," "Gimme sweet drink," "Hurry sweet drink," "Please hurry sweet drink," "Please gimme sweet drink," and all these combinations were possible. For the toilet chair she had a sign "Dirty good."

At the University of California at Santa Barbara, David and Ann Premack have also been teaching language to a chimpanzee called Sarah. In their case, instead of using sign language, they have used a number of pieces of plastic of different shapes and colors, each representing a word. These pieces of plastic are backed with metal and

324

are used with a magnetic board so that they will stick to the board in any position required. Dr. Premack and his wife have succeeded in teaching Saraha a vocabulary of 130 terms. They say that she uses them with a reliability of between 75 percent and 80 percent. They asked the question in an article in 1972, in *Scientific American*, "Why try to teach the human language to an ape?" They felt that in their case the motive was to try to discover the fundamental nature of language. They pointed out that languages are said to be unique to the human species, and it is well known that other animals have fairly elaborate communication systems, although it is doubtful that they actually have anything equivalent to human language. Human language is a form of communication which is very highly refined, although it is possible that some aspects of it which are thought to be unique to the human actually belong to a more general system of communication. They believe that if "an ape could be taught the rudiments of human language, it should clarify the dividing line between the general system and the human one."

This is the way they work with the chimpanzee Sarah. First, they would take a slice of banana and put it between the trainer and Sarah. This was done several times. Then a pink plastic square was placed near the chimpanzee, and the piece of banana was moved out of her reach so that to get the banana, it was necessary for Sarah to place the piece of plastic on what they described as a "language board"—the magnetic board on the side of her cage. Once she had learned that, all she needed to do to get the banana was to put the pink plastic square on the board; then the fruit was changed, and the same activity was repeated with a piece of apple. On a number of occasions she was given the opportunity to pick up and eat the apple slice; it was then moved out of her reach, and instead, she had to place a blue plastic sign on the board. Eventually she was able to associate the blue plastic form with the apple and the pink plastic square with the banana. This was repeated with a number of other fruits. Another plastic sign which was equivalent to the word "give" was added, and she was forced to put this common plastic sign in front of the fruit that she wanted.

To help her distinguish the significance of the plastic sign which meant "give" from other verbs such as "wash," "cut," and "insert," the following procedure was adopted: When she wanted to be given the apple, she had to put "give" in front of the apple. If she put the "wash" sign in front of the apple, the apple was immediately placed in a bowl of water and washed. After a period of time, she learned that a particular action was associated with each of the plastic verb symbols.

Communicating with Sarah in Santa Barbara from article by David Premack in Scientific American, Oct., 1972.

There were a number of trainers involved in her activities, and so that she could learn the name of each of them, certain procedures were carried out. The Premacks give the following details of this:

> To facilitate the teaching of personal names, both the chimpanzees and the trainers wore their plastic word names on a string necklace. Sara learned the names of some of the recipients the hard way. Once she wrote "Give apple Gussie" and the trainer promptly gave the apple to another chimpanzee named Gussie. Sarah never repeated the sentence. At every stage she was required to observe the proper word sequence. "Give apple" was accepted but "apple give" was not. When donors were to be named, Sarah had to identify all the members of the social transaction: for instance, "Mary give apple Sarah."

Sarah was also taught the concepts of *same* and *different*. For example, she would be given two objects of the same nature and also sets of three objects of which two were similar and one was different

326

Communicating with Sarah in Santa Barbara from article by David Premack in Scientific American, Oct., 1972.

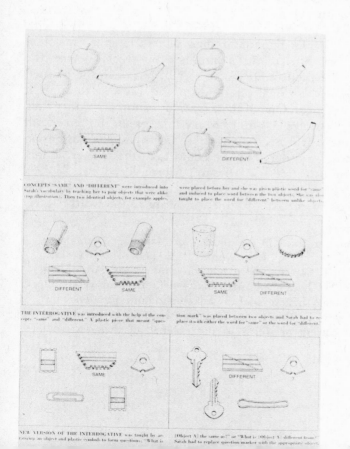

CONCEPTS "SAME" AND "DIFFERENT" were introduced into Sarah's vocabulary by teaching her to pair objects that were alike (top illustration). Then two identical objects, for example apples, were placed before her and she was given plastic word for "same" and induced to place word between the two objects. She was also taught to place the word for "different" between unlike objects.

THE INTERROGATIVE was introduced with the help of the concepts "same" and "different." A plastic piece that meant "question mark" was placed between two objects and Sarah had to replace it with either the word for "same" or the word for "different."

NEW VERSION OF THE INTERROGATIVE was taught by arranging an object and plastic symbols to form questions. "What is [Object A] the same as?" or "What is [Object A] different from?" Sarah had to replace question marker with the appropriate object.

from the other two. She was required to put together the two objects that were like each other. Once she learned this, she was given a plastic word which meant "same" and was asked to place the plastic word "same" between the two like objects. There was another plastic word for "different," and she would take two unlike objects and be required to place the plastic word "different" next to them.

She was given one very interesting item to teach her the significance of the conditional sense. The Premacks used a special sign for conditional which was an "if/then" sign. She was taught to use that in association with the rest of a sentence. She was first trained in the use of the conditional, for example, of being given the choice between an apple and a banana. If she chose the apple, and only if she chose the apple, she was given some chocolate, of which she was extremely fond as well. So she learned the conditional relation between words. She was given a written construction such as "Sarah take apple? Mary give chocolate Sarah," and she was then provided with just one plastic word which was the conditional word. She had to take away the question mark and put the conditional symbol in its place if she was to get the apple and the chocolate. Then she was presented with the sign that said, "Sarah take banana? Mary no give chocolate Sarah." Then she was asked to use the conditional symbol in this sentence in place of the question mark. When she performed this, she was given a banana but was not given any chocolate. Then she was given "Sarah take apple if/then Mary give chocolate Sarah." That was associated with the sentence "Sarah take banana if/then Mary no give chocolate Sarah," and so on. She made a number of errors before the significance of these sentences got home to her, but once she learned the significance of a conditional relationship, she was able to apply it to other types of sentences. One of these was, for example, "Mary take red if/then Sarah take apple," and "Mary take green if/then Sarah take banana." Here it was necessary for Sarah to pay close attention to what Mary chose in order that she could take the proper action. She did not get confused and was able to perform very well.

The Premacks have a very interesting comment about the significance of the ability of Sarah to think in terms of the words:

> Was Sarah able to think in the plastic word language? Could she store information using the plastic words or use them to solve certain kinds of problems that she could not solve otherwise? Additional research is needed before we shall have definitive answers, but Sarah's performance suggests that the answers to both questions may be a qualified yes. To think with language requires being able to generate the meaning of words in the absence of their external

representation. For Sarah to be able to match apple to the actual apple, or Mary to a picture of Mary indicates that she knows the meaning of these words. It does not prove, however, that when she is given the word apple and no apple is present, she can think apple; that is, mentally represent the meaning of the word to herself. The ability to achieve such mental representation is of major importance because it frees language from simple dependence on the outside world. It involves displacement; the ability to talk about things that are not actually there is a critical feature of language.

That Sarah may have had at least some elementary ability to conceive of a mental image of something she could not see is suggested by the following experiment. Early in her training, she was often given a piece of fruit and two plastic words, one of which was the correct one for a fruit and one which represented the word for another type of fruit. She was required to place the appropriate word on the board before she was allowed to have a piece of fruit. However, she made very many wrong choices, and for a time the investigators thought that this was simply poor performance, but then it occurred to them that perhaps she was actually trying to express her preference for fruit. When they were offering apples and she put up the word for banana, what she was doing was really saying that she preferred banana to apple. The Premacks conducted a number of tests to try to see what her fruit preferences were. In one test they used actual fruits, and in another they used the name of the fruit. Her choices between the words followed almost exactly the same sequence as her choices of the actual fruit; this means that she could understand what the plastic symbols actually meant.

In another test that suggested that she understood the significance of these plastic disks and was not using them in some kind of a trick, she was given an actual apple and had to describe it; in other words, she had to say whether it was red or green or whether it was round or square and so on. She picked out the descriptive features which were characteristic of the apple. The Premacks say this is "evidence of the interesting fact that the chimpanzee is capable of decomposing a complex object into features."

Then, instead of the apple, a blue plastic triangle which represented the word for apple was placed in front of her, and again she was given a comparison test; in other words, she had to pick between red and green, round and square, and so on. The properties that she picked were those of an apple which indicated that the blue triangle did in fact mean "apple" to her.

It is obvious that the Premacks have been successful in teaching a

Lana, chimpanzee, and Biji, orangutan, taking part in the language project at Yerkes Primate Center, are told about the computer they will be working with.

simple language to their chimpanzee Sarah, and it has been done in such a way that a complex subject has been reduced to a series of simple steps which were highly accessible to the chimpanzee. Although the experiment was not specifically designed for any other purpose, it does, in fact, have some important spin-offs since the program that was designed to teach Sarah to communicate has now been successfully used with people who have language difficulties caused by brain damage. It could be used with advantage in teaching the autistic child, which is yet another example of how a fundamental study carried out with any real practical intent will nearly always have some kind of practical application.

The Yerkes Center's study on language with great apes, which is under the supervision of Dr. Duane Rambaugh and Professor Harold Warner with the assistance of Dr. Ernst von Glaserfeld and Dr. Leo Pisani from the University of Georgia, will be carried out on a chimpanzee, a gorilla and an orangutan. The design of the computer

330

Lana, chimpanzee in the Yerkes language project. She is tapping out a message to the computer, using various keys which have symbols on them. This particular message was "Please machine give Lana water." The computer immediately obliged.

equipment used was the work of Professor Harold Warner of the Yerkes Center. Chimpanzees, as we have noticed, have been used for studies in communication in the past, but the orangutan and gorilla have not, so it will be interesting to see how they compare. The idea is to have the signaling or the communication carried out by a computer rather than a human. There are a number of reasons for this. One of them is, of course, that cues in addition to those the animals are supposed to translate into language can be conveyed by a human observer without his realizing it. In the Yerkes' experiment, there is a keyboard with lighted square keys which are about $1\frac{1}{4}$ inches by 1 inch; they burn at half brilliance when the machine is on, and whenever one is depressed, they light up to full brilliance. There are five banks of keys. The first bank contains agent words and pronouns; the second bank, activity words; the third bank, prepositions; the fourth bank, object words; the fifth bank, other words such as "yes," "no," "not," etc. Initially the animal is asked simply to request things by a single word. For instance, if it wants a piece of apple, it depresses the key which has the sign for apple. But later on it is asked to attach words. Eventually it will be expected to say, "Give me apple," or "John, turn off noise," or "Take Mike away," or "Judy, take me to playroom," and so on. The animal always receives a food reward, at least in the beginning, in order to teach it the correct symbols for various expressions. Of course, it works with a human in the beginning until it understands what it is supposed to do; then the communication is taken over by the computer. The experiment has actually started, and in a relatively short time the chimpanzee used in the project, a female called Lana, learned quite a number of symbols, including those for apple, monkey chow, water, etc. She can also ask for toys, for someone to come in and play with her and can ask the computer to show her a movie or play her some music. These last two she usually asks for in the evening when everyone has gone home. The orang which is being used in this experiment has not yet learned to use the machine; the reason is that she is still frightened of it. It is an interesting fact that the orangs are much more sensitive than the chimpanzees and much more scared of strange objects that they do not understand. On one occasion I went around our great ape wing with a child's colored book of pictures opened to a picture of an alligator. I showed this to several chimpanzees, and they immediately ran up to the front of the cage and touched the picture with their fingers. When I showed it to a gorilla, it took a look at it and then stood back several feet and looked at it from a safe distance. Most of the orangs I showed it to took one look at it and then disappeared into their inner quarters and refused to look at it anymore.

The Premacks feel that Sarah, has a language ability very comparable to that of a two-year-old human. In fact, some of the demands made of Sarah are those that would not actually be made of a two-year-old child. All of which goes to demonstrate that we have grossly underestimated the chimpanzee and almost certainly the other great apes and makes one think very soberly about individuals who keep them locked up in tiny cages and use them like guinea pigs in their various forms of medical research. We should also ponder whether it is a reasonable life for them even to be cooped up in a zoo cage.

A discovery by Drs. Davenport and Rogers at the Yerkes Primate Center that chimpanzees possess the ability of cross-modal transfer of information also demonstrates a previously unsuspected mental ability of these animals. For example, they can be shown an object or even a photograph of an object and subsequently are able to select it entirely by touch, without seeing it, from a number of other objects. This action had previously been regarded as impossible for chimpanzees to carry out.

In an experiment recently carried out by the Yerkes Center four chimpanzees—three females and a male—were liberated on Bear Island, off the coast of Georgia. Bear Island is part of the Ossabaw Island complex, owned by the West Foundation. By the courtesy of Mr. Clifford West and Mrs. Eleanor West, their island, about 100 acres in extent, surrounded by water and marshes and very well wooded, was made available to the center to establish our colony of chimpanzees. The four chimpanzees had been in a compound 100 feet square with a number of other chimpanzees at our field station. Three of them, two of the females and the male, were wild-born, and one, Joni, the youngest, was restricted reared—in other words, an animal that has been taken from its mother at birth and brought up in total isolation for twenty months, not permitted to see another animal or human being for this time. Joni was mentally disturbed and had developed a number of stereotyped actions such as standing and rocking from side to side and sitting and rocking backward and forward. The other females were Girlie and Saki, and the male was called Jiggs.

Jiggs had been the dominant animal in a group of sixteen chimpanzees in a compound at our field station. He had an oil drum which he would roll around the compound, and this was probably a symbol of his authority. He would stand at one end of the compound,

Author stands by the sign designed to keep unauthorized persons off the island.

Jiggs, chimpanzee, lends added note of warning to the sign.

The chimpanzees are released onto Bear Island (Ossabaw Island complex) and make for the woods. Below: What are those strange creatures in the cage? Chimpanzee just released on Bear Island (Ossabaw Island complex).

and perhaps the other chimpanzees would be at the other end watching some human observers just outside the wire. Jiggs would start rolling this barrel with great velocity across the compound until it crashed into the wire at the opposite end with great force. The chimpanzees that had been in the way in the beginning would, in a kind of bored fashion, move out of the way at the last moment. It is incredible that none of them ever got injured. They always seemed just to be able to avoid the barrel. Jiggs was a big, rough, rambunctious animal. We thought he would be an ideal type to put out in the wild with the other animals.

The great day came, and the animals were loaded on a truck and driven south of Savannah where a boat waited to take them to Bear Island. Bear Island had in the meantime been prepared for them. There was a landing area protected with chain-link fencing so that a boat could land to provide them with food without the chimpanzees boarding it. Then there was an enclosed area in which the food and water were placed so that the animals would always eat and drink in there; in this way it would always be possible to catch them again when required.

They were duly liberated into this fenced enclosure; the door was opened, and they all ran out to look around and to see what was going on. Girlie and Saki took off at high speed and were not seen again for the rest of the day. Joni hung around the landing area and climbed into one of two little A-frame houses (with plexiglass fronts and backs so that we could see through them) standing on short stilts with an entrance through the floor. These had been provided so that in the cold weather the animals would have somewhere to go where they could be safe from the cold wind, and in a confined space their combined body heat would help them warm each other.

The big surprise was Jiggs. He hated to lose contact with his human colleagues, and far from behaving like a big tough Tarzan and rushing into the trees, he hung around the humans even more closely than Joni, who stood off in the distance and rocked back and forth. He would keep coming back into the feeding area and would press himself up against the chainlike fence so that he could be groomed through the wire. When, at the end of the day, the boat stood off from the island and prepared to leave for the mainland, Jiggs came rushing forward to the water's edge, metaphorically wringing his hands; one could almost see tears rolling down his face. He did not want to be left alone in the wild under these circumstances without the comfort and the security of having humans around. It was obvious that the act that he had been putting on at the field station all the time was nothing

336

WØEXD/4

Project Mercury chimpanzee astronaut Ham (wearing medal) is shown on left. Ham operator WQEXD, Air Force pilot, and veterinarian Major Richard J. Brown provided medical consultation in selection, training, and utilization of all animal space travelers. WQXED (right) gives another future chimpanaut a checkup.

but a big bluff. We went off for the night and came back early the next morning. As we arrived, we called out, "Jiggs!" There was a bellow, and Jiggs came rushing from the bushes where he had apparently been holed up; he could not get back to see his human friends fast enough. At the end of the first day, Girlie and Saki had come back at the last minute because by now they were thirsty and they wanted a drink from the water fountain which had been installed for them. Time has passed, and the animals have now been there for a good seven months. They have started climbing trees, making nests as they would have in the wild, have been progressively eating more and more of the leaves and other material found in the environment, and are beginning to look as if they have put on weight and have grown much longer and more luxuriant coats.

For the first few days after they were placed on the island, when they came up to get their food, they could be seen to be absolutely covered with ticks. These ticks caused us some worry; however, after

Jiggs and John Haynie have a tête-à-tête through the wire (Haynie is in the cage).

the chimpanzees began to integrate themselves into some sort of a social group and began to groom each other more, the ticks more or less disappeared. It became obvious that they were pulling the ticks from each other. They were not completely free of ticks, a few could be found, but the numbers were relatively small in comparison with what they had in the first few days. Which brings up the conjecture whether the procedure of grooming among Primates, together with its offshoot of petting and lovemaking that we see in humans, may not have really originated, from an evolutionary point of view, from the necessity of keeping each other free of ectoparasites in the wild.

The latest report from the island is that Saki is probably pregnant. Girlie was also pregnant, but one day it was observed that she was missing. When days passed and she did not appear, we thought she might have got lost or drowned in the marshes. An expedition went ashore on the island to look for her, but not a sign of her could be seen. Then one day, after Girlie had been missing for about two weeks, Jiggs was seen running to the edge of the cleared area near the landing area and whimpering. When this was investigated, Girlie was found lying dead on the edge of the clearing—she had obviously been dead some

Jiggs, chimpanzee, surveys his new domain—a desert island and three female companions.

days. An autopsy showed that she had been pregnant, had aborted, and then probably got an infection which had killed her. When the group of humans had gone ashore to look for her, she was probably hiding somewhere sick. Later on she dragged herself, or Jiggs dragged her, back to the clearing to die. It is a sad story, especially since, if we could have got to her when she was sick, we could probably have cured her with antibiotics. This incident gives an indication of the hazards that are present when animals, or indeed humans, return to the natural state. Humans on a desert island would be faced with exactly the same problem.

Unfortunately Joni, too, was lost. We received reports that she was not very well, and Dr. Keeling, our chief veterinarian, made a visit to the island to see what her problem was. He captured her and found that she apparently had been injured in some way on the side of the face and the head, which were very much swollen and bruised. He wondered if perhaps she had fallen out of a tree. She was kept isolated in a cage overnight, but the next day she died. At her autopsy it was not possible to be sure of the precise cause of death. Among the possibilities we thought of was snakebite, but the pathology was not

339

Jiggs relaxes on his new island.

compatible with death from this cause. She may have died as a result of eating some poisonous plant.

The other two animals are doing extremely well, and we are delighted that they survived the cold spells without any difficulty, even though the temperatures have been below freezing. In the wild chimpanzees can be found at altitudes as high as 10,000 feet. At that altitude, even in the tropics, nights can be extremely cold.

Partly to keep intruders away and partly to keep some supervision over the whole project, we have hired a young man from Pennsylvania State University, Mark Wilson, who is spending a year at Ossabaw taking part in a very interesting educational program set up by the West Foundation called "Genesis" in which a number of city students are given the opportunity to come each summer and spend some months living in a very primitive situation. They live, as far as possible, off the land where there are all kinds of wildlife, but they are not, of course, completely deprived of modern-day food. They are expected also to learn something of the ecology of the island. In many cases some of them use this opportunity to write theses. Mark Wilson is

The author confers with Yerkes and Emory personnel about the release of the chimpanzees.

spending part of his time helping supervise this project and part looking after our animals. One of the other Genesis boys, Jim Erliker, is helping him, and the project in the Savannah area is overseen by Jim Brown, a local architect, who has been connected with Ossabaw Island projects for some time and who provides the boat that enables us to keep in touch with and supervise this particular project.

The Pygmy Chimpanzee

On October 13, 1928, at a meeting of the Cercle Zoologique Congolais in Léopoldville (now Kinshasa), the capital of the Belgian Congo (now the Republic of Zaire), the presence of a small chimpanzee on the south bank of the Congo River was described. That this was a new race of chimpanzee was demonstrated by an article in the following year in the *Review of Zoology and Botany of Africa* by Dr. Ernst Schwarz. Later on the director of the Congo Museum at Tervueren, Dr. H. Schouteden, published a detailed article on this new race of chimpanzee which was called *Pan paniscus*. The animal

341

Oona—pygmy chimpanzee
from Dania, Florida.

Oona—pygmy chimpanzee
from Dania, Florida.

was found on the south bank of the Congo and presumably has been isolated there over the millennia by the Congo River. In 1930, Dr. James P. Chapin of the American Museum was able to obtain an adult female chimpanzee. Dr. Harold J. Coolidge, who described the anatomy of the pygmy chimpanzee in great detail in 1933, says that it is remarkable that this animal should have escaped the attention of scientists until as recent a date as 1928.

Some of the anatomical features of *Pan paniscus,* the pygmy chimpanzee, differ from the other varieties of chimpanzee. The forehead of the *paniscus* is higher and is more convex, the back of the head is more rounded, and the face does not stick out as much as a chimpanzee's (it is less prognathic). It has much smaller ears than the chimpanzee; they are much closer to human ears in size. The face is completely black, whereas most of the other chimpanzees have light-skinned faces. The pygmy chimpanzee also appears to be more agile, and there is evidence that it is more intelligent than other chimpanzees. The standard chimpanzee cry in the wild is *hou-hou-hou,* whereas the *Pan paniscus* has a call which is much more like *hi-hi-hi.*

343

The voice is neither as loud nor as shrill as a chimpanzee's but is higher-pitched. The pygmy chimpanzee also has smaller eyebrow ridges than the ordinary chimpanzee. In other words, in many respects the pygmy chimpanzee is much closer physically to the human than it is to other chimpanzees. Dr. Elwyn Simons and Dr. David Pilbeam, from the Peabody Museum at Yale University, believe that it is much more closely related to the thin, gracile human animal which eventually became *Australopithecus,* and it may well be that this animal is much more closely related to man's ancestors than any other living ape.

The animal has fine glossy black hair over its body but has a white tuft of hair around the anus. Another interesting thing is that the second and the third toes on the feet are connected by a web of skin.

In December, 1923, Dr. Robert M. Yerkes purchased a young male chimpanzee, which he called Prince Chim. This was a black-faced chimpanzee, and Yerkes became dedicated to him, even to the extent of writing a eulogy for him. He commented upon the great intelligence of this animal compared with his other, white-faced, female chimpanzee, Panzee, that he got from British West Africa. The animal was not only more intelligent but more agile. Dr. Coolidge, in an analysis of the information about Dr. Yerkes' Prince Chim, has come to the conclusion that Dr. Yerkes actually had a *Pan paniscus.* Evidence for this comes from the teeth, which indicated that the animal was twice as old as Dr. Yerkes thought it was, from its coloring, and many other anatomical characteristics. In particular it had black skin, small ears, and webbing between the second and third toes. The animal also was caught in an area where *Pan paniscus* is found. Dr. Yerkes speaks of how bold and aggressive the animal was, how constantly alert and eager he was for new experiences, and he says, "Seldom daunted, he treated the mysteries of life as philosophically as any man. Never have I seen man or beast take greater satisfaction in showing off than did little Chim."

Finally he says of him, "Doubtless there are geniuses even among the anthropoid apes. Prince Chim seems to have been an intellectual genius. His remarkable alertness and quickness to learn were associated with cheerful and happy disposition which makes him the favorite of all, and gave him a place of distinction not only in their regard but in their memories."

At this point I should mention that the Yerkes Research Center, in association with the National Academy of Sciences and with the cooperation of various officials of the Republic of Zaire, is endeavoring to organize an expedition to the area south of the Congo where the pygmy chimpanzees live. Incidentally, they are a rare and endangered species. They may not be exported from the Congo or imported into the United States. It is planned that a study will be made of the approximate number of these animals still existing in the wild. If possible, a number of them will be established in a breeding colony on islands on Lake Tumba, a large lake which opens into the Congo River, where the animals can be preserved for further study by scientists from all over the world.

The Orangutan

Marco Polo in his account of his visits to the Orient refers to animals that were obviously apes of some sort—it is not certain whether they were gibbons or orangutans, although the latter is the more probable. However, there is some confusion on when the animal we now call the orangutan was first known because Dr. Tulp used the name "orangutan" to describe the ape which he talked and wrote about and sketched. Although he calls it an orangutan, it is said to have come from Angola and was probably a chimpanzee. This was in 1641. But the picture of Tulp's chimpanzee does in fact look to me to be more like an orangutan than a chimpanzee. The first apparently unequivocal record of an orangutan in scientific literature was in 1658. At this time Jacobus Bontius, a Dutch physician working in Java (Indonesia), gave a description of it. Dr. Jacob Le Comte described it in 1697, and Dr. Tyson and a number of others gave further details during the eighteenth century. One of the later authors, Hoppius, who wrote in 1760, describes the color of the coat of the orangutan as being not too different from that of a well baked brick.

Dr. Bontius was the first person to give the name "orangutan" to the ape which is found in Borneo and Sumatra. Part of his description is as follows: "A female who seemed to have an idea of modesty, covering herself with her hand on the appearance of men with whom she was not acquainted; who sighed, cried, and did a number of other actions, so like the human race that she wanted nothing of humans but the gift of speech."

The author and conservationist Tom Harrisson makes the rather dry comment that "one uneasily feels that in those early days, sharp Malays passed off cretinous step sisters on credulous scientists, in this thirst for sensation and data."

Bontius said that the Javanese claimed that these animals have the ability to speak but will not do so because they are frightened of being made to work if they do.

In the latter part of the nineteenth century Restif de la Bretonne wrote a fantasy about the discovery of Australia, but it was actually based largely on the reports of the exploration and the discovery of Australia by Captain Cook. In it he includes a fictional letter addressed by an ape to the members of his own species. He points out that this ape, which was an orangutan, had an intelligence very close to that of a human's and further that male apes are capable of having sexual intercourse with native women if they get the opportunity. This myth has been discussed in detail in my book *The Ape People*.

Literature, particularly early literature, is full of references to the sexual behavior of apes and their desire to have sexual contact with human females. But any woman who believes this would be a thrill would undoubtedly be disappointed by the penises of these animals. A woman ravished by an ape would scarcely feel a thing. The penis of a chimpanzee, for instance, is long but very, very slender, a little thicker for most of its length than a pencil. The gorilla has a very short penis and also not a very large one. It is no more than a couple of inches long.

There were actually rumors in Borneo that male orangs carry off the native women for sexual purposes. The famous naturalist Georges Buffon, in 1766, accepted this fact. Gustav Flaubert, in his *Quid Quid Voluaris*, reputed to have been published in 1837 (he was only sixteen at the time), described an ape known as Djalioh who had been produced by sexual intercourse with a male orangutan and a Brazilian black woman. Flaubert's story involves a complicated and rather violent relationship between this ape-man and a young married couple. Tom Harrisson, formerly curator of the museum in Kuching, Sarawak, Borneo, has recounted a story written some years ago by an administrative officer who was stationed in Borneo, of the rescue of a Dyak lady who had been captured by an orangutan.

In view of all the stories about apes and human women, one might ask if there have ever been any hybrids between a man and ape or have there ever been any hybrids between the apes themselves. To this we cannot give a definite answer. There is no real evidence that a successful cross has ever been made between a human and an ape. Restel de la Bretonne has produced a fictional one, and this is the creature that was supposed to have written a letter to his species. According to him, the writer of the letter was called César de la Malaca, and he is described as the product of an incident between an orangutan and a native woman. He had been brought up and educated in China, and he was then given to an Australian who was reputed to have presented him to Restif de la Bretonne, who, still in

Portrait of an orang.

this fictional account, put him in the care of what was described as "a respectable lady." He was brought up under her tutelage and was educated in the manners of Western people. He then became disillusioned with humanity and decided to write a letter to his own species which would be a warning and a consolation to those who felt they wanted to imitate man either as they were living free in the forest or while enslaved by man. He says that ignorance is an imperfection, "but knowledge has inconveniences that frighten me . . . . To ignore the pains of the spirit; to ignore death, as you do; to see only the moment, that is to be immortal. The more we know, the more we become aware of dangers, the less happy we are."

Then he finally points out that man actually causes more pain to his fellowmen than he does to other members of the animal world. He believes that this is the reason that he is such an unhappy person. César then gives a catalogue of the various types of human vices that shows the extent to which man pays no attention either to the laws of nature or to the Christian religion. He finishes up by saying, "Thank God that I am an ape and not subject to human laws and prejudices," although of course he is only half an ape.

Captain Daniel Beeckman came to the southern part of Borneo in 1712 and described his visit in a book which he published in 1714. He mentioned the "oran-ootans" and points out that in the Malay language this means "men of the woods." He said that they grow as high as six feet, that they walk erect, that they have much longer arms than men do and have quite good faces, that they are very strong and that they throw stones, sticks, and branches at people who offend them. He points out that the natives believe that these people had previously been men but had been metamorphosed into beasts by some kind of god as a result of their blasphemy. Beeckman was also able to purchase a live orang and paid a sum of one English pound sterling for it.

About this animal he says the following:

> I bought one out of curiosity, for six Spanish dollars, it lived with me for seven Months, but then died of a Flux; he was too young to show me many Pranks, therefore I shall only tell you that he was a great Thief, and loved strong Liquors; for if our Backs were turned, he would be at the Punchbowl, and very often would open the brandycase, take out a bottle, drink plentifully, and put it very carefully into its place again. He slept lying along in a human Posture with one Hand under his Head. He could not swim, but I know not whether he might not be capable of being taught.

Linnaeus, in his *System of Nature* in 1758, showed some confusion about the orangutan and actually put it in the same genus as the human. He called it *Homo nocturnus* or alternatively *Homo sylvestris orang-outang*. Of it he says:

It lives within the boundaries of Ethiopia (Pliny), in the caves of Java, Amboina, Ternate. Body white, walks erect, less than half our size, hair white, frizzled. Eyes orbicular; iris and pupils golden. Vision lateral, nocturnal. Life span twenty-five years. By day hides; by night it sees, goes out, forages. Speaks in a hiss. Thinks, believes that the earth was made for it, and that sometime it will be master again, if we may believe the travellers.

Most of this, of course, is completely inaccurate as far as the description of the orangutan is concerned. Tom Harrisson points out dryly that "this supposed orang ambition has fortunately not since been noticed by anxious mankind. Not that any additional excuses have been necessary, since 1758, to incite people to treat orangs as somehow basely inimical to human mastery."

In the nineteenth century there was a great interest in orangs, and there were many visitors to Borneo, mostly dedicated to the murder of orangutans. Evolutionists such as Alfred Russel Wallace and various others, including Dr. William Temple Hornaday, who was chief taxidermist of the United States National Museum, visited Borneo and were responsible for the demise of a number of these fine animals. In the title page of his work Hornaday quotes Byron as follows: "There is a pleasure in the pathless woods/ There is a rapture in the lonely shore." He certainly did nothing to give any rapture or pleasure to the orangutans that he came in contact with. He killed forty-three orangs in one year, and here are some quotations from his book *Two Years in the Jungle* (published in 1929) on some of his activities.

His native helpers had located a huge male orang for him.

The men were all greatly excited, but I knew that the old fellow was old and waited for a good shot. When the moment the opportunity came, and I fired twice in quick succession at the orang's breast. He stopped short, hung for a moment by his hands, then his hold gave way and he came tearing down, snapping off a large dead branch as he fell, and landed broadside in the water, which went flying all over us, he fell within ten feet of our boat and we secured him without getting out. As we seized the arms and pulled the massive head up to the surface of the water, the monster gave a great gasp, and looked reproachfully at us out his half-closed eyes. I can never forget the strange and even awful sensation with which I regarded the face of the dying animal. There was nothing in it in the least suggestive of anything human, but I felt as if I had shot some grim and terrible gnome or river guard, a satyr indeed.

Three miles farther on I spied a baby orang up in a treetop, hanging to the small limbs with outstretched arms and legs, looking like a big red spider. It gazed down at us in stupid, childish wonder, and I was just aiming for it, when Mr. Eng Quee called my attention to the mother of the infant, who was concealed in the top of the same tree. As soon as I fired at her, she climbed with all haste to her little one, which quickly clasped her around the body, holding on by grasping her hair, and, with the little one clinging to her the mother started to climb rapidly away.

Fortunately, we were able to get the boat in amongst the trees without much trouble, and all immediately went overboard. We had scarcely done so when a third orang, a young male about two years old, was discovered looking down from a nest overhead, which he immediately left and started to follow the old mother. As he went swinging along underneath a limb, with his body well drawn up I gave him a shot which dropped him instantly, and then we turned our attention to the female. She was resting on a couple of branches, badly wounded, with her baby still clinging to her body in great fright. Seeing that she was not likely to die for some minutes I gave her another shot to promptly end her suffering, then she came crashing down through the top of the small trees and fell into the water, which was waist deep.

We strained to secure the baby, but it was under water fully a minute before we found it, quite unable to swim and very nearly drowned. We managed to resuscitate it, however, then the other two were lifted into the boat and we drew out into the stream.

That day Mr. Hornaday had seven orangs to skin and cut up, and he says that many wondered how he was able to spend all day skinning and cutting up the seven orangs. So he concludes his comments about this by saying:

> Well, tastes differ, that's all. As for myself, I would not have exchanged the pleasures of that day, when we had those seven orangs to dissect, for a box at the opera the whole season through.
>
> It is a pity that the men who "don't see how you can do it" could not have been there on that memorable occasion. When we finished, there was a mountain of orang flesh, a long row of ghastly, grinning skeletons, and big, red-haired skins enough to have carpeted a good size room.

It is hard in these days of conservation of animal life to appreciate his point of view, and one can only sit back in horror at the slaughter and the fact that anyone could massacre these beautiful intelligent animals for any purpose even when this was done with the object of simply getting a large number of skins and skeletons for museums.

Alfred Russel Wallace, the famous naturalist who could also be described with Darwin as being the co-originator of the theory of evolution, also had several field days in Borneo, having made one of his visits under the auspices of the white Rajah of Sarawak, who was Sir James Brooke. Wallace shot a number of orangs and admitted that he never killed one with a single shot.

A British naval captain, Rodney Mundy, also was a hunter of orangs and has some experiences in killing them which are distressing to anyone who reads them today.

Tom Harrisson in his introduction to Barbara Harrisson's book *Orang Utan* has this to say about the twentieth century as compared with the nineteenth as far as orangs are concerned:

> Broadly, then in the first half of the twentieth century there was less crude slaughter of orangs and other apes under the pretext of natural science, Christian sport, or the protection of indigenous rights. But the poor apes were now increasingly caged for human delight. In order to cage one, invariably at least one other had to be killed—unless special methods (not at first known or generally practicable) were employed. Those taken captive frequently died before they reached their intended zoos. Of the survivors, less than one in five survived more than three years, although an orang's normal expectation of life is at least twenty-five years.

In actual fact the orang's normal expectation of life is more likely to be that of the chimpanzee, which we now know is around forty-five to fifty; the Philadelphia Zoo has an orang with an estimated age of over fifty. The trade in orangs has been going on right up to the present. The animals were being collected both in Borneo and in Sumatra by the process of shooting the mother to get the young one and exporting the young one to Europe. The United States has now attempted to play its part in suppressing this trade by forbidding the further import of orangutans into the United States. Any that come in illegally are seized and are placed in the National Zoo in Washington so that whoever paid the cost of bringing it in is unable to recoup his cost.

Most of the great apes seem to have been derived from an animal which existed between 10,000,000 and 20,000,000 years ago called *Dryopithecus*. Studies and excavations of fossils in the Eurasian area showed that the earliest *Dryopithecus* there was about 16,000,000 years

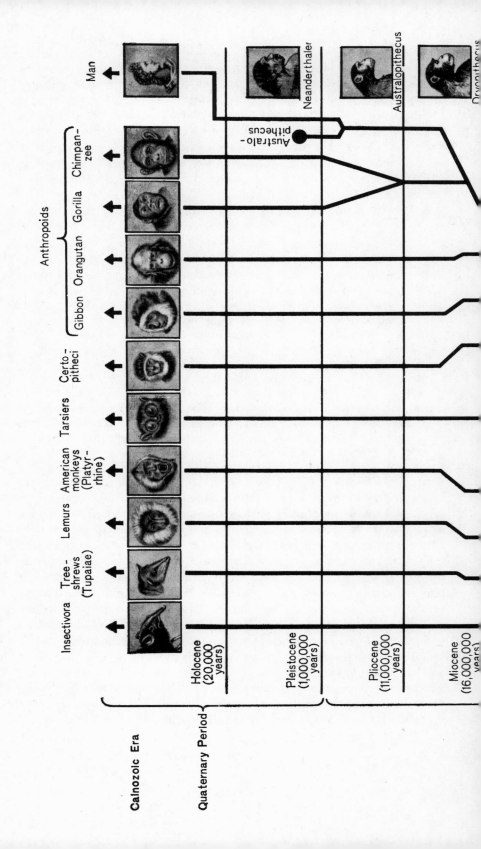

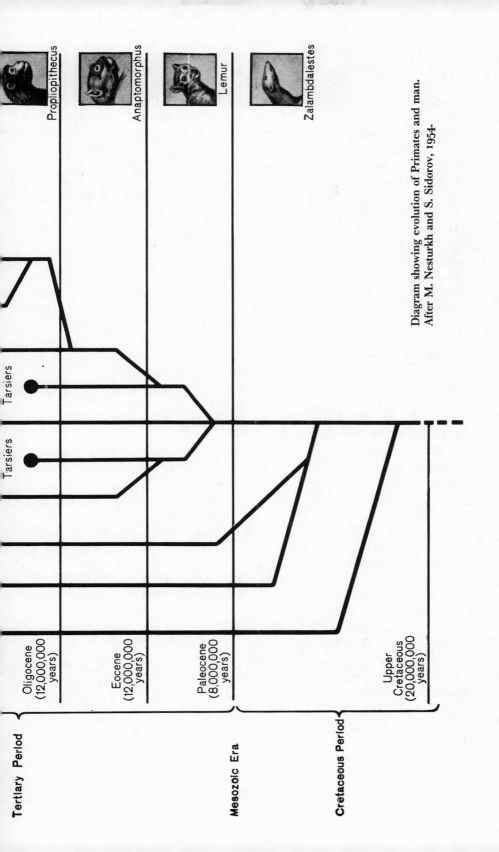

Diagram showing evolution of Primates and man.
After M. Nesturkh and S. Sidorov, 1954.

old. It is possible that from one of these Eurasian forms of *Dryopithecus* the orangutan originally derived, although he probably came from the *Dryopithecus* stock earlier than the chimpanzee or the gorilla; at least that is the view of David Pilbeam and Elwyn Simons of the Peabody Museum at Yale.

A number of scientists who have been studying the biochemistry of Primates believe that these various apes came much later in prehistory than the fossils indicate. At the moment, however, the fossil evidence seems more reliable. It seems pretty certain that by the time *Pithecanthropus* (the Java man, now said to belong to a group known as *Homo erectus*) was living in Southeast Asia and those large apes called *Gigantopithecus** were living in China, the orangs were already well developed and were widespread in the Southeast Asia region. Dr. Eugène Dubois, who discovered *Pithecanthropus,* found a number of orang teeth, which were not fossilized, in the highlands in the central part of Sumatra. Such teeth have also been found in a cave in Niah and in other parts of west Borneo. In the great caves of Niah, Tom Harrisson describes having dug out an orang bone which had been cooked; the time scale to which this belongs is about 35,000 B.C., which represents the early Stone Age period. The cavemen of those days may have kept orangs as pets. He comments: "It is clear, then, that Asians were intimate with the orang thousands of years ago; and it is probable that long before this ape was known to western science, men were killing or keeping it in semi-captivity."

The jaw of an orangutan forms part of one of the most fantastic scientific hoaxes that has ever been perpetrated, and perhaps it would be of interest to the general reader to know something about it.

Toward the end of the nineteenth century and the beginning of the twentieth century, there had been a number of interesting fossil finds which were shedding light on the origin of man. Dr. Dubois, mentioned earlier, had, in 1891 and 1892, found the bones of the Java man, *Pithecanthropus erectus,* who was said to have lived about half a million years ago and to have been an ape-man who walked erect. In 1907, a jawbone was found in Heidelberg, Germany, which was thought to have belonged to an ancestor of Neanderthal man. This stimulated scientists into hoping that it might be possible to find some even more ancient human fossil material in Europe.

At the time the Heidelberg jaw was discovered, Charles Dawson, an amateur geologist, was living in the south of England and was very interested in the geology of the area in which he lived. He had done a

---

* Specimens of *Gigantopithecus* which survived up to a few thousand years B.C. or even up to historical times may have been the origin of the legend of the Abominable Snowman.

good deal of digging and actually discovered a number of animal fossils. He lived near the town of Piltdown in Sussex. Close to the town there was a gravel pit in which various local workmen used to dig out gravel that they could use for mending roads. This gravel really belonged to an old river bed which had probably been formed half a million years before, during the early part of the Ice Age. It seemed a good spot to find fossils, and Dawson began to dig there.

In 1908 he began to find some human material. In 1911 he wrote to Sir Arthur Smith-Woodward of the British Museum that he had found part of the top of a skull and a jawbone with two teeth; near them he had discovered some tools made of flint and various bones belonging to other animals. Sir Arthur Smith-Woodward made a detailed study of the bones and the jaw, and so did Sir Arthur Keith, the famous anthropologist, and both of them accepted these specimens as genuine and said they belonged to an ape-man who lived about half a million years earlier. The apelike appearance of the jaw was the main reason for suggesting that the remains were those of an ape-man. The teeth in the skull were very flat, as human teeth are, and the bones which formed the skull looked about the right size for a man. So there was a curious mixture of ape and human characteristics in these remains. At that time scientists had pictured an ape man as being one which had a large brain, with a humanlike skull but a jaw that still remained apelike, so that the Piltdown skull fitted very well into this conception. Later on, with other findings of fossil skulls, it became obvious that this was not the way the ape-man developed; in general, the lower jaw became more humanlike faster than the skull itself.

In 1916, Charles Dawson died. Dr. Smith-Woodward continued to dig in the Piltdown gravel for a number of years to come and was never able to find any more fossils. Eventually the various discoveries of fossil humanlike skulls gave a fairly good picture of the origin of man through *Australopithecus* (South African man) *Pithecanthropus,* which is now *Homo erectus,* Neanderthal man, and modern man. In this picture, Piltdown man did not find a place at all. In 1954, Dr. Smith-Woodward suggested that there were two lines of human evolution which were quite separate; one line developed into Piltdown man which occurred in Europe and another line in Africa and Asia went through *Australopithecus* and *Pithecanthropus.*

Towards the end of the 1940's, Dr. Kenneth Oakley, of the British Museum, together with Dr. Joe Weiner and Professor Le Gros Clark, of Oxford University, decided to apply to the study of the Piltdown fossils new techniques which had recently been developed for measuring the age of fossils. One of these was to measure the amount of fluorine in the fossil bones. Fluorine is present in soil, and if bones are buried for a long period of time, they tend to absorb it. The longer they are buried, the more fluorine they absorb. By measuring the amount of fluorine, it is possible to get some conception of the age of the fossils.

More modern methods involve the use of measuring the amount of radioactive material found in bones. But it was the fluorine estimation method which was used first in the investigation of the Piltdown bones which were supposed to be half a millions years old. The fluorine tests showed that the Piltdown fossil material was certainly not older than 50,000 years, so it certainly could not have been in the line of ancestry of modern man because modern man was already in existence 50,000 years ago. At the same time, the dating of the bones of animals that were found around this specimen (they were elephant and rhinoceros bones) showed them to be half a million years old.

In 1953, Dr. Joe Weiner was considering all the possibilities concerning this discovery. The presence of the skull and the apelike jaw did not make sense, but also the fact that they belonged to different fossils was not very logical because this would mean that 50,000 years ago large apes were living in England. This was not only unlikely but was actually impossible because at that time there was an ice age and there were certainly no apes in Europe then. Weiner was driven to the conclusion that perhaps a human being had put all these pieces together. He considered first of all the jaw which was definitely like that of an ape, in fact, rather like that of an orangutan. Yet it had teeth that had been worn flat and in this respect were like the teeth of a man. However, X rays showed that the roots of these teeth were long roots like those of an ape, not like those of a man. Also, when he drilled into them to get a sample for analysis, he found that the brown stain on these teeth which is characteristic of all the fossils was only on the surface and that the teeth were quite white under that stain.

When these specimens had been discovered, Dr. Smith-Woodward had analyzed some of the bone from the brain case in the skull for the presence of organic matter, which normal bone contains. For example, expressed in terms of nitrogen, fresh bones have about 4 percent nitrogen; it was found that the brain case had only about 1.4 percent, which meant that it had very little organic matter. However, at the time these analyses had been done, the jaw had not been tested for organic matter. When this was done, it was found to have 3.9 percent nitrogen, which is very close to that of fresh bone. So it was obvious that this orangutanlike jaw was certainly not a fossil.

Further study of the ages of these materials was carried out by an

356

analysis of the radioactive materials in them because the longer these materials lie in the ground, the more radioactive material they absorb and so become progressively more radioactive. These studies confirmed the relative newness of these so-called ancient fossils.

There were a number of queries that had to be answered in this puzzle. Were the jaw and the skull of the same age? Could the flatness of the surfaces of the teeth be due to a natural wear? Was the color of the bones in the skull and the jaw a natural stain absorbed from the local soil? There was a bone clublike instrument found near the skull, and it also needed to be investigated. Finally, did the chemical content of the other fossil materials found in the same place match that of the Piltdown soil?

The radioactive studies finally demonstrated, and confirmed the studies of organic material, that the skull was actually some thousands of years older than the jaw, and therefore, they could not have possibly been from the skeleton of the same animal. The teeth were studied in some detail. When teeth wear normally, the edges are beveled, but the Piltdown teeth had flat, sharp cut edges. Also, the first molar would normally have shown many more signs of wear because it erupts in the jaw first and has a longer time to be used. But in the Piltdown jaw the two molars showed identical degrees of wear. There were a number of other reasons which suggested this wear was not normal, but eventually a detailed study under the microscope showed scratch marks on the flat surfaces of these teeth, so it was obvious that they had been filed flat by whoever had perpetrated this hoax.

The brown color of these bones which presumably should have been due to iron absorption in the soil, was, in fact, due to an agent commonly used in paints called Van Dyke brown. This chemical contains chromium, but it could not have come from Piltdown because there was no chromium in the soil there. The club was of elephant bone and was presumably carved by the ancient man to whom the skull was supposed to have belonged. It would have been a fresh bone when it was carved, but you cannot whittle a fresh bone in this way; you can do it only with a dried-up fossil bone. It was obvious then that this club had been carved after it had become a fossil and was, therefore, also a fake. The other fossils found at the site did not match the soil in any way; in fact, some of the fossils looked like those which had been found and recorded in Tunisia in North Africa. Some of the teeth were hippopotamus teeth and were similar to those that had been found in caves on the island of Malta; other fossils were similar to those found in other places of England. When all these facts were added together, it was obvious that the whole Piltdown project was a forgery, and Professor Weiner investigated the possibilities of who might be the forger. All the evidence pointed to Charles Dawson. So was laid to rest one of the most puzzling findings in the history of paleontology. The lower jaw was, in fact, that of an orangutan. It was probably 400 or 500 years old, and it had been obtained from Chinese gold washings in a cave in Sarawak in Borneo.

The size of the brain of the orang is about 400 to 450 cubic centimeters, which is comparable to the size of the brains of the gorilla and the chimpanzee. The eyes are very close together, much closer than in the chimpanzee or in the gorilla, and lie in big round orbits; the chimpanzee and the gorilla have squarer orbits than the orang.

The orang is now restricted to parts of Sumatra and Borneo. It is happiest in the lowlands and is usually found below 2,500 feet. Nests have been found as high as 5,000 feet, and on Mount Kinabalu in Borneo, orangs have been seen as high as 8,000 feet. Although it has a very restricted habitat now, in the past the orang obviously had a much wider distribution. Almost certainly it occupied the Malay Peninsula and undoubtedly extended through Burma into India and certainly into the Siwalik Hills of India, which are near Chandigarh in the north of India. The remains of orangs have been found in three of the provinces of China, indicating that it extended right over the Asian area. There is also evidence that it existed in most of Borneo and Sumatra, not only in the restricted areas where it is now found. There is a color difference and one or two other minor differences between the type found in Sumatra and that found in Borneo. The Sumatran is usually a brighter red with longer hair while the Borneo is a darkish, duller color. The largest population of orangs is now found in North Borneo in a region known as Sabah and extends also into Kalantan, which is in Indonesian Borneo. They are also found in the state of Sarawak, but none is in the neighboring state of Brunei.

Before 1959, there were no real estimates of the population of orangs left in the wild, but since then, some attempts have been made to take a census. It appears that certainly no more than 5,000, and possibly less than this number, are left in their natural state.

It has been said that in the wild the female orangs only produce young once every four years and that there is a high infant mortality—almost 40 percent. At the center, where we have been extremely successful in raising orangs, we are able to get our female orangs to reproduce much faster than that. We have one who has had three babies within a period of four or five years. With the care that we give the orang babies, most of which we take from the mother and

358

A baby orang at the Fort Worth Zoo poses for the camera.

bring up in a nursery, we have not found the orangs any more difficult to raise than our one gorilla baby or our many chimpanzee babies.

The male orang reaches a height of four feet or a little bit over, and some earlier records of animals of giant height are obviously inaccurate. It has a very wide arm span, however, which has been measured from seven to eight feet. Probably one of the tallest animals was recorded by Hornaday, who shot one that measured four feet six inches. The weight of adult orangs seems to be grossly underestimated in most books. Most say that orangs do not grow beyond 250 pounds, and some have even claimed that they are not any more than about 120 pounds. But in actual fact, we know that in captivity, at any rate, males can weigh as much as 450 pounds, and, according to G. S. de Silva, they also achieve this weight in the wild.

The baby orang looks very much more like a human than does the adult, particularly since the male, when it becomes mature, develops tremendous fleshy pads on either side of the face. The function of these cheek pads is unknown. It may simply be a secondary sexual characteristic, presumably attractive to the female, but this is the only conceivable function that they have, unless they also serve to protect

359

the face either in fighting or from thorns and broken sticks in the wild.

I mentioned that the hair, particularly in the Sumatran orang, is very red. It is also very long; some of the individual hairs, particularly those around the arms and on the head, have been measured at a foot or more in length, and this is why the animal looks so hairy. The actual amount of hair on the head of an orang is relatively small compared with some apes. The orang has 158 hairs to a square centimeter on its scalp—far less than man, the gorilla, and even the chimpanzee. The body is fairly sparsely covered with hair, too. The face, the ears, the palms, and the soles are free of hair. The hair on the head is darker than that on the rest of the body and may be dark brown. As the animal gets older and more mature, the whole coat becomes dark. Orangs have very expressive brown eyes and a not very prominent nose.

Hanging down beneath the chin, and particularly obvious in the adult, is a great pouch; this is the air sac. It is connected to the animal's larynx (the tube that carries air down to the lungs). The orang can fill his air sac with air so that it expands like a great balloon under his chin and makes him look as though he had an enormous goiter. It is not certain what this structure is intended for, but it seems most likely to be a resonating organ when the animal calls in the forest, probably enabling the call to be heard over a much greater distance. Our animals in captivity rarely inflate it, but occasionally they do. The noises which orangs make are more limited in variety than those of the chimpanzee. Our orangs make a whimpering, whining kind of a noise which has been described by a number of authors. They have a kind of soft bark on occasions, and I have also heard them roar.

Another sound which is very characteristic is a very loud, sharp click. I have noticed this noise, particularly when a human an orang dislikes for some reason or another comes near it. We recently had an orang that developed a skin disease requiring it to be isolated from the other animals. It was being treated for this condition, which meant that from time to time it had to be put to sleep for examination and application of the treatment. It resented the veterinarians who did this. Every time one came near, it would give this sharp, clicking sound. Presumably orangs use it for other things that they also do not like.

In addition, they will suck their lips to make a kissing sort of sound. Schaller has also talked about a gulping sound they make and what he describes as a two-tone burp. All these sounds were given by an orang which apparently had been annoyed by Schaller. We have noticed the

Courtesy of Paul Steinemann of the Zoological Gardens, Basel, Switzerland

Male Sumatran orang.

Orang and her baby. Orangs often place their babies on their heads.

kissing sound but have not seen it used in relationship to annoyance. The grunt that Barbara Harrisson describes we have also heard, but I would not describe it as a bark, as some authors have done.

The roar of the orang has been described by Dr. David Horr as a segmented vocalization, beginning with a sound made low in the throat and rising rapidly in a crescendo till it became a high-pitched squeal; then the pitch dropped a little before the call was repeated a second time. There may be a dozen of these segments repeated in a single trumpeting. Orangs have been seen making this call when their air sacs were inflated, but they also have been seen making it when the air sacs were not inflated.

Young adult males in the wild may make this call a number of times in a day, but the older males use it very rarely; they usually do not vocalize more than once a day. A mature male often gives the call when he is resting in the early morning or in the late afternoon, but the call has been heard at night in the wild. Dr. Horr suggests that this type of vocalization is a classic locator call which announces the existence and the location of an adult male and serves to warn other

Dr. Ronald Nadler at Yerkes Primate Center tests ability of newborn orang to support itself by one hand.

males in the area of his presence so that they stay spaced apart from each other. At the same time it also informs females who are in heat where the male may be found. The males have a much wider range in the jungle than the females do.

Females who are not in heat either pay no attention to the male's call or move away from that direction. Dr. Horr cites one case of a female with a clinging infant who, when she heard the call of an adult male, responded by giving him a threat vocalization, though the male was not even near her. Adult females that were in heat, however, moved in the direction from which the call came with "considerable enthusiasm."

The orang is a large, heavy animal, and therefore, its movement in the trees is not likely to be rapid, nor is it likely to jump from branch to branch as do some of the lighter animals, like the gibbon and the siamang. The animals climb up the trunks of trees, using both arms and legs, and climb along branches in the same way. Dr. Yerkes was very impressed with the locomotion of orangs. The way an orang climbs a tree is reminiscent of a man. It uses its long arms to hold the

363

Dr. Ed David, President Nixon's former science adviser, visits the Yerkes Center and makes friends with a baby orang. Also in the picture, Dr. Sanford Atwood, president of Emory University (center), John Lannon, administrative assistant to Dr. David, and the author.

trunk of a tree or to reach up and grab branches and pull its body up to them. At times, during its climbing, it has to support the weight of its body completely on its legs, which also function actively in the climbing process. This fact alone distinguishes very clearly the orang from the gibbon in the trees, for the gibbon moves by using its arms alone. The orang has none of the gibbon's flightiness in the branches, and it is prudent and sure as it climbs, even when it is closely pursued. It depends, not on speed for its safety in the treetops, but on the cunning and caution it uses to camouflage itself.

When in Borneo, I noted that one of Barbara Harrisson's semiwild orangs used the following method to get about. The orang would get to the top of a tall, thin tree and sway back and forth on the tree until it swayed over close to another tree; then it would grab the second tree and let the first one go. This was all to the good if the second tree would bear his weight. In some cases the second tree would not, and the animal would come crashing down to the ground with some force; but it did not appear to do him any damage. This animal progressed through several trees in this way, crashing some to the ground and in some cases crashing down only with the branch that he had grabbed.

Yerkes orang exercising on a jungle gym.

Orangs seem to be able to drop as far as 20 to 30 feet to the ground, sometimes for no apparent reason but possibly because they have been frightened by a snake.

The earlier authors writing on the orang engaged in an argument on whether it was capable of walking erect. There is no question that it is able to do so. It has been seen walking erect on the ground in the wild on occasions, and I have seen it walking on the ground and then running quickly in the erect position to the base of a tree it wanted to climb. Orangs in captivity are often seen to adopt the erect posture. Our own animals pull themselves up onto their feet and then walk across the floor of the cage to visit with humans on the other side of the wire of their cages. Anyone who has seen orangs in circus acts knows that these animals are trained to stand erect for significant periods of time without any difficulty or fatigue at all.

There were similarly some controversies on whether orangutans can swim. Dr. Garner quotes a report of an orang swimming as follows: "On one occasion, this ape was induced to go aboard a steamer which lay in the harbor. The purpose was to kidnap him and carry him to

365

Europe. Either through fear, or instinct, or reason, or some other cause, this ape jumped overboard and swam ashore, although he was naturally afraid of water. . . ."

A number of other researchers, however, have found that the orangutan is unable to swim. They have put young orangs of various ages in pools to see if the animals would take to the water at all, but they simply became panic-stricken. We have lost a Yerkes orang from drowning.

There was also an argument among the early researchers about the branches that fall to the ground when an orang is moving through the trees. Some authors claim that they are thrown by the animal at his pursuers, and others say they are just branches that have broken off in the course of the animal's progress. As in many other controversies, both explanations are at least partly correct. There is no doubt, as I have seen, that orangs moving through the trees will knock off branches simply because of the animal's weight, and it is able to grasp another branch in time to prevent itself from falling with the branch that is broken off. On the other hand, the animals do sit up on trees and break off branches and throw them at people below. Wallace, whom we mentioned earlier, describes the throwing of branches at humans in this way:

> It is true he does not throw them at a person, but casts them down vertically; for it is evident that a bough cannot be thrown to any distance from a top of a lofty tree. In one case, a female mias [this is the name often used in the early days by the natives in Borneo for orangutans] on a durian tree, kept up for at least ten minutes a continuous shower of branches and of the heavy spined fruits as large as thirty-two pounders, which most effectually kept us clear of the tree she was on. She could be seen breaking them off and throwing them down with every appearance of rage, uttering at intervals a loud pumping grunt, and evidently meaning mischief.

Dr. Richard Davenport of the Yerkes Center, during his expedition to Borneo to study orangs, has also described how isolated males would sit up in the tree and break off branches and deliberately throw them in his direction.

At night orangs always make nests to sleep in. One of the ways of estimating whether there are many orangs in a particular area is count the number of nests that can be seen in the trees. In the literature there is some confusion about the height and type of trees for nesting. Both Wallace and Hornaday mention that the nests are usually made in small trees or saplings and that when the animal is asleep in its nest, the tree often sways considerably because of its weight. Hornaday says that they never build their nests higher than 25 feet, but he also admits that they often build them in trees that are in very awkward positions to approach, for example, on the sides of swamps. Authorities in Malaysia believe that orangs, as a general rule, build their nests in tall trees and well away from the ground, but that sometimes they do construct nests in small trees. Nests have been seen in Borneo which were only 10 to 12 feet off the ground. The building of nests in tall trees may be a later development, perhaps there has been a change in the numbers or habitats of the predators, especially the clouded leopard which is the chief enemy of the orang. Also, more frequent penetration of man into the forests may affect the height at which the animals now build their nests. There is a report that orangs that get so old that they have difficulty in climbing trees will sleep on the ground, but nobody has ever been able to confirm this.

When an orang is about to make a nest, usually as darkness begins to fall, it climbs up its tree, stands up, and, using one arm as a support, grabs hold with the other arm of branches that are a little distant. It pulls them in, breaks them, and piles them up in a crosswise fashion. This is not done deliberately, but they become crosswise automatically as the animal works around in a circle pulling the branches in. Eventually it is surrounded by a circle of broken limbs, usually about three feet across. Then it has to go through the procedure of making the floor. It does this by breaking off smaller branches and putting them in the middle of the nest. Finally it grabs a branch with a number of leaves and runs it through its hand, pulling the leaves off; these it lays down over the twigs. Then it has a complete nest for the night and lies down. The nests are used more than once; it is said that the animals continue to use them until the leaves become dry.

When there is rain or inclement weather, the orangs will put branches or sometimes leaves over themselves. Perhaps it is this instinct that causes them in captivity to put any gift, such as a bag or a piece of cloth or paper or a book, on their heads. Dr. Richard Davenport, during his expedition into Borneo, actually observed an orang weaving a type of cover over its nest, presumably to protect itself from the elements. It was building the nest from underneath, using the same building technique, so that it was a kind of upside-down nest. The construction of roofs is, however, not usual.

The orang seems to be particularly keen on fruit as an article of diet. When any of the trees in the forest are fruiting, there are certain to be orangs in the area. Among the fruits they eat are the durian, the rambutan, and the mangosteen. They also eat the fruits and the leaves of a number of other species.

The durian tree has an interesting fruit. It is very large and can be up to 30 pounds in weight. The rind is very thick and very hard and has a number of strong spines. Its odor, which can be smelled from a long distance, is very disagreeable and is said by some to resemble the smell of a civet. However, if you can hold your nose or do not sniff while you are cutting it, the taste is actually delicious once you get it in your mouth. The pulp has rather the consistency of custard with a number of large seeds. The natives also use the seeds for food; they roast them, break them open, and then pound them into flour. The orang is very wasteful in the use of this fruit and probably eats only the center pulp.

The rambutan is a small fruit the size of a walnut covered on the outside with long, soft structures which look like prickles but actually are not sharp. When the skin, which is fairly thin, is removed, there is a rather large stone covered with a thin layer of a white, semitranslucent, juicy material which is delicious.

The mangosteen is about the size of a medium apple. The dark-purple skin is thick and easily broken. The fruit inside is divided into segments and looks just like a peeled orange, except that it is white and semitranslucent. The taste bears no relationship to the orange at all and is absolutely delicious. I can quite understand why the orangs are very fond of it.

They also eat young, fleshy leaves, the inner bark from trees, shoots of the pandanus palm, and terminal bamboo shoots. Orangs have been known to eat orchids occasionally and a number of species of insects, particularly termites. Sometimes they eat the dirt from termite mounds. There are also reports that they have been seen to fish for termites in the same way that some chimpanzees do. They have not been seen in the wild to eat meat or eggs but will eat both of these in captivity.

Most authors agree that they will eat pretty well anything in captivity, and their diet at the Yerkes Center may be of interest. In the morning they are given several cupfuls of milk mixed with a special dried yeast. Later in the morning they get a hard, biscuitlike material called "chimcrackers," which was originally designed by Dr. Yerkes with the help of the leading American nutritionists of the day. It consists of a variety of cereals, dried milk, and cod liver oil. In the afternoon they get various fruits, vegetables, and soybean cake which

Exhibit of Impressionistic paintings of ape faces by the young French painter from Roquebrune-Cap-Martin, Lucien Tessarolo.

By courtesy of His Serene Highness Prince Rainier of Monaco

is made of ground-up chimcrackers mixed with cooked soybeans, put into trays, and placed in the refrigerator so that it sets into a soft cake which can be cut into segments.

They get at least one orange a day to be sure they get their vitamin C. They also get cabbage, carrots, and sweet potatoes or Irish potatoes. They appear to enjoy all these foods. We have, on occasion, tried them with hamburgers, which they have eaten without asking for ketchup.

The orang, because of the variety of things it eats in the wild, can always get some sort of food somewhere in the forest, but Dr. Horr points out they have to hunt for fruit because there are no large concentrations of this favored item of the diet anywhere in the jungle.

Orangs are big eaters, and a single adult male can strip a tree of most of its fruit in a couple of days. The availability of fruit must be a limiting factor to the number of orangs that can occupy a particular area. Fortunately, adult males range over a very wide area so that they do not overburden the food supply in any particular area. Since

Orangs at the Center for Acclimatization of Animals in Monaco. Apollo—resident male.

By courtesy of His Serene Highness Prince Rainier of Monaco

By courtesy of His Serene Highness Prince Rainier of Monaco

Orangs at the Center for Acclimatization of Animals in Monaco. Antu—female orang formerly of the Yerkes Center.

370

they are solitary for most of their lives, the presence of an individual in a particular section of the jungle does not constitute too serious a threat to the food supply.

The range of an adult male orang is about one to two square miles. Dr. Horr describes how in an eight-day period he and his party followed an adult male for a distance, in a straight line, of about two miles, but the circuitous route that the orang took was much longer than that. The ranges of the adult males overlap the ranges of the females. The only social unit that exists among the orangs is the adult female with her baby which is still dependent on her. Sometimes she has two offspring with her, and more rarely a third is found nearby. Although it is not known for certain, it is said that in the wild a mother orang has a baby about once every three or four years. The female with her offspring has a very small range; according to Dr. Horr, it amounts to a little more than a quarter of a square mile. Her range may also overlap that of another adult female.

The developing juvenile which is big enough to leave its mother and be independent represents another population unit. It starts off by being partly dependent on its mother, begins to forage farther and farther from her until eventually it sets up housekeeping for itself. In the wild, for the first year of life, the baby orang clings continuously to its mother; it takes mostly milk during this time. In the second year it consumes solid food and spends progressively greater amounts of time away from contact with the mother's body on its own. By the third year it is fairly independent. It even constructs its own sleeping nest but may from time to time sleep in its mother's nest, and it does keep contact with her. After the third year it is off but may keep near its mother for about four years until it sets up its own range. In the wild orangs start breeding at six or seven years. The experience at Yerkes is that in captivity the animals are more like eight, nine, or ten before they produce infants.

When male orangs meet in the forest, they threaten each other. Dr. Davenport describes what he calls a "dive" which he thinks is part of the threat pattern. In this particular activity the orang stands up on all fours on a branch, then suddenly raises his hands and throws himself forward, still keeping hold of the branch with his feet. He hangs on and finishes, hanging upside down by his feet, and then scrambles back onto the branch again—this is a spectacular sight.

Menstruation is not very obvious in orangutans. The menses are scanty and in most cases cannot be seen, but there is a sexual cycle of about thirty days. Although we have seen copulation many times in orangs in the center, it has not been seen very often in the wild. In captivity orangs use a variety of methods for copulation. On two occasions I have seen a male orang at our center sitting with a female's legs drawn across his thighs and his penis in the vagina, and unlike the chimpanzee, he was thrusting in and out in long, slow strokes for very long periods of time, giving every evidence of enjoying himself. In the first copulation between a pair of orangs, the ejaculation of the male is quite fast. After a few days it may take twenty minutes, including a certain amount of resting but not much fast, active pelvic movement, only slow movements. I have also seen orangs copulating at the center with both of them hanging from the wire on the roof of the cage.

In new pairings between orangs—that is to say, when new males and females come together—the female never presents or offers herself to the male. The male actually chases her and forces copulation, and so this is really a type of rape. On one occasion, however, we did see a female that appeared to initiate copulation. The male lay on his back, and the female squatted on top of him with his penis inside her. In the cage type of hanging copulation, the female hangs by all fours from the wire roof, and the male approaches her hand over hand, also hanging from the roof. In this hanging technique the couple can try all kinds of copulatory positions.

The geologist Leong Khee Meng, who was camping in the north of Borneo in 1968, had the opportunity to see a pair of wild orangutans copulating. The time was about seven thirty in the morning, and he heard a number of sharp, piercing screams behind his camp. He sent one of the native helpers out to investigate, who reported that a pair of orangs were copulating in the trees. The male was only a small orang, presumably a partly grown or just mature juvenile. The female was about twice his size. She had her arms outstretched and was holding onto the nearby branches to support herself. Her ventral part was directed toward the floor of the forest. She was actually resting at an angle between five and ten degrees. The male had got on top of her back and had both his hands around her middle. His head was resting sideways on her back. At the beginning of each mating session, the female called out a sharp, piercing cry, which died away as the male began to thrust into her. Each period of thrusting lasted between half a minute to a minute, and the animal thrusts about ten or fifteen times during this time. After each session the animals would rest for another half minute to a minute before they would start again. In each case the female heralded the beginning of the next period of penetration of the penis with the same piercing cry. According to the observers, the female's lips were open, giving the impression she was breathing through her mouth, during the period of copulation. She also seemed to be the more active partner.

On one occasion when the animals climbed higher in the tree, she was the one who did the climbing and the male simply clung to her. When the animals saw that they were being observed, the female actually grasped a branch with plenty of leaves, broke it, and bent it down in such a way that it blocked them from the view of the humans on the ground. The latter moved to another spot so that they could continue to observe the activities of the orangs, and the animals then climbed higher up into the tree, where they began mating again. Since the view of the observers was progressively obstructed as the animals continued to move around in the tree, they also shifted to different points of observation. This seemed to irritate the orangs, and eventually they stopped copulating. One of the native helpers ran down to the bottom of the tree and began shouting at the animals, and the female then began to break off branches and throw them down at him in a very hostile manner. The male appeared not to be particularly interested and simply climbed higher up the tree, finally climbing across to another one. The female threw branches at the native helper for five minutes or so and then climbed higher in the tree, crossed over to her mate, and commenced copulating again. The observers left the scene at that time, which was about half past eight, and the animals had been mating intermittently for about an hour by that time. They could still hear her piercing cries for some time after they left the area. Eventually they moved out of earshot so they were not able to estimate precisely how long the procedure went on. When they returned to their camp at two o'clock in the afternoon, the apes had gone.

When a male attempts to mate with a female that is still carrying an infant clinging to her, she usually reacts violently in her attempts to avoid him. The females try to avoid breeding until each of their offspring has reached a level of independence which would be adequate to permit the mother to care for a new infant which, of course, is highly dependent on her. The period of pregnancy is about 275 days. The Yerkes Center has been particularly successful in breeding orangutans. Our original group of between thirty and forty animals was acquired back in 1963, and they were immature animals at that time. We had to wait until about 1965 before any were old enough to produce infants. Our first orang, Seriba, was born on May 20, 1966. Since that time we have produced twenty-nine orangs, which is an average of four or five a year. These babies are doing extremely well, and many of them are getting quite big.

Yerkes newborn orang suspends himself by his fingers.

Governor Jimmy Carter of Georgia and Mrs. Carter visit the Yerkes Center and befriend an infant orang. The author is in the center.

Most of our orangs have been taken from the mothers at birth to ensure their survival. Their rate of survival is much lower if they are left with the mother than if they are taken away and brought up in a nursery situation. They are kept, to begin with, in human incubators of the type used for premature human infants and kept there for a time, complete with diapers, until they are ready to go into the nursery, where they are kept in large cages. In good weather they are taken outside for an hour or two under the supervision of an animal caretaker. They are placed in a wire-enclosed area equipped with a jungle gym, and they are able to climb and swing about and interact with each other to their hearts' content.

What we would like to do, but have not yet been able to find the funds for, is to have a large outdoor enclosure at our field station where a group of more adult animals could be placed together as a family group. The best that we have been able to do in this respect has been in the Atlanta Zoo where we have a family group of eight. This

375

Dr. Boris Lapin, director of the Russian Primate Center, poses with a Yerkes orang during a recent visit.

includes three mothers and their juveniles, a male, and a new baby which has been born to one of the mothers since she has been in the zoo. It is a very interesting group of animals. One special item of interest is that two of the females picked continuously on one of the other females. They forced her into corners, nipped at her, sometimes made it difficult for her to get to her food and so on; but as soon as a male was introduced, he acted as a stabilizing factor for the family, and the three ladies were kept under control.

Barbara Harrisson, whose work has been described in my book *The Ape People*, was one of the first to try to educate back into the wild young orangs that had been taken from their parents and kept in domestic captivity for varying periods of time. Barbara was given the use of a reservation in the Bako National Park in Borneo, not far from Kuching. At the time I visited her, she had three orangs—Arthur, Cynthia, and George—that had been transferred to the jungle. There was a special enclosure nearby where they were fed every day since that particular area of the jungle did not have much fruit. However, Arthur, who was the oldest animal, quickly began to roam and

376

Young orang poses with the author.

developed a two-mile-radius territory as his range. He got his water from rocky pools by sucking it between his lips or by scooping the water into his mouth. He actually went into shallow water, which is of interest. Dr. Vernon Reynolds, who visited the Bako orangs, describes how on one occasion Arthur explored a cave, collected some leaves together, and made them into a bed and lay down on them. Our orangs at the center will do the same thing if they are given any material that can be manipulated into a bed in a corner of their cages.

Arthur also built nests in the jungle in the evenings but usually built in a region where he could still keep within calling distance of Cynthia and George, who were at that time too timid to go into the jungle for any great distance. Arthur not only built a nest at night, but also built one before his midday siesta. Barbara Harrisson says that she has seen three young wild orangs do the same thing. Arthur was also observed hanging under his nest when it rained.

Dr. Reynolds made some observations on the reactions of the animals to each other's vocalizations. For example, if Arthur heard Cynthia or George whining in the distance, he would stop, watch for a while, and then would proceed toward them. When he returned, he would greet the other youngsters with what Reynolds describes as a "loud raspberry blowing."

One extremely interesting observation about Arthur was his reaction when he met a snake. He picked up a stick and chased the snake and hit it before it managed to get down a hole. Reynolds also points out that when Cynthia saw a snake up on the tree, she fell out of it in shock.

I have described my personal contact with Arthur in my book *The Ape People*: how he regarded me as a good type of character to wrestle with and how I tried to keep him off with a stick, which he pulled out of my hands; he then put his teeth around my ankle and began to bite into it until Barbara Harrisson had to beat him to get him to let me go.

The animals of Bako are no longer there. Arthur was transferred to the province of Sabah in the north of Borneo, where a rehabilitation project for errant orangs was getting under way under the supervision of the office of the chief game warden. The Game Branch of the government in Malaysia was formed in 1964, and at that time it was decided that no more orangs would be exported to zoos and that animals in captivity should be returned to the wild; accordingly, an experiment for this purpose was set up in the Sepilok Forest Reserve. This is about 10,000 acres in extent and was financed by the state. G. S. de Silva recorded this experiment in the *Malayan Nature Journal*

(1971). The 10,000 acres includes quite a number of the fruit trees which the orang likes so much. Altogether, since 1964, forty-one animals have been involved in this project. At the time De Silva wrote in 1969, twenty-four animals were being maintained in the reservation. Their ages varied from one to fifteen years. Most of them were in pretty poor condition when they arrived; some of the animals had been in captivity for several months, and some had been in captivity for years.

They had been molded both psychologically and physiologically by the conditions of their captivity. For example, one sub-adult female called Winnie, prior to joining the project, had been brought up to acquire a taste for beer and cigarettes and even tossed off an occasional whiskey. These commodities, of course, were not provided for her in the reservation; but she had developed a craving for tobacco, and if she got an opportunity, she would try to snatch cigarettes from the staff. Another animal, an eighteen-month-old male, started to masturbate every time he saw a partly dressed human. This peculiar behavior was investigated, and it was found out that his previous owner had trained him to perform this act for the entertainment of fellow workers at a timber camp. Another animal had been kept in a metal cage with bars on all sides, none of them closer together than two inches. The animal had become flat-footed and was not able to flex its toes.

When an animal arrives at Sepilok, it joins other animals in its own age group. It is fed and exercised, and its reactions are observed. If, after two weeks, it has become used to the company of the other orangs and the attendants, it is allowed to exercise on its own, climb trees, and feed with the other orangs. The animals are checked every morning for any signs of sickness; then they are fed and allowed to roam at will in the trees. At 10 A.M. they are given a drink of milk, fortified with proteins and minerals; when the milk arrives, those that are in the vicinity up in the trees immediately come down and get it, while the others respond to a call. They are also fed again at midday, at 3 P.M., and at 5 P.M. After 5 P.M., they are allowed to roam about, sleep on trees, or sleep in cages, and the doors of the cages are open twenty-four hours a day.

When fruit is available in abundance in the wild, the food supplied in the camp is cut down; sometimes it is cut off completely for one or two days. Some of the animals go off quite a distance and sometimes disappear and sometimes return. One adult male, called Moses, gradually went farther and farther away until one day he went off into the forest and did not return. This was in 1966, but in 1968 one of the

379

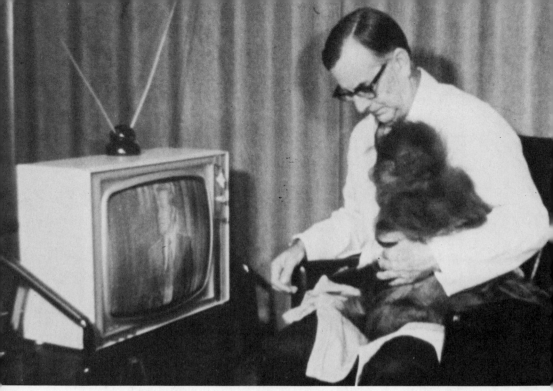

The author and Yakut, the orang that appeared on the *Dick Cavett Show*, watch Johnny Carson together.

people at Sepilok saw an adult male that turned his head whenever the name Moses was called out. So presumably after this extended period away, he had returned back to visit his friends at Sepilok. He did not respond when other names were called.

A typical animal of the type that populated Sepilok was a female about eight years old called Joan. When she was found, she was chained by her neck underneath a house near the river. She had been badly treated and was severely emaciated. Apparently she had been captured when she was very young and had been raised as a pet for a number of years. She was fairly well treated by her original owner but, when he died, was neglected, eventually became unmanageable, and was taken to the forest and abandoned. She was rediscovered in the forest by someone who offered her some food, and she then followed him back to his kampong (village), where she was given only partial freedom because she was likely to raid cultivated fields if she was completely free.

At the time she was discovered and taken to Sepilok, she was obviously undernourished and was unfriendly and suspicious of

After appearing on the *Dick Cavett Show*, Yakut decided he was no longer an orang and was a member of the human race. A T-shirt testifying to this fact was purchased for him.

strangers. She was taken by boat overnight to Sandakan. She still had the chain around her neck, and during the night, she kept trying to tie the chain to the bars of the iron cage in which she was restrained. The crew could not sleep because of the clanking noise this made. Finally, in desperation, they fastened the chain to the bar. The moment they did that she apparently felt secure, crept into her sack, and slept soundly. While she was in captivity, she had been chained up every night by her owners, and now she was so used to it that if this were not done, she could not sleep.

At Sepilok she was allowed a good deal of freedom and would feed on leaves, fruit, and grubs. She spent a week there staying close to her cage. Eventually she went into the forest and disappeared for three days. From that time she paid regular visits to the forest, and no attempt was made to stop her from going there. Her periods away grew longer, and at one time she was absent for more than ten days. On one occasion, after she had been away for a week, she was found at a timber camp, where, according to the men, she had come in one evening and had been very friendly and docile with them. She had been six months at Sepilok when she was taken off into the forest some two miles away and let go. But she came back to the base camp two weeks later. In fruiting season she disappeared into the forest for a period of six weeks. Now she goes and comes or stays in the forest for various periods of time, and no one worries about her.

One of the game rangers in this project, James Wong, said that under these semiwild conditions the orangs seem to have a great tendency to imitate the actions of their keepers or their attendants. He noticed orangs intently watching laborers cutting firewood and carrying it to the camp on their backs. As soon as they departed, the orangs picked up pieces of firewood left by the laborers, put them on their shoulders, and walked around with them.

Sometimes individual orangs would form very close relationships with members of the staff. For example, Paul, a juvenile male, became very attached to De Silva and would fly into fits of rage—screaming and rolling on the ground—if he tended to the needs of any of the sick orangs.

In 1964, Arthur, the animal from Barbara Harrisson's group in Bako, was introduced to Sepilok. By now he was an adult male. This was the animal that had bitten my ankle in Bako. As soon as he arrived, he attacked one of the game rangers without any provocation at all. About a week later he entered a bungalow and turned it upside down, throwing a heavy cage several feet away from the house. On another occasion he tried to attack the ranger and his staff. Eventually Arthur had to be destroyed. He had previously attacked an assistant game ranger, pinning his arms to his side and biting him on the face. This had been done in the presence of his keepers; for some reason the animal had made a deliberate attack and afterward had to be physically restrained.

Winnie, the female who had a taste for whiskey and cigarettes, got very irritable when the craving for tobacco became very strong and sometimes wrestled with members of the staff and sometimes tried to bite them in an attempt to get their cigarettes.

A number of the animals were gentle, seizing the hand of a member of the staff and putting it in their mouths. They pressed the hand slightly with their teeth, but they liked mainly to lick the palm. They may do this because of the salty flavor of sweat on hands and palms.

Joan eventually became pregnant and had a baby. The birth is described by De Silva, based on the observations of game ranger S. Hong. The description is very similar to what we have observed in our twenty-two orang births at the Yerkes Primate Center, but it is a very interesting description and so is repeated verbatim here.

382

At about 8:30 a.m. on the twelfth of May, 1967, Joan was observed behaving in an abnormal manner. She looked uncomfortable and attempted to lie down in various positions and seemed to be straining. Although the cage in which she slept in the night was immediately opened, she refused to leave it, and grasped the iron bars and sacks with her hands and feet. Joan even attempted to bite—something she had not done before. As she preferred to stay in the cage, no attempt was made to take her out. But the door of the cage was left open. At about 11:30 a.m. when the animal was under observation, a sticky, colorless discharge started to drip from her vagina. Progressively, the discharge became thicker and profuse and contained blood tinges. [The first stage of labor should have occurred during the morning. The gripping of the iron bars and sacks was an indication of the uterine contractions.]

At about 11:35 a.m. a clear transparent sack protruded from the vulva. When this appeared Joan was in a semi-standing position, her knees were slightly bent and her legs were wide apart. As the sack came out, she held it in her hands but no attempt was made to manipulate it. About four to five seconds later the vulva aperture widened and the chamber burst. When this occurred, Joan grasped the sides of the cage with her hands and adopted a squatting position. Then she began to whimper and cry and tears were visible. The baby was then expelled, and its head appeared with its face upwards. As the delivery progressed, it was noticed that the hands of the baby were flat against its sides. The whole process took about a minute and as the baby came out, Joan took it in her hands. The baby was covered with mucous [sic] and blood, its eyes closed and the body pale. Immediately after the baby was born, Joan licked its face clean and pressed its cheeks so that it opened its mouth, then placing her mouth on that of her baby she blew three or four times [Hong presumes that air was being forced into the baby's mouth. De Silva thinks that prior to this act a normal respiratory rhythm actually had already been established]. After this the baby's fingers and toes began to move and Joan commenced to lick its body. While this was being done, its eyes and mouth opened and shut several times. Joan then chewed the umbilical cord and severed it well away from the baby's umbilicus. Holding the baby with her left hand Joan started straining and pulling the umbilical cord. After some manipulation it was expelled with the placenta. All this was first sucked and then eaten without hesitation. By this time the baby was attempting to position itself on the mother. It gripped Joan's fur and went off to sleep. The baby cried soon after it was born. When the writer observed Joan and her baby four hours later, she appeared to be frightened of strangers. The same attitude was displayed toward the staff even though they had been with her for several months. Under normal conditions she never displayed the least sign of fear. She carried the baby, which was asleep, close to her breast and at times she nibbled at the umbilical cord. At birth the baby was about a foot long. No attempt was made to weigh the animal. When bedding was placed into the cage, Joan promptly covered herself and the baby with it. As it was apparent that she wished to sleep, the animal was left to its own devices. Prior to sleeping, Joan drank about a bottle of milk fortified with a protein additive. The next day, at about 3:00 p.m., Joan got out of her cage with her baby and went off in the rain towards the edge of the forest. She halted near the bole of a tree, placed the baby on her back near her right shoulder, covered its head for a few seconds with the palm of her hands as if to ward off the rain. Then steadying the baby on her shoulder with one hand she climbed the tree and took refuge among the branches. Clutching the baby to her breast, she remained on the tree, even though it was raining heavily, and made no attempt to come down. Prior to climbing the tree, the infant was transferred from her breast to her shoulder, so that it would not get hurt while she climbed. The next day at about 6:00 a.m. the animal returned to Sepilok holding the sleeping baby to her breast. Further observations revealed that the remaining portion of the infant's umbilical cord was missing. Apparently Joan had chewed it off an inch or so from the umbilicus. Five or six months after the baby was born, Joan would not tolerate any human or animal interference. Gradually, she appeared to realize that the staff meant no harm to her baby. She would, if her confidence was gained, permit members of the staff to come near and stroke her offspring. On one occasion, the hand of a visitor was taken by Joan and placed on the baby's head. On the nineteenth of June, 1968, Fisher observed Coco, a sub-adult female, playing with the baby while Joan looked on. Joan has up to now nursed and weaned her baby without any assistance. She has been observed to chew leaves and put them into the baby's mouth. From its absence, it is assumed that the father does not assist in looking after its young. Although the writer [De Silva] has observed wild family groups comprising male, female, and baby for lengthy periods, at no time is the male observed to look after the baby which clung to its mother or exercised near by.

A beautiful color picture of Joan eating a banana with her baby sitting alongside her appeared as a two-page spread in *Life* magazine on March 28, 1969.

Not many orangs have been kept by humans as pets outside Southeast Asia. Very few have been kept anywhere by people who are able to write about them. Probably the only book I know of which deals with the orangutan is that by Barbara Harrisson, and some details of her work were given in *The Ape People*.

384

Chimpanzees in the Bobby Berosini monkey act at Las Vegas.

Young orang, Manis, is an instantaneous success in the Bobby Berosini monkey act in Las Vegas.

Back in 1948, Winifred Felce described her work with apes, particularly the baby apes, in the Munich Animal Park, a zoological garden. One of the animals she had under her charge was a little orangutan called Tony who had been transported from Sumatra on a ship with another orang. He had been consigned to a firm of animal importers in Europe. On board the ship the man who was responsible for the orang babies, knowing that the boredom and lack of exercise on a long ship journey are not very beneficial to the animals, let them out to have the run of the ship for the greater part of the day. They had a wonderful time climbing all over the cages and boxes of the other animals, peering in at them, teasing monkeys, knocking things over, and stealing the fruit and the rice that was put near the boxes as food for the captive birds being taken to Europe. The little orang had a number of accidents, including drinking a substantial quantity of iodine which nearly killed him. While he was quite young, he learned to walk upright and even to run on his feet without using his hands. In doing so, he learned to put the soles of his feet flat on the ground and extend the toes fully instead of curling them inward and walking on the side of the feet as orangs tend to do. He was also able to jump onto his legs without losing his balance or having to touch the ground with his hands.

When he was in the zoo in Munich, he showed an interesting orang reaction to rain. His keeper and some other apes had been playing in the playground when there was a sudden rainstorm and they rushed for the house. But when they got into it, they could not find the little orang, Tony. They searched all over the place. Then they called him. Out in the yard there was an overturned washtub which was used for bathing on warm days; suddenly this rose, and Tony peered out. As soon as the rain stopped, he climbed out of it. This tendency of orangs to protect themselves from the rain has been observed in the wild; in addition to getting under their nests, they tend to cover themselves with branches and leaves. On one occasion the keeper described how she took Tony into town to visit a fruit and vegetable market. He was sitting alongside her in the car as she drove when a sudden storm pattered on the roof of the car. Tony, even though he was not getting wet, put a rug over his head and shoulders.

He seemed to have a very good sense of humor. When he was sitting at the ape tea parties, at which a keeper would sit with a number of apes, he would, without anyone knowing, put pieces of unwanted food on his neighbor's plate or even slip them under his chair, and he was as good at stealing food as he was at giving it away. Sometimes in the kitchen, when he was being led past a plate of raw eggs, he would shoot out his arm and grab an egg before anyone realized what he had done. Then he would hide the egg in his mouth without its affecting his facial expression at all. At the tea parties, when all the apes and the keeper were seated and eating milk pudding and stewed fruit with spoons, he would spoon the food into his mouth and then, without looking to the right or to the left, would remove a few spoonfuls from his neighbor's plate. It was almost impossible to notice this unless you were looking straight at him. If he did this to a neighboring chimpanzee, the chimpanzee would usually complain by stretching its arms out or complaining vocally to the keeper. At some of these parties there was a rope attached to the roof above the table, and Tony was known to leave his seat at the table and grab the rope and swing on it, grabbing the keeper's hair as he swung past him. The visiting public who were watching the tea party thought this was tremendous fun, but the keeper had to put up with it. If he left the table to go catch Tony, all the other apes would leave the table too, so he was really on the spot.

Mrs. Felce records a good deal of interesting information about other apes as well. One that appealed to me particularly was her story of a four-and-a-half-year-old chimpanzee called Peter who escaped from the zoo and ran into a church on a hill near the zoo. He strolled down the aisle in the middle of an afternoon service. As he did so, many of the congregation, thinking it was the devil in person, sprang onto the pews with horror. Peter was dismayed by this treatment and by the strange surroundings and started to whimper. He showed great relief when the keeper suddenly appeared at the doorway of the church to fetch him and take him home.

The Gorilla

There is a good deal of doubt about when Western man first knew of the existence of the gorilla. It was obviously known to African natives ever since they began to share the same environment. Probably the earliest possible record of the gorilla is contained in a report given by a Carthaginian explorer called Hanno. During the fifth century B.C. he made an exploration down the west coast of Africa. It is not certain how far down the coast he went, but some authors believe that he reached as far as Sierra Leone. He refers to a number of wild people whom he called "gorillas." There were many women among them, and the women had very hairy bodies. When an attempt was made to capture some of them, the males got away by climbing up on rocks and throwing stones at the party, but Hanno and his group were able to capture three females, "who bit and scratched and resisted their captors." Hanno's party killed these animals, skinned them, and brought their skins back to Carthage, where they were nailed up for people to see. The fact that they were called gorillas by Hanno does not necessarily mean that this is what they were. The modern-day range of gorillas is something like a 1,000 miles south of Sierra Leone.

Hanno's gorillas were, in fact, interpreted by his people as being really the mythical Gorgons, probably because they were described as hairy women. H. W. Janson points out that this confusion by classical authors of the hybrid types of creatures that occurred in ancient fables—such as satyrs, sphinxes, and cynocephali—with the different types of simians of which they heard makes it almost impossible to find out precisely what monkeys, especially apes, they had the opportunity of seeing. Pliny described an animal which he called a "cephos" which had feet and hands like humans and which appeared at the Roman games on one occasion. It may have been a gorilla. It apparently fought well, but there is no certainty that it really was a gorilla.

The date of Hanno's expedition was 470 B.C., and he set out from Carthage with about fifty galleys. When he returned to Carthage, he wrote up his report on a tablet and hung it in the Temple of Moloch. Hanno's story was recorded in Greek in a document called *Periplus*. The gorilla skins were deposited in the temple of Juno. They stayed there until 146 B.C., when the Romans attacked the city and destroyed it. Pliny has stated that at the time of the Roman invasion two of these skins were still preserved in the Temple of Astarte.

It was really 2,000 years after Hanno before it was certain that Western man had seen gorillas. This was in 1559, and the explorer was Andrew Battell, who had been captured by the Portuguese. He met two kinds of apes, one of which I mentioned earlier. In 1625, his account appeared in Samuel Purchas' *Purchas His Pilgrimes, A History of the World in Sea Voyages and Land Travell by Englishman and Others*. He described one of the two sorts of apes that he met as a "Pongo" and there seems little doubt that this was a gorilla. (The description appears on p. 280.)

Lord Monboddo, also mentioned earlier, received a letter in 1774, concerning an ape from a sea captain. The letter described this animal as follows: "This wonderful and frightful production of nature walks upright like a man; is from seven to nine feet high, when at maturity, thick in proportion and amazingly strong. . . ."

In 1819 Thomas Bowdich published *Mission from Cape Coast Castle to Ashanti*. He talks about a number of apes in the area of Africa known as Gabon. He called them "ingenu" and described them as being five feet tall and four feet broad. But the discovery of an animal which was unquestionably the gorilla was made by two missionaries, the Reverend Wilson and the Reverend Savage in 1846. The Reverend Mr. Savage visiting his colleague at the Gabon River in that year, was intrigued by the presence in the house of a large skull said to be that of an animal like a monkey but very big and very ferocious. The two of them collected a number of skulls and sent them off to Sir Richard Owen, the distinguished English anatomist, as well as to the anatomist Jeffries Wyman.

Savage described the gorillas as follows:

> They are exceedingly ferocious, and always offensive in their habits, never running from man as does the chimpanzee. . . . It is said that when the male is first seen he gives a terrific yell that resounds far and wide through the forest, something like Kah-ah! prolonged and shrill. . . . The females and young at the first cry quickly disappear; then he approaches the enemy in great fury, pouring out his cries in quick succession. The hunter awaits his approach with gun extended; if his aim is not sure he permits the animal to grasp the barrel, and as he carries it to his mouth he fires; should the gun fail to go off, the barrel is crushed between his teeth, and the encounter soon proves fatal to the hunter.

A fanciful conception of the gorilla is also provided by Sir Richard Owen, who in 1859 said: "Negroes when stealing through shades of the tropical forest become sometimes aware of the proximity of one of these frightfully formidable apes by the sudden disappearance of one of their companions who is hoisted up into the tree uttering perhaps, a short choking cry. In a few minutes he falls to the ground a strangled corpse."

Savage wrote about his experiences in the *Boston Journal of Natural History* in 1844, and his fellow missionary Wilson had the following to say about the gorilla:

> It is almost impossible to give a correct idea either of the hideousness of its looks, or the amazing muscular power which it possesses. Its intensely black face not only reveals features greatly exaggerated, but the whole countenance is one expression of savage ferocity. Large eyeballs, a crest of long hair, which falls over the forehead when it is angry, a mouth of immense capacity, revealing a set of terrible teeth, and large protruding ears, altogether making one of the most frightful animals in the world.

389

Another author, Henry A. Ford, had this to say about the gorilla in the jungle:

> When he hears, sees, or scents a man he immediately utters his characteristic cry, prepares for an attack, and always acts on the offensive. . . . Instantly, unless he is disabled by a well directed shot, he makes an onset, and, striking his antagonist with the palm of his hands, or seizing him with a grasp from which there is no escape, he dashes him upon the ground, and lacerates him with his tusks.

These descriptions completely misled people. Belief in the ferocity of this animal has persisted almost universally down to the present day. Even in my childhood my father would often tell me stories of savage gorillas, but the story that I was always told and which used to frighten me and lead me sometimes to have nightmares was that a gorilla would rush up to you, enfold you in his arms, and crush you to death against his breast.

The first gorilla to reach Europe came to London in 1855. She was a young female, whose name was Jenny. Her owners planned to exhibit her all over England in a traveling menagerie, but unfortunately the poor animal died before the show had gone very far. Zoologists examined her and were eventually convinced that she was a gorilla. She is now known in scientific literature as Wombewell's gorilla. Actually a scientist named Walter Waterman first described this animal, but he was under the impression that she was a chimpanzee. According to his account of what happened, the menagerie had reached Warrington in Cheshire, where, "without any previous symptoms of decay, Jenny fell sick and breathed her last." Later on she was examined by other distinguished scientists, and there was little doubt that she was, in fact, a gorilla and that she was the first to reach Europe.

The second specimen came to Europe in 1876. Called the Falkenstein gorilla, it had been purchased from a Dr. Falkenstein by the authorities of the Berlin Aquarium. It was a male and was only 30 pounds when it arrived in Berlin. The poor creature, however, only survived until November, 1877, and the total time that he spent in civilization was only a year and four months.

Another gorilla arrived in France in 1883. The famous zoologist Alphonse Milne-Edwards described a male gorilla about three years of age which, like all his predecessors, survived only a short time. Milne-Edwards studied the animal while it was alive and claimed that it was a brutal creature—also savage and morose.

Up to 1903, only one live gorilla had arrived in the United States, and it apparently lived only a few days. W. Reid Blair, director of the New York Zoological Park, described this animal in a letter he wrote to Dr. Yerkes on April 7, 1928:

Sometime in 1898, there was a report of the arrival of a baby gorilla in Boston but it is my impression that this animal died on shipboard. We have no record of it having reached any zoological garden or circus.

Our first gorilla was a female about 2½ years of age, which was brought to the zoological park [New York] by R. L. Garner in 1911. It was obtained in Equatorial West Africa. When it reached here on September 23, 1911, it was in such poor condition that it died on October 5 of malnutrition and starvation. It was 34 inches in height but weighed only 25 pounds. Our second gorilla was the female Dinah. She reached us on August 21, 1914, and died on July 31, 1915, after having been on exhibition in the Zoological Park for eleven months and ten days. She had, however, been in captivity about two years, one year of this time being spent in Africa where she had been under the care of the Society's agent, the late Richard L. Garner. The cause of death in this case was malnutrition and rickets. Dinah weighed, on November 15, 40½ pounds. Her standing height was 3½ feet and the extreme spread of her arms and hands between the tips of the middle fingers was 4 feet, 2½ inches. At the time of her death she was somewhat emaciated and weighed somewhat less than 40 pounds. We judged she was four years old at the time of death.

Gorillas continued to be imported into various countries and to survive for very short periods of time. In 1925, Ben Burbridge, an American hunter, brought the first specimen of mountain gorilla to the United States. This species of gorilla has the scientific name of *Gorilla gorilla beringei*. In 1922, he had delivered a mountain gorilla to the Belgium Zoological Garden in Antwerp as well. Dr. Yerkes has this to say about the American mountain gorilla:

In 1925, on his way to the United States with the specimen named Congo, he [Ben Burbridge] left in Antwerp a second young male. We are informed that the Antwerp specimens died within a few months of arrival. Congo, a female, whose age when captured was estimated as approximately four years, was kept near Jacksonville, Florida, where she has been kept by Mr. John Ringling. At the date of writing, she has been in this country for about two years, and throughout that period has continued in excellent health and has grown steadily and rapidly.

During the years 1927 and 1928, at least five infant gorillas from West Africa were brought into the United States. One . . . was purchased for the Philadelphia Zoological Garden. He has been exhibited for more than a year, at the date of writing, and continues to thrive. The National Zoological Park has a specimen, the New York Park another and two individuals are known to have died.

The gorillas referred to in this last paragraph were lowland gorillas.

In 1855 a twenty-year-old reporter, Paul Belloni du Chaillu, proposed to the Philadelphia Academy of Natural Sciences that he go to Africa to look for gorillas and that they finance the expedition. He was a good storyteller and had a good deal of curiosity, but no scientific training. Certainly he had the energy necessary for this expedition. Furthermore, the customs and the language of the country were already known to him, for he had grown up in Gabon in West Africa and had come back to New Orleans in 1852. At age seventeen, he had already written articles for a number of newspapers in which he talked about the jungle in the Gabon and the kinds of animals he had seen in it. He pointed out that some African tribes were under the impression that gorillas were actually men who had done something bad, whom the witch doctors had then turned into hairy manlike creatures. Other tribes thought that they were men who had become lazy and run away from the work of the village, adopted the ways of living of animals, and slowly deteriorated until they simply were manlike animals. There were also some tribes under the impression that gorillas were able to talk, but would not do so because they thought they would then be caught and sold into slavery.

Armed with the money that the Philadelphia Academy of Natural Sciences provided, Du Chaillu set off for Africa and landed in 1856. He collected a veritable caravan of helpers; some of them were to act as beaters to startle the gorillas out of their cover. He spent four years in Africa on his expedition and claims to have walked 8,000 miles during that period—2,000 miles a year, or about 7 miles a day every day for four years, most of it through heavy jungle. The figure is difficult to accept.

For some time he had difficulty in catching up with any gorillas; he found only footprints and occasionally fragments of wild sugarcane broken up and left on the ground. Eventually in a wild cane field Du Chaillu did, in fact, see his first gorillas. They were mostly young, and frightened by the beaters, they ran back into the forest. Du Chaillu said they were like "hairy men running for their lives." In fact, he said it would be like murder to shoot them. However, the native helpers told him that if he met an adult male gorilla, he had to shoot it, and

From Paul du Chaillu *Africa* (London, John Murray, 1861)

Gorilla hunting. A hunter spots a mother gorilla playing with her baby.

From Paul du Chaillu *Africa*, (London, John Murray, 1861)

Paul du Chaillu kills his first gorilla.

his first shot would need to be on target since if he missed, he would have no time to reload his gun.

Here is Du Chaillu's account of how he disposed of a gorilla:

Suddenly Miengai uttered a little cluck with his tongue, which is a native's way of showing that something is stirring, and that a sharp lookout is necessary. And presently I noticed, ahead of us seemingly, a noise of someone breaking down branches or twigs of trees. This was the gorilla, I knew at once, by the eager and satisfied looks of the men. They looked once more carefully at their guns, to see if by any chance the powder had fallen out of the pans; I also examined mine, to make sure that all was right; and we marched on cautiously. The singular noise of the breaking of tree branches continued. We walked with the greatest care, making no noise at all. The countenances of the men showed that they thought themselves engaged in a very serious undertaking; but we pushed on, until finally we thought we saw through the thick woods the moving of the branches and small trees which the great beast was tearing down, probably to get from them the berries and fruits he lives on.

Suddenly as we were yet creeping along in a silence which made our heavy breath seem loud and distinct, the woods were at once filled with the tremendous loud barking roar of the gorilla.

Then the underbrush swayed rapidly just ahead and presently before us stood an immense male gorilla. He had gone through the jungle on his all fours; but when he saw our party he erected himself and looked us boldly in the face. He stood about a dozen yards from us, and was a sight I think I shall never forget. Nearly six feet high (he proved four inches shorter), with immense body, huge chest, and great muscular arms, with fiercely glaring large deep grey eyes, and a hellish expression of face which seemed to me some nightmare vision: thus stood before us this king of the African forest.

He was not afraid of us. He stood there, and beat his breast with huge fists till it resounded like an immense bass drum, which is their mode of offering defiance; meantime giving vent to roar after roar. The roar of the gorilla is the most singular and awful noise heard in these African woods. Begun with a sharp bark, like an angry dog, it then glides into a deep bass roll, which literally and closely resembles the roll of distant thunder along the sky, for which I have sometimes been tempted to take it where I did not see the animal. So deep is it that it seems to proceed less from the mouth and the throat than from the deep chest and vast paunch.

His eyes began to flash spears of fire as he stood motionless on the defensive and the crest of short hair which stands on his forehead began to twitch rapidly up and down, while his powerful fangs were shown as he again sent forth a thunderous roar. And now truly he

394

reminded me nothing of some hellish dream creature—being of that hideous order, half-man and half-beast, which we find pictured by older artists in some representations of the infernal regions. He advanced a few steps—then stopped to utter that hideous roar again—advanced again, and finally stopped when only a distance of about six yards from us. And here, just as he began another of his roars, beating his breasts in rage we fired, and killed him. With a groan which had something terribly human in it, and yet was full of brutishness, he fell forward on his face. The body shook convulsively for a few minutes, the limbs moved about in a struggling way, and then all was quiet—death had done its work, and I had leisure to examine the huge body. It proved to be five feet, eight inches high, and the muscular development of the arms and breasts showed what extreme strength it had possessed.

I must say that when I look at our own beautiful gorillas, my heart bleeds to read descriptions of this sort. Du Chaillu, incidentally, was the first white man to kill a gorilla.

Du Chaillu was a good writer and took every advantage of the dramatic parts of his trip, and as a result, at first he was a great success; but subsequently scientists became very skeptical of his reports. When he got back to the United States in 1859, he was received almost like a hero. The Royal Geographical Society of England invited him to give a talk. His audience was crowded with scientists, and he received a standing ovation. He claimed to have brought home, in addition to his specimens of gorillas and mammals, 2,000 birds, many of which he claimed to be new species. He had also captured a young gorilla and wrote the first description of a young gorilla in captivity:

It was a young male gorilla, evidently not yet three years old, fully able to walk alone, and possessed for its age, a most extraordinary strength and muscular development. Its greatest length proved to be, afterwards, two feet, six inches. Its face and hands were very black, eyes not as much sunken as in the adult, the hair began just at the eyebrows and rose at the crown, where it was of a reddish-brown. It came down the sides of the face in lines to the lower jaw much as our beards grow. The upper lip was covered with short coarse hairs; the lower lip had longer hair. The eyelids very slight and thin. The eyebrows straight and three-quarters of an inch long.

When Du Chaillu's claims and his publications were investigated, he did not fare very well. Apparently the drawing he published of the skeleton of his gorilla was one he had copied from a drawing in the

British Museum. The frontispiece, which enlivened his book and which is reproduced in this one, was actually copied from a drawing in the Paris Museum. Some of his new chimpanzee species were also copied from illustrations in the Paris Museum, and neither of the new species was actually new. His claims were challenged, in fact, by John Edward Grey in 1861.

William Winwood Reade in his book *Savage Africa*, published in London in 1863, describes his expedition to West Africa. He was not a naturalist and described himself as "a young man about town," but can be more accurately labeled a nineteenth-century controversialist. He describes an encounter with a gorilla in which he crawled cautiously through the undergrowth, trembled all over, clenched his teeth, and then trod on a dry branch which broke. The net result of this was that the gorilla he had been stalking turned and fled. He had the following comment about Du Chaillu:

> After five months careful investigation I found that the gorilla neither beats his breast like a drum, nor attacks man in the above manner; M. Du Chaillu has written much of the gorilla which is true but which is not new, and a little which is new, but which is very far from being true. I am compelled to put aside as worthless the evidence of M. Du Chaillu; who has had better opportunities than any of us of learning the real nature of the animal, but who has unhappily, been induced to sacrifice truth to effect, and the esteem of scientific men for a short-lived popularity.

Because of the dramatic style and perhaps some overexaggeration, Du Chaillu was judged rather harshly. Gorillas do beat their chests. It is well known that the big male gorilla will, in fact, make a charge, which is called a bluffing charge, and if the individual stands his ground, the gorilla will stop before he actually comes into physical contact. Of course, one needs a certain amount of nerve to believe this and especially to act it out. However, the gorilla, despite this bluff, is in general not a violent animal and is, in fact, peace-loving and amiable. Du Chaillu completely misled people about the ferocity and the brutishness of the gorilla.

R. L. Garner published *Apes and Monkeys* in 1896, wherein he described his experiences on an expedition to West Africa. Intimidated by Du Chaillu's stories of the ferocious nature of the gorilla, he built himself a cage in the forest and sat in it day after day waiting for the animals to pay him a visit. In actual fact, he saw relatively little from this. With all his faults, Garner was the first individual to make

396

Courtesy of the Cheyenne Mountain Park Zoo

Male gorilla.

an attempt to carry out a field study of the gorilla and to find out how the animal lived in the wild.

Because he saw so little, Garner actually depended a great deal on information that he got from the natives about gorillas. He talks about the gorilla being polygamous, that he chooses a wife and remains married to her from then on, showing some marital fidelity. However, he may from time to time adopt a new wife but still keeps the old one. He gathers around him a family consisting of his wives and their offspring. These family groups, he thought, were kept well separated from each other and can be easily composed of ten to twelve animals. He also believed that they never made any nests. He made this claim because he had never seen any nests. Any solitary gorillas, he said, were actually young males looking for a wife.

Carl Akeley, who went to Africa after Garner, disputed Garner about the fact that gorillas made no nests. He said that they did, but that they made them only on the ground. He was incorrect as well. Akeley was a taxidermist and a collector for the Museum of Natural History in New York. He shot five gorillas on his expedition to take back to the museum, but more important, he actually took 300 feet of film of wild gorillas, the first film of gorillas in the wild that had ever been made. He was so impressed with the animals that he was concerned about the fact that he had shot some of them; he expressed the fear that if shooting were not controlled, this animal might be eliminated. So even at that time, in the early part of the twentieth century, the idea of conservation of the great apes had got home to at least one person. He asked that a gorilla sanctuary be instituted and that arrangements be made for scientific studies of the great apes.

George Schaller in *The Year of the Gorilla* (1964) has this to say about lowland gorillas: "Today the habits of lowland gorillas are still largely unknown. Hunters shoot them, zoo collectors catch them, and explorers take random notes on them in passing, but only one scientist has made a definite long term attempt to study the ape."

The gorillas that live in the west of Africa are the lowland gorillas. In the central and eastern parts of Africa are their cousins the mountain gorillas.

As 1900 dawned, the studies on the gorilla shifted away from West Africa over to the mountainous parts of the eastern Congo and of western Uganda. In this area there is a vast valley which stretches to the southern part of Lake Tanganyika from the upper White Nile. This is the giant Rift Valley, which is about 30 miles long. From the valley rise two masses of mountains. They rise to more than 16,000 feet on one side and to 10,000 feet on the other.

The fact that there was a large ape in this area was rumored to early explorers. John Hanning Speke and James Augustus Grant, in 1861, had been told when they traveled in this area (they were the first Europeans to search for the origin of the Nile) that there were manlike monsters in the mountains in that area that were not able to talk with men. Then in 1866, David Livingstone made his famous walk through this area, and he talked about seeing natives fighting with creatures which he labelled gorillas. We do not know for certain that they were. In 1890 Henry Stanley said that he believed that the gorilla lived in the northeastern part of the Congo.

In 1902 a German officer, Captain Oscar von Beringe, traveled near the northern part of Lake Tanganyika. Climbing the mountains in that area at an altitude of 9,300 feet, he and his party saw above them a number of apes, and this is what Von Beringe has to say about them:

> We spotted from our camp a group of black, large apes which attempted to climb to the highest peak of the volcano. Of these apes we managed to shoot two, which fell with much noise into a canyon opening to the northeast of us. After five hours of hard work, we managed to haul up one of these animals with ropes. It was a large manlike ape, a male, about one and a half metres high and weighing over 200 pounds. The chest without hair, the hands and feet of huge size. I could unfortunately not determine the genus of the ape. He was of a previously unknown size for a chimpanzee, and the presence of gorillas in the lake region has as yet not been determined.

This gorilla shot by Von Beringe was, in fact, the mountain gorilla, and it was named *Gorilla gorilla beringei* in his honor. The mountain gorilla resembles the lowland gorilla very closely. It takes an expert to separate the two, although the distinguished primatologist Adolph Schultz has listed some thirty-four structural differences between the two animals. The mountain gorilla has, as one might expect, longer hair than the lowland gorilla since he finds it harder to keep warm up in the mountains.

After Von Beringe's discovery, there emerged the usual pattern following the finding of a new animal. First of all, museum collectors swarm all over the area, eager to collect as many specimens as they can. The mountain gorillas were no exception. Schaller has listed the following mortality figures. In a small 150-square-mile area in the region of the Virunga volcanoes, the last stronghold of mountain gorillas, fifty-four gorillas were shot between 1902 and 1925. Prince Wilhelm of Sweden led an expedition in 1921 which eliminated sixteen gorillas, and between 1922 and 1925 the American hunter Burbridge accounted for nine. The American Museum of Natural History got five gorillas—shot by the naturalist-sculptor Carl Akeley. Akeley was, however, concerned about the future of the mountain gorilla and strongly advised the Belgian government to establish the area as a permanent sanctuary for the animals where their lives would be peaceful and safe and where scientists could study them. As a result, the Albert National Park was set up on April 25, 1925. In 1929 it was extended to include the whole of the Virunga volcano chain.

After Dr. Yerkes established the Yerkes Laboratories for Primate Biology in Orange Park, Florida, he sent Dr. Henry Nissen off to study chimpanzees in the wild in Guinea in West Africa and suggested that Harold C. Bingham go to Africa to look for the mountain gorilla. Bingham carried a rifle and was assisted by a large group of helpers. He was able to see the gorillas, and he spent some time following them. He also did a small amount of relatively unsuccessful photography. Then came a dramatic situation in 1939. The party had trailed a group of gorillas and had made the decision to return to camp. Suddenly a large gorilla charged straight at the party, heading for the guide and Mrs. Bingham. The guide simply took to his heels. Mrs. Bingham stepped to one side, and Bingham fired at the gorilla. The animal still ran on and then dropped in its tracks—dead. Bingham claimed that he saved his wife by that shot, but we still do not know whether this is a fact or whether the gorilla was only indulging in one of its bluffing runs. However, no one can blame Dr. Bingham for not taking chances on this occasion.

During the 1930's some interesting observations on the gorillas were made by Charles Pitman, a game warden in Guinea. He found gorilla nests in the trees, showing that the animal was still arboreal to some extent, and on another occasion he made an expedition with the governor of Uganda to see the apes in the wild. The group of people followed the group of apes some distance, watched them, and then decided that they had seen enough and would return back to their camp. However, the gorillas, not to be outdone by the humans, decided that they wanted to do a little observing of their own. As the party withdrew, they were followed by the group of gorillas. While watching them, a big male gorilla barked at the party. Pitman had this comment to make about this: "For a game warden it was an unenviable position. On the one hand the sacred person of the governor, on the other the almost sacred and strictly protected gorilla. . . ." However, Pitman, the governor and the gorilla maintained their sangfroid and everything ended satisfactorily on both sides. The entertaining part, of course, is the gorillas returning the compliment to the humans by having their own expedition to observe them.

The mountain gorilla in the wild has now been extensively studied. One of the really comprehensive works on the naturalistic behavior of a wild ape is *The Year of the Gorilla* by George Schaller—both in a popular and in a highly technical version. Prior to that time two young women, Rosalie Osborne and Jewel Donisthorpe, studied the gorillas in Virunga although they did not make very extensive observations. In 1955, Walter Baumgartel bought a hotel at the foot of the Virunga volcanoes in Uganda and called it the Traveller's Rest Hotel. Because there were gorillas in the neighborhood, he hoped to establish the hotel as a tourist attraction. This is where former secretary Rosalie Osborne, and former journalist Jewel Donisthorpe spent a year studying these animals. They described nests, they saw food remains, and they heard the male animals roar at intruders, but very little more.

It was not until Schaller really tackled the job properly that accurate information began to emerge. In his fascinating account of the mountain gorilla, Schaller too describes an occasion when he, the observer, became the observed. He came across some gorillas that were feeding on a steep slope, about 100 yards above him. He sat down near the base of the tree and he picked out the leader of the group. He called this animal Big Daddy, because he had two silver areas in the gray hair on his back. Usually the old males develop silver hair, and they are often called silver backs. The leader saw Dr. Schaller and gave two grunts, whereupon a number of females and youngsters looked in the same direction and then went to his side in case there was danger. There was another male whom Schaller called D. J., a striving executive type that had not yet reached the top—the second-in-command. Then there was a third male, even bigger than the leader, whom he called the Outsider, who roamed the periphery of the group. He was estimated to weigh between 450 and 480 pounds.* Finally, there was another silver back male whom he called Split-Nose because of a cut on his left nostril. He was a young animal, and his back had only just developed a silver tinge; but he made up for his relative youth with considerable vocalizing when he saw Dr. Schaller.

D. J. started the movement which ended with Schaller under

---

* It is often claimed that male gorillas reach 600 pounds or more, and they often do in zoos, where they usually become obese. Of ten adult male mountain gorillas that Schaller knew of, which had been killed and subsequently weighed by hunters, the heaviest animal had only reached 482 pounds, and the average was 375 pounds.

observation. He left the place where he was resting and moved uphill. Then he slowly angled toward Schaller, always keeping a screen of vegetation between them. However, he had to stand up every now and then to see where he was. The moment Schaller looked toward him he would duck down and then sit without any motion for a time before continuing his stalking activities. He finally approached within 30 feet of Schaller, roaring loudly and beating his chest. This demonstration completed, he peered out to see if it had had any effect. Schaller said he was never able to get used to the roar of a silver back male because it was usually very sudden and shattering. The effect that it had on him was invariably that he wanted to run, but at least he got satisfaction from the fact that the other gorillas in the group jumped when the roar started just as much as he did. At this point Schaller became nervous and began to climb the tree; he climbed to about 10 feet. Then one of the females came to within about 70 feet from him, sat on a stump, propped her chin on folded arms, and regarded him. Slowly the entire group came toward his tree. He was a little worried about this behavior because it had never happened to him before. Females carrying infants came over to look at him, and two juvenile gorillas climbed another tree to get a better look at him through the branches and vines. One juvenile, estimated to be about four years old, climbed into a small tree only 15 feet from the tree Schaller was in. They both sat there, as he says, "nervously glancing at each other." One of the other gorillas, the only black back male in the group, came to within 10 feet of the tree, eating a blackberry leaf as he did so. He stood near the base of the tree and looked up at Schaller with his mouth open. And so it occurred that the observer was treed, completely surrounded by gorillas, who sat by the hour observing him. After a while the sunshine turned into hail, and the hail into a drizzle of rain, and at last the gorillas ceased their observations and set out to forage for food.

About this time Schaller heard a very peculiar sound, and after looking around for a time, he was aware that D. J. was copulating with a female on the steep slope. She was on her knees and elbows, and he had mounted her from behind and was holding onto her hips. Since they were both on a steep slope, every time he thrust, they slid downhill. Within about fifteen minutes, they had slid 40 feet and only stopped sliding because they finished up against a tree trunk. Then, Schaller says, "a hoarse trembling sound, almost a roar, escaped from D. J.'s parted lips, interrupted by sharp intakes of breath." Apparently he had completed his orgasm, and the female lay there without

Gorillas copulating. This is only one of the positions gorillas may utilize for this purpose.

moving for about ten seconds before she got up and walked uphill while the male stayed behind still panting.

With the gorillas' attention so directed, Schaller was able to descend his tree and go about his business.

On another occasion, Schaller was also able to witness copulation. On this occasion it was initiated by the female. The Outsider, the big silver back male that was not really a member of the group but that skimmed the periphery, seemed to be looking into the distance when a female gorilla appeared behind him, put her arms around his waist, and thrust against him a number of times. It took a while for the significance to sink in, but when it did, the male turned around, grabbed the female by the waist, pulled her down so that she was sitting on his lap, and then started to thrust. The leader of the group, who was lying down only 15 feet away, got up and approached them. The Outsider then retreated a few feet, and the leading male, Big Daddy, and the female sat down for a while together. Then he walked away, and the Outsider returned. Schaller says:

She gazed into his eyes, and there must have been something in that look, for he did not tarry. With the female in his lap, he thrust rapidly, about twice each second, and soon emitted the peculiar call which I had heard during previous copulations. The female twisted sideways and squatted beside the male who rolled onto his belly to rest in this position for ten minutes. Suddenly he sat up and pulled the female onto his lap, but she broke away. They both then rested for half an hour. After a long rest, the female got up again and stood once more by the back of the male and he repeated the whole process again this time apparently bringing it to a climax.

Schaller says that he was surprised at the magnanimity with which Big Daddy shared his female with the Outsider and apparently with other males and even temporary visitors. He thought that perhaps this tended to make the group more peaceful. In any case the gorillas seemed to like each other and were very nonaggressive and relaxed animals. He also draws attention to the fact that Eskimos and also some other native peoples share their wives with visitors in just the same way Big Daddy permits other animals to share his.

Schaller has an interesting comment to make about the reaction of the animals observed in the wild and the observer:

> Animals are better observers and far more accurate interpreters of gestures than man. I felt certain that if I moved around calmly and alone near the gorillas, obviously without dangerous intent toward them, they would soon realize that I was harmless. It is really not easy for man to shed all his arrogance and aggressiveness before an animal, to approach it in utter humility with the knowledge of being in many ways inferior. Casual actions are often sufficient to alert the gorillas and make them uneasy. For example, I believe that even the possession of a firearm is sufficient to imbue one's behavior with a certain unconscious aggressiveness, the feeling of being superior, which an animal can detect. When meeting a gorilla face to face, I reasoned, an attack would be more likely if I carried a gun than if I simply showed my apprehension and uncertainty. Among some creatures—the dog, rhesus monkey, gorilla, and man—a direct unwavering stare is a form of threat. Even while watching gorillas from a distance I had to be careful not to look at them too long without averting my head, for they become uneasy under my steady scrutiny. Similarly they can consider the unblinking stare of binoculars and cameras as a threat, and I had to use these instruments sparingly.

Schaller describes the gorilla day as follows: Between six and seven they wake up, stretch their arms, yawn, reach over, and break off pieces of wild celery or other material conveniently close, and eat them. They also eat thistles and nettles, bamboo shoots, various fruits, sometimes the bark off several trees, and a variety of leaves. Most of the plants they eat taste bitter to humans. Gorillas defecate in their nests in the wild; presumably because of the amount of vegetable material they eat, their feces are quite solid and rather like horse dung in consistency. As a result, they do not worry about defecating in their nests because they can still sit on the feces and it does not stick to them or make them messy. They defecate pretty much wherever they happen to be at any particular time.

Schaller noted that he never saw gorillas eat any animals in the wild. He never saw them eat birds' eggs, insects, or mice, even though he noticed that there were opportunities when they could have done so. He saw them pass dead animals without handling their remains even though they were fresh, did not disturb a nearby pigeon's nest on another occasion. They will readily eat meat in captivity; in fact, our Yerkes gorillas have enjoyed cooked hamburgers. They do not seem to kill wild animals and eat them in the way chimpanzees will with young baboons. For breakfast the animals sit up in their nests, grab any food they can reach, move very little to get a piece of food if it is not convenient to them, and then go back to the nest and sit. As Schaller says, the only sounds that you can hear are "snapping of branches, the smacking of lips, and an occasional belch."

The infants apparently watch what their mothers eat and learn what to eat and what not to eat, so that food habits are handed down from one generation to another.

They move around in a small area eating; then as morning gets on, they eat less but wander about, choosing titbits here and there. They feed about two hours in the morning and worry very little about any other activity. Once they have got over the early-morning feeding, they move farther from each other and may actually move out of each other's sight. By about ten o'clock the feeding is usually stopped. Schaller also points out that he has never seen gorillas drink water in the wild state. In the area that he worked there was very little water, and when it rained, the water was rapidly absorbed by the soil and ran into the cracks of the rocks. However, the vegetation is succulent, and since it rains frequently, the leaves often have moisture on them and there is a heavy dew on the foliage in the morning. These probably contribute enough moisture for a gorilla.

From midmorning the animals begin a siesta that usually lasts until midafternoon. They lie around, usually near the silver back male that is the leader, and soak up the sun. Some lie on their backs, some on their bellies, some on their sides, and others sit and lean their backs against the trunk of a tree. Sometimes they construct a nest to have their siesta in. They can produce a serviceable nest in as short a time as five minutes.

Describing this rest period, Schaller says, referring to the silver back male:

> These males are also tolerant and gentle, and this is especially evident in the periods of rest. The females and the youngsters in the group genuinely seem to like their leader, not because he is dominant, but because they enjoy his company. Sometimes a female rested her head in his silver saddle or leaned heavily against his side. As many as five youngsters occasionally congregated by the male, sitting by his legs or in his lap, climbing up on his rump, and generally making a nuisance of themselves. The male ignored them completely, unless their behavior became too uninhibited, then a mere glance was sufficient to discipline them.

Not only do the animals sleep and doze and sit during this siesta period, but some of them also engage in toilet, mostly grooming themselves. Females are more active in grooming themselves than males and, in fact, groom twice as much. The juveniles are even more active self-groomers than the females. Infants, on the other hand, hardly ever groom since the mothers handle this chore. When the baby is still small, the mother will put it on her lap or over one arm and groom it by putting her fingers into its hair and pulling the hairs to one side. She seems to be especially careful of the grooming in the area of the anus which bears a little tuft of white hair. The youngsters seem to dislike this procedure; perhaps it is just being turned upside down that they dislike, but at any rate, they do a good deal of kicking while it is going on. A young gorilla is never physically chastised by its mother.

Adults do not seem to groom each other unless one of them requests it. For example, one of two females sitting near each other may lean over and tap the other animal on the arm and then rise and turn her rear end toward the sitting female, who would then groom the area that had been indicated. When adults groom each other, they seem to confine their grooming primarily to the rear end, back, and other parts of the body an animal cannot itself get at easily.

Young gorillas, of course, use the siesta period as an extended period of play and an opportunity to explore the environment without fear of separation. They do a fair bit of climbing into the trees, and in this they differ markedly from the adult animals. In fact, the silver back males only climb about a quarter as much as the juveniles. Gorillas spend about 80 to 90 percent of their hours awake on the ground, and when they climb, they climb in a very deliberate way. Their climbing ability can be compared with that of a ten-year-old boy. They get hold of a branch, hold it while they get a safe foothold, and then pull themselves up onto the next foot- and handhold. If they are not too sure whether a branch will support them, they sometimes pull it sharply to test it before they trust it with the weight of their body. They get down the trunk very easily simply by sliding down feetfirst, using the soles of their feet as brakes.

There is usually very little aggressiveness within the group. Most of the quarreling is done by the females. They will sometimes sit and scream at each other and do a certain amount of wrestling and biting. When the silver back male gets fed up with this, he goes over to the females and grunts at them, and they stop screaming immediately. They never seem to injure themselves during their quarreling. The role of the male is of interest in view of what we saw, mentioned earlier, with our orangutan group in the zoo in Atlanta. There was quarreling among the females until the male was brought into the group, which then stabilized under his influence and the females ceased to quarrel in his presence.

Gorillas may rest up to three hours in the middle of the day; then the silver back male decides when they are going to break it up and when they are going to start to travel and how far they will go. Once the group starts to move, the babies all rush to their mothers and climb on their backs. They tend to move in single file when they are traveling in the forest, with the silver back male at the head and often a young black back male acting as a rear guard. They usually travel in the middle of the afternoon, and they do not go very far—a half mile would be a long journey.

Once they have made their journey, feeding is again the activity, and they feed until dusk. This is a rather leisurely period of feeding, sometimes with little short periods of travel. As darkness falls, the animals become more and more relaxed in their movements and close in near the silver back male. Then at five or six o'clock the leader of the group starts to break branches in order to build his nest; this is the signal for the other members of the group to do the same thing. So after a ten- or eleven-hour day, they are all bedded again for another night's sleep.

Gil Clayton, when working in the Virunga range in Rwanda, was observing wild mountain gorillas with Dr. George Frazer. While a group of gorillas was under observation, there were sudden terrifying

407

roars from the males, and everyone rushed for cover. A leopard had invaded the area. A leopard is rare at that altitude, about 9,000 feet, and it followed the gorillas. Eventually the males formed together, and roaring at the tops of their voices, they charged the leopard. It turned and ran. Later one of the big male silver backs appeared to become sick. He left the group accompanied by a black back male that the researchers called Junior. Junior seemed to be looking after the sick silver back. The two humans, however, felt that since this animal was sick and had begun to show signs that he might die, there was a good chance that the leopard would come back.

The next morning the humans returned to the group to find that the old silver back and Junior were not with them. Clayton and Dr. Frazer tracked them and eventually found the silver back dead and Junior on all fours alongside him, presumably on guard. He suddenly showed excitement, reared onto his legs, and roared; Clayton turned to see that the leopard was slinking toward him. The leopard continued to advance despite Junior's intimidating roar, and Clayton expected Junior to turn tail and run, particularly since there were no other gorillas in the immediate vicinity to help him. However, just as Clayton had lifted his rifle to shoot the leopard, Junior charged. Shortly after that the leopard leaped at the gorilla, and they became locked together in a fight to the death. The leopard was attempting to dig one claw into the gorilla's eyes while his rear claws were raking Junior's abdomen. They fell to the ground twice and rolled over; neither would let go. Blood ran down the gorilla's face, chest, and arms. He struggled to his feet, still screaming, and grabbed the leopard's rear leg, wrenched him away, and threw him by the leg into the bushes. The leopard turned to attack again, and as he did, Clayton fired two shots at him. Junior had fallen back on the ground and was bleeding freely from his many wounds. He and Clayton gazed into each other's eyes for a moment, and then the leopard, merely wounded by the bullets, leaped onto Clayton's back, knocked him down, and sent his rifle flying into the bushes. The leopard was all over him. Clayton managed to get his left hand on its throat, pushed the animal back, and tried to get his knife out of its sheath with his right hand. He brought his knees up and put them against the leopard's chest in an attempt to hold him off. He finally managed to roll the leopard over, sit above it, and drive his knife into the thorax just below the rib cage. It took several stabs of the knife before the animal finally died.

When Clayton got to his feet, badly bleeding, he found that Junior had crashed off through the bushes to return to his group. Clayton said it took about a month for his own wounds to heal. Junior was now in charge of the group, and his wounds healed completely too. He was in charge because the old silver back male had been the number one man and he was now dead.

Dian Fossey was an occupational therapist in California who became interested in the possibility of studying gorillas in the wild. She was inspired by Jane Goodall's work and received financial assistance from the National Geographic Society on the recommendation of Dr. Louis Leakey. She was thirty-seven years old, and she set out for the gorilla country in the east of the Congo. She finally reached the Virunga range, which consists of eight volcanoes reaching higher than 14,000 feet. Part of this range is in Rwanda, part in Uganda, and part in Zaire.

Dian Fossey studied and formed a closer relationship with the mountain gorillas at even closer quarters than Schaller had done. In the course of her studies she managed to establish herself with a number of groups of gorillas almost as a member. In some of them, the relationship was so close that juveniles and young adults would come and examine the buckle on her knapsack, play with the laces on her boots, and pick up and examine the strap of her camera. She used a special technique in achieving this rapport. Instead of going to them and following them and staring at them continuously with binoculars, she tried to gain their confidence by acting like a gorilla. For instance, she imitated their feeding habits and the way they groomed, and eventually she picked up some of their vocalizations, particularly some of the deeper belches. She said these methods were effective but not very dignified, for example, "Thumping one's chest rhythmically, or sitting about pretending to munch on a stalk of wild celery as though it were the most delectable morsel in the world."

She found that when she was approached by the big males, folded arms were accepted by the gorillas as a gesture of submission, the same kind of gesture they used for that purpose. She came across one group of gorillas which had an unusual composition, consisting of a number of males and an elderly doddering female gorilla that was wrinkled and graying. Dian Fossey estimated her age at about fifty years. This group, unlike those observed by Schaller, did a good deal of mutual grooming; but eventually the old lady died, and the group did not continue this mutual grooming habit.

She found that the character of a group often changed when the leader changed. For example, one group she was observing was run by a silver back male she called Whinny. He got sick and remained sick for some months and then finally died. Another silver back took over as leader of the group, and she named him Uncle Bert. He immediately restricted the activities of the group. Under the other leader the gorillas had accepted the presence of Dian Fossey with great calmness, and she had no problems in observing them. Under the new leader, the same group did a great deal of chest beating, banging the foliage, and other alarm activities. This leader also changed their behavior and the route they followed in the jungle. However, she does note one very touching incident:

> Perhaps too quickly I labeled Uncle Bert a cantankerous old goat. One day, as their rest period was breaking up, a small infant approached him and leaned against his back. I was about to predict an unhappy fate for this baby, but Uncle Bert surprised me. He picked up a long-stemmed Helichrysum flower and tickled the baby with it. Soon the infant was scampering about like a puppy, and Uncle Bert was lying on the slope, tickle switch in hand and the most idiotic grin on his face.

On one occasion when she was watching the group, a juvenile called Icarus had climbed up a small tree and was swinging by his arms when the branch suddenly broke and came crashing down. The group seemed to blame Dian for the accident. The whole group charged her, led by the males with the females following, and they stopped only 5 or 10 feet away, apparently when they saw that the juvenile was perfectly all right and climbing another tree. However, the males remained very tense and continued to give barks of alarm. Then she was dismayed to see a small infant gorilla climb the same small tree which had broken and also began a series of acrobatic activities on it. The silver back males looked at the baby and looked back at her; whenever their glances met, the silver back males roared their disapproval. Finally, Icarus climbed up onto the tree and started a game with the baby which eventually led him back to the group. With a sigh of relief, Dian found that the crisis was over.

Dian Fossey also stresses the gorillas' lack of aggressiveness and says that in 2,000 hours of observation, she only saw about five minutes of behavior which could be called aggressive. Most of this was bluffing behavior. The closest call she had was when five large males charged, roaring at her. The leader was only three feet away when she spread her arms wide and shouted "Whoa" and the whole group stopped. She described mountain gorillas as nothing but "introverted, peaceful vegetarians," a judgment which is probably correct.

Dian Fossey was even able to make physical contact with one of the animals in the groups which she examined. An animal she called

Peanuts on one occasion approached within a few feet of her while she was sitting and began to show off in front of her by beating his chest and by strutting around. Then he moved to within a couple of feet of her. She gestured in a way she believed the gorillas find reassuring and began to lift one arm and scratch herself under the arm with the other. She was delighted to see that Peanuts imitated her. She then offered her hand, palm up, which she thought the animal would recognize as being like his own palm. Peanuts stood up, was uncertain for a moment, and then finally touched her hand with his fingers. This is probably the first recorded friendly physical contact between human and wild gorilla. Such an act helps make the brave hunters of the past look foolish with their stories of the ferocious beasts that attacked them.

Mountain gorillas have three vocalizations that are very clear. One is called the belch vocalization. If it is given, it usually brings responses from many other animals nearby. Dian Fossey managed to crawl into a feeding group of gorillas and start the belch vocalization herself and the animals around her answered it. She said that this vocalization expresses a feeling of comfort and well-being; the same vocalization with the same message also figures in the vocal repertoire of human primates.

Another type of vocalization is called a pig grunt. The pig grunt is a harsh, staccato sound, and it is used when discipline is required—for instance, when a silver back male has to stop a group of squabbling females or if he wants to move on and warn the group to get up and follow him. Females use a softer version in the discipline of their babies.

The alarm bark is a vocalization used when the animal is curious. If the silver back gives this alarm bark, the whole group is immediately attentive to what is going on.

Toward the end of 1971, Miss Fossey started a census of these animals in the area where she had been working, partly with the object of finding out just how many there were and also to stimulate conservation activities.

The distribution of the gorilla in Africa is much more restricted than that of a chimpanzee. The chimpanzee is widespread and will adapt to a variety of conditions, although it is fundamentally a dweller of the forest. The lowland gorilla is found in the Cameroon-Gabon area of West Africa. It occurs in that area from the Cross River in Nigeria to the region of the lower Congo. The forests of central Africa, although they have chimpanzees, have no gorillas at all. However, there are gorillas in the relatively low-lying areas of the east and in the

Snowflake. The only known white gorilla, Snowflake has blue eyes.

mountains, such as in the Virunga area where Dian Fossey worked.

All these eastern gorillas were originally thought to belong to the subspecies *Gorilla gorilla beringei*. Colin Groves of Cambridge University, in his study of the gorillas of Africa, states that only the animals that live up in the highlands, particularly those of Virunga, are true mountain gorillas and that those living in the lowlands are different subspecies *Gorilla gorilla graueri*. Scientists have discussed why the distribution of gorillas is so discontinuous; why, for example, is it

412

Paki gorilla and her newborn baby Kishina. First gorilla to be born at Yerkes.

absent from the central forests of Africa? It is believed that this is probably because in the past, as they developed, they were unable to cross several large rivers which carved up that part of Africa.

A few years ago, a completely white gorilla was found in the wild in West Africa and was taken to the Barcelona Zoo. It was called Snowflake and has been studied with great interest. Although it is white, it is not a true albino, because it has blue eyes and a true albino would have pink eyes. Its photo appears in the illustrations in this book, p. 412.

Not many gorillas have been kept and studied in captivity. One of the most famous was Toto, kept by Mrs. Hoyt, and their relationship is described in detail in my book *The Ape People*.

The Yerkes Center, which originally bought fifteen lowland gorillas when they were tiny babies weighing only 20 or 25 pounds, has now raised many of them to maturity, and recently we have successfully bred them. A baby gorilla called Kishina was born on August 7, 1972; the name means "the source" or "origin" in Swahili, and the word is also used for a type of dance.

413

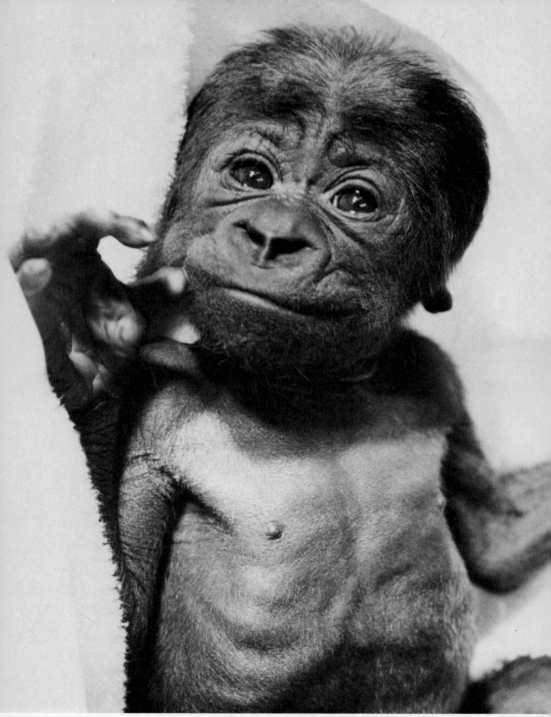

Portrait of Kishina. Yerkes Center's firstborn gorilla. Center now has sixteen gorillas.

Two young gorillas talk things over in the Yerkes nursery playroom.

Dr. Ronald Nadler, at the Yerkes Center, has described the birth of Kishina. About four hours before the birth, the mother, Paki, became restless and agitated and could not maintain one position for very long. During this period she seemed to be straining and was obviously having labor contractions. The birth sac began to protrude from the vagina. At first Paki touched it rather gently; but eventually she tore it, and the fluid spilled to the floor. At this stage Paki actually tore off a bit of the sac and ate it and also licked up some of the fluid. Then the head of the infant appeared in the vulva. Paki kept touching the head and her vulva and licking her fingers. Simultaneously, she moved around the cage in a squatting position. Eventually, she walked to one corner of her cage, lay on her side, and the infant was delivered. The baby gave a few gasps and then began to breathe normally. At first Paki paid no attention to the baby, instead licking blood and fluids from her hands. After a few minutes, however, she began to handle the new baby, Kishina, and became attentive to it,

Two young Yerkes gorillas in a play bout.

licking its body and especially its head. She did not attempt to bite through the umbilical cord but pulled it and the attached placenta until it finally parted from the umbilicus.

After an hour of licking, Paki picked Kishina up and pressed her to her chest, and she did this more often as the time progressed. Eventually Kishina got a nipple in her mouth and twenty-four hours after delivery began to nurse for long periods.

Paki behaved like a good mother for a few days and then, for some peculiar reason, began to reject her baby. We could not risk anything happening to our firstborn gorilla, so she was taken away and put in our ape nursery, where she received expert infant care, and today she is a thriving gorilla toddler.

A gorilla was born in a zoo in Columbus, Ohio, on December 22, 1956. The birth was not observed, but the baby was found on the floor of the cage and required artificial respiration. This little animal was a female. The Columbus gorilla baby was the first proof that it was possible for gorillas to reproduce in captivity.

One of the most famous gorillas born in captivity is Goma, born at

First gorilla in second zoo generation, Lowland gorilla Goma, born in 1959, with her newborn son.

Courtesy of Dr. E. Lang and Jorg Hess of the Basel Zoo, Switzerland

Stephi, male gorilla, and Achilla, female, with her baby daughter.

the Basel Zoo. In view of the Basel Zoo's interest and experience in keeping anthropoids, it is not surprising that it was in their zoo that the first gorilla was born in Europe. Goma is the Swahili word meaning "dance of joy"; it is also the name of a town situated on the banks of Lake Kivu in the Congo. Dr. Ernst M. Lang, the director of the Basel Zoo, explains that the animal had to have a name which began with *G* because in any particular year all the animals born in the zoo have names beginning with a single letter, and they were fortunate to find both an appropriate letter and an apt name.

Goma's mother was Achilla. The zoo had received her on October 23, 1948. She had come to Paris with a shipment of animals that had journeyed from the Cameroons in West Africa. She weighed only 17½ pounds when she came to Basel and was raised with a young chimpanzee. They played together a lot and, according to Dr. Lang, gave every appearance of having a happy childhood.

In 1952, Achilla complicated life for everybody. She was drawing

418

From E. M. Lang, R. Schenkel, and E. Siegrist, *Gorilla—Mutter und Kind* (Basel, Basilius Press, 1965)

Gorilla baby, Goma, and her mother, Achilla, at the Basel Zoo.

with a four-color ball-point pen, got frightened by something, put the pen in her mouth and swallowed it. A foreign object of this size could not be permitted to remain in the stomach for very long, but gorillas had never been operated on before, so the veterinarians were concerned about the anesthetic. The animal was too strong to be held down, so vets first gave her a drink spiked with a soporific drug. This made her behave like a drunk within half an hour, but she was still not unconscious. By this time a pediatrician, a children's surgeon, and a physician from the Basel hospital had arrived; in addition, there was the zoo veterinarian and Dr. Lang himself. Dr. Lang then gave Achilla an injection of morphine that would, as he said, "send a man straight to dreamland," but they had to give her a number of shots before they were able to make her somnolent and get her on the operating table. Once there she was attached to an anesthetic machine. It was not long before the stomach was opened and the ball-point pen removed. When they finally got her back in her cage,

419

Dr. Tuttle working with Inaki. Inaki has small electrodes inserted in the muscles of her arm, which are connected by a lead to scientific equipment that records the electrical changes in the muscles as Inaki walks on her knuckles. The electrodes cause no pain, and she plays happily with Dr. Tuttle.

however, she did not wake up. Her pulse had slowed, and her respiration was scarcely detectable. Dr. Lang almost gave up hope; but finally she was given a heart stimulant and heavy massage, and she opened her eyes. From then on the wound healed very well, and she recovered fully. When the scar healed, she removed the stitches herself without bothering to have the doctor do it.

In 1953 the zoo was able to get a 74-pound male gorilla from the Columbus Zoo in Ohio. They put Achilla and the new male, whose name was Stephi, in neighboring cages. They were able to smell each other through the bars but were suspicious of each other in the beginning. After a time they began to put their fingers through and play with each other. Eventually they were well enough acquainted that they could be put together in the same cage. Dr. Lang was thrilled because the Basel Zoo was at that time the only zoo on the European continent that had a breeding pair of gorillas, and the

Dr. Russell Tuttle and Yerkes gorilla Inaki, who is working with him.

zoological profession in Europe was eager to see if breeding would be successful. Achilla had reached 154 pounds in weight and was about four feet in height; Stephi had reached about 310 pounds and was over five feet.

By March, 1959, they found that Achilla could not tolerate the male in the same cage except for a very brief time. As she got bigger and bigger, it was obvious that there was a baby on the way. The night of September 22 and 23, 1959, Goma was born. The zoo staff did not see the evening birth, but when the keeper came to her cage early in the morning, Achilla already had the tiny gorilla baby in her arms. Its face was pale with a rosy tint, and it had a good cap of black hair. She was not holding the baby to her breast but had it on her left arm with the face turned so that it was away from the breast and so that the baby was not able to seek for milk itself. After thirty-six hours she had still not nursed it, so they decided to give Achilla a little sedative. When it took effect, she put the baby down and went off to lie down, and they were able to go in and get Goma. Goma was then taken home to the Lang household to be brought up by them. She demanded food about every two hours and drank up to 33 cubic centimeters (a little over an ounce) at a time.

Goma's birth was a great triumph for Dr. Lang and the Basel Zoo. She was soon the focus of reporters, photographers, and television cameramen. Books were written about her, were translated into a number of languages, and a number of scientific papers were written as well.

The details of Goma's upbringing make fascinating reading and were published in a book called *Goma, the Gorilla Baby*, written by Ernst Lang and profusely illustrated with a fascinating series of photographs. An English translation was prepared by Fisher and Schaeffer and published by Doubleday.

The zoo in Frankfurt has also been very successful in breeding baby gorillas. It has had five now, four of which were offspring of a single female—Makula. The fifth animal was called Dorle, and his birth was very fast, taking only a few seconds. The mother had been seen playing with a group in an outside enclosure, and the birth took place there. The mother seemed to be as surprised as the keeper when the baby was born. Although she looked at the baby, which was shouting and whimpering, she did not touch it or pick it up. So the keepers had to go into the compound and retrieve it and take it immediately to the nursery. The animal weighed two kilograms (about five pounds), and he was bottle-fed.

Two gorillas have been born at zoos in Japan. The first was born in the Kyoto Zoo on October 29, 1970. Then on February 2, 1971, in the Ritsurin Zoo in Takamatsu, another gorilla was born.

The Lincoln Park Zoo in Chicago has also had a gorilla, and other zoos, including the famous San Diego Zoo and the equally famous National Zoo in Washington, D.C., have been successful in breeding gorillas; but only about fifty have been born in captivity throughout the world to date.

One striking anatomical fact about the gorilla is that its ears are very much smaller than the ears of any other anthropoids. The ears of the chimpanzee are very big and stand out very obviously. The ears of the gorilla, however, are quite small, are situated close to the head, and are very much more similar to human ears.

Dr. Duane Rumbaugh found that in some intelligence tests the gorilla, like the orangutan, was a little superior to the chimpanzee. The gorilla appears to those who come in contact with it to be less intelligent than the chimpanzee because the latter is much more motivated and is a much more active animal, whereas the gorilla is much more retiring and introverted.

"What's this funny-looking stuff?"
Gorilla in San Diego Zoo examines a bunch of grass.

Courtesy of Dr. Duane Rumbaugh

Courtesy of Dr. Duane Rumbaugh

Everyone to his own grass!! Gorilla marches off purposefully with his bunch of grass.

423

Beauty and the beast. Joan Berosini of the Bobby Berosini monkey act.

Courtesy of Bobby and Joan Berosini

In the Gabon, in West Africa, a few years ago the De Meyers', a French couple, decided to adopt a baby gorilla and, having enjoyed the experience, adopted another and another until they now have nine. It all started four years ago when some natives brought them an orphan gorilla whose mother and father had been speared. Later they were brought other orphan gorillas. Their weights, medical histories, and other details have been recorded. Gorilla babies, like human babies, are virtually helpless when they are first born. Like the chimpanzee, they grow much faster physically than the human baby, but mentally they grow at about the same rate as the human, up to about a year or eighteen months. Arthur, one of the gorillas of the group, who is now aged about four, has to be caged during the day; otherwise, he would tear the place to pieces. Another baby, three-year-old Zozotte, specializes in scaring Africans who visit the house because it likes to see them run. Another of the three-year-olds likes to chew on a rug before it drops off to sleep. On one occasion a circus offered Henri de Meyers $16,000 for Arthur but was turned down flat. Henri's answer was: "We do not sell our children."

424

Courtesy of Bobby and Joan Berosini

Bobby Berosini monkey act in Las Vegas. Gorilla, orang, and chimpanzees. Apes love to be in music hall acts.

Joan Berosini of the Bobby Berosini monkey act and friend.

Courtesy of Bobby and Joan Berosini

This section can be most appropriately closed with a hilarious story about our own gorillas. Recently, a wild animal park called Lion Country Safari opened up south of Atlanta. A lake with three islands was constructed on the grounds. Since we have been rather pressed for accommodations for our apes, we offered to put some of them on the islands—a pair of different apes on each one.

One day three workmen on a raft paddled over to the gorilla island to carry out some minor maintenance; one of them went ashore, and the other two stayed on the raft a few feet away. One of the gorillas, called Rann, saw the workman on the island and charged him; the workman looked up, saw the animal bearing down on him, and promptly dived into the lake. By the time Rann got to the edge of the lake, he was going so fast he could not stop and went on into the water too. However, he managed to flounder to the raft, and as he grabbed it, the two men on it dived into the lake. Rann climbed onto the raft and gravely surveyed the humans milling around in the water. Eventually the humans pushed the raft back against the island and Rann stepped ashore; the workmen clambered back on the raft, and all was well.

XII. Man and His Future

THERE WERE TWO primary changes that led to the development of our ancestors into man instead of into an ape. These were the adoption of the erect posture and the development of a larger brain.

Erect Posture

The erect posture performed a number of valuable functions, but one of the most important was the freeing of the hands from locomotion. This permitted the hands and the arms to be used for carrying things, for holding and throwing weapons, and for manipulating objects so they could be more minutely examined and understood and eventually be changed and fashioned to become tools. In this way, the adaptable grasping hand, liberated from the earth, became a major factor in the development of man.

The ancestor of man split off from the rest of the Primate stock during Miocene times—a period of 35,000,000 to 25,000,000 years ago. The Miocene was a period of great geological disturbance. During this time the great mountains of the world were being formed—the Himalayas, the Alps, the Andes, and the Rockies. At the same time, in many areas, grasslands or savannas were replacing the forests. This meant that animals forced to inhabit the grasslands had to adapt themselves to a new type of countryside which, because of its tall grass, favored for Primates the bipedal walking (the erect posture and walking on two legs) more than it favored any other type of locomotion.

Man may have come down into the savannas primarily as a four-footed animal, walking more or less like the gorillas and chimps of today and then, by process of adaptation, come to assume an erect posture. On the other hand, man may have arrived already at least partly erect, because he had been conditioned to this posture by the process of brachiating, which is the type of progression well exemplified by the gibbons.

In 1971 the Englishman O. J. Lewis showed that the structure of the wrists of the ancestors of man and apes indicates that they were animals capable of brachiation, so that man and apes must have come from an animal which was originally swinging in the trees. So at least the originators of man were already adapted to a vertical position. We know that gibbons not only use brachiation in the trees but also walk about on the branches of trees on their hind legs about 10 percent of their time. Of course, gorillas and chimpanzees can adopt an erect posture on the ground, at least for short periods of time, and so can the orangutan, which is the most arboreal of the great apes.

Jane Goodall found that chimpanzees became erect if they were trying to look over tall grass. This has also been observed by Schaller in gorillas, which is precisely why it is believed that the savanna stimulates an animal to adopt the erect posture. If apes are looking for another animal, they will often stand up and look around. If they are observing a human who is observing them, they will often stand up to get a better look at him.

They will also stand erect if they are carrying anything in their hands. The Dutch zoologist Dr. Adrion Kortlandt has seen them walking bipedally in a banana and papaya plantation with their arms full of fruit. Both chimpanzees and gorillas will stand up on their legs

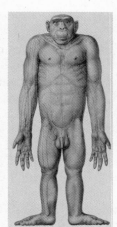
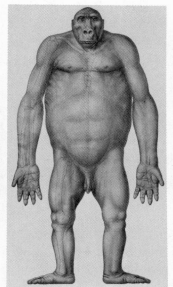
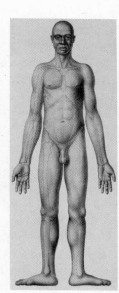

Original drawing by Dr. Adolph Schultz (Courtesy of Dr. Schultz)

Relative proportions of orangutan, chimpanzee, gorilla and man.

when they are making threats, as well as in greeting and during courtship.

It is of interest that Goodall's chimpanzees had in their environment a good deal of open, savannalike country, whereas, the English behaviorist Vernon Reynolds and his wife, who confined their studies of chimpanzees to those that lived in dense forests, found that the animals very rarely adopted the erect posture. Sometimes they would do so walking along the branches of trees, but even then they held onto other branches with their hands to support themselves. Dr. Nissen, who was sent out by Dr. Yerkes to study chimpanzees in West Africa, also studied chimps in the rain forest. Although he saw animals that would stand up on their hind legs, he never saw an animal walking erect.

Kortlandt also studied chimpanzees in the same kind of environment as those studied by Jane Goodall. He found that they were very terrestrial animals that did not take to the trees very often. He estimated that about 10 to 15 percent of the progression of the animal was done upright on two legs. He said that they also assumed the erect posture if they wanted to jump across a small stream. Of course, the adoption of the upright posture for a short period to enable an animal to carry food is not at all unusual among Primates. Stump-tail

429

macaques have often been seen to do that. Capuchins do it, and rhesus monkeys do it. There are some beautiful films of Japanese macaques walking erect, carrying an armful of potatoes which they take to sea to wash before eating them.

Dr. B. A. Sigmon, of the Department of Anthropology at the University of Toronto, Canada, has stressed that bipedalism is used by animals wherever it improves the chances of that animal's survival. Obviously standing up in the long grass to look for an enemy or an unusual object or to find the rest of one's group is important to survival, and so is standing up in order to carry food away.

Animals that developed anatomical and physiological changes which favored the erect posture obviously had a genetic factor which would favor their survival. Changes in the body which were necessary were changes in the foot, the pelvis, the vertebral column, and the skull, to mention the main items in the skeleton, and of course, with those skeletal changes there had to be changes in the muscles which served them. One of the important muscular changes was the development of the big gluteal muscles which form the backside of humans and which are very minor in nature in the apes. This muscle not only helps maintain the erect posture, but also permits humans to move the leg in the upright position with much greater speed than it can be moved in the apes. Speed, of course, would be very important to an animal in the savanna dependent on its ability to see its enemies and to flee them if necessary.

In the erect posture the vertebral column needed to form an S-shaped curve, which relocated the center of gravity in the animal and helped it stand erect. In quadrupedal animals the hole in the skull called the foramen magnum, through which the spinal cord passes and connects with the brain, is more toward the back of the skull. In an erect animal if the foramen magnum were in this position, the animal would be looking up at the sky all the time. Therefore, in erect animals the foramen magnum has shifted underneath the skull, so that it can be balanced properly in a horizontal position on top of the spinal column. Of course, the neck muscles had to be much stronger in a quadrupedal animal to hold up the head, which had a much bigger face and snout than the human. When the skull became balanced on the top of the vertebral column, it was not necessary to have such big neck muscles, so these became reduced in size. Some of the bony developments on the skull which were necessary for the attachment of the big powerful neck muscles of the quadrupedal animals, therefore, became reduced. Thus, the skull of a human is smoother and more streamlined than that of an ape.

430

As the human strides along, the bipedal nature of walking requires the hip to swing with each step to obtain a reasonable length of stride and to maintain the center of gravity in motion. This attracts little attention in males, but in females, partly because there is a much greater deposit of subcutaneous fat in the gluteal region, the appearance of this region of the body during walking has a powerful sexual attraction for males.

The necessity for walking in the upright position means that a number of strains have been imposed upon the body, and since not everyone's anatomical parts have been appropriately modified to take up these extra strains, a number of medical problems result. Back pain, for instance, is a very common complaint in doctors' offices. Slipped intervertebral discs, osteoarthritis in the hips, knee problems, flat feet, and so on are all associated with the lack of proper anatomical adaptation to the erect posture.

Man is also subject to frequent falls. He can be very easily thrown off-balance, particularly older people whose muscles cannot always compensate for a sudden change of position or alteration of the center of gravity. Even the gibbon, which is the most erect of the apes, takes a number of falls, although most of these occur during brachiation.

There are two major Primates that have more or less adopted the erect posture on a permanent or semipermanent basis—the gibbon and man. The gibbon's adoption of the erect posture, however, has not led to the results it has in man. The gibbon's erect posture is dependent on the use of his arms and hands for swinging, so that a lot of the time when he is erect, his hands cannot be used for any other purpose. His feet are, of course, available for carrying things, and gibbons do use them this way. However, when a man is erect, he is dependent only on the action of his legs to stand upon his feet, and so his hands are free and are available for all kinds of manipulation. Because the gibbon tied his hands to locomotion, he has not progressed much beyond what he was doing millions of years ago.

The raccoon also has hands of great sensitivity. He does a great deal of manipulation and handling of external objects and touching them with his fingers and has an enlargement of the area of the cerebral cortex which is concerned with the sensory appreciation of the hand. The tactile pad and the prehensile ability of the tail of the spider monkey also have specialized regions in the cortex.

An important feature of the Primate hand, particularly of the human hand, is the very fine control over the fingers and the thumb, as well as the existence of the opposable thumb—that is, the tip of the thumb can be used to touch the tips of all the other fingers. The fine muscular control of the thumb and fingers permits a precision grip, the type of grip, for example, that is required to hold a needle in one hand and pass a thread through it with the other. This is a degree of control beyond any other Primate.

Professor Raymond Dart has made the following comment on the use of Primate hands:

> The performance by men of various skilled feats of body that do not lie within the compass of anthropoids is, of course, due to human bipedalism, that is the divergent uses to which human hands and feet are put as compared with those of anthropoids. This divergence as used by human beings and by anthropoids of their limbs, was dependent not only upon relative profundity of their ancestor's mental cogitations but upon the divergence of their ancestor's diet. The forest loving vegetarian anthropoids clung to their four-handed climbing and fruit while the terrestrial predacious Australopithecines, depending on their speed of foot and deftness of hand, lusted after flesh! Elliot Smith (1927) showed how a simple change in habit from terrestrial to arboreal life in one of the same family of insectivora was competent to revolutionize the entire structure of the forebrain by altering the relative proportions of its constituent cortical areas without causing any increase in body size. Thereby he demonstrated how cerebral improvements in vision, tactual discrimination and hearing respectively brought about, by divergent employment of the animal's body, conditioned primate evolution. Just as arboreal life transformed the insectivore brain, and was the prime factor in the emergence of the primate brain from the insectivore brain; so predacious terrestrial life, by adopting a divergent posture and employment of the limbs, inevitably transformed the anthropoid brain and was the prime factor in causing the australopithecine to differ in its path from that of the anthropoids and to resemble more closely that of mankind.

Drs. J. T. Robinson, L. Freedman, and B. A. Sigmon, the first from the United States, the second from Western Australia, and the third from Canada, have studied a number of fossil remains and have concluded that *Paranthropas* (an apelike, humanlike creature which lived at least 4,000,000 or 5,000,000 years ago and which is out of the mainstream of human evolution) and *Australopithecus* (which lived about the same time and is thought to be in the mainstream of human evolution) walked in the erect posture. They both had the S shape in the spinal column and changes in the pelvis which are characteristic of man as contrasted with the apes. In *Paranthropus* the lower limbs were shorter in proportion to the body than they were in *Australopithecus*.

432

The former had a more mobile foot—that is to say, a foot that probably still had some grasping ability compared with the rather stiff and solid foot of the human. These findings suggest that *Paranthropus,* in its bipedal position, still only had the capacity for slow movement. *Australopithecus,* however, must have been able to run rings around *Paranthropus* because he had all the changes in the skeleton necessary to suggest that he was not only able to adopt the erect posture, but he also had the appropriate skeletal modifications as well which would permit him to move his legs with considerable speed. The evolution of the erect posture, therefore, probably resulted first in the ability to walk predominantly in the erect position; the development of speed came later in the evolutionary process.

A very interesting theory was put forward in 1960 by Sir Alister Hardy of Oxford University. Sir Alister suggested, ingeniously, that the development of the erect posture by man occurred because he came down from the trees and took up an aquatic environment. So some animal, whose wrists were anatomically adapted for swinging in the trees, was eventually, according to Sir Alister Hardy, forced from the trees to the seashore or the lake shore to look for food. He found plenty of food there in the form of shellfish and sea urchins, and he used his hands to find stones which he could use to crack the shells of these animals to obtain the succulent meat within. Presumably when man first came out of the trees, his legs were not strong enough to support his body for any length of time, and he would then go into the water for support. By walking in water, he was more or less kept in the erect position by the support of the water. He could then dive down, swim down, or lean to put his hands down to pick up shells of crabs or sea urchins or whatever was there which served him as food. Japanese macaques will walk for a long period bipedally in the water, if the water reaches high enough up the body to support their weight.

It is noteworthy that if you put a baby into water when it is very young, in fact even before it has learned to sit up and certainly before it learns to walk, it will automatically go through motions of swimming. There are in existence two or three aquababes—babies who have been taught to swim and even to dive and swim underwater, which they do very efficiently, before they are able to walk.

There are two or three things about man which are difficult to explain but seem to be more understandable if one thinks of this possibility of a marine period for him. One of these is the loss of hair. To compensate for this loss of hair, man has developed a very thick layer of fat under the skin. This thick layer of fat is not found under the skin of any of the great apes, although, of course, they have hair.

433

The only mammals which have this comparable thick layer of fat are the marine mammals. Think of the blubber that underlies the skin of the whales, for which they have been mercilessly slaughtered over the last hundred years or so. This fat gives a smoother, more streamlined appearance to the body of the human compared with apes. Streamlining, of course, would be helpful in swimming. The fact is that the apes cannot swim at all, and man, in general, can swim with great facility; he is capable of a very long swim at speed, such as the twenty-mile crossing of the English Channel.

Since man has this very thick fatty material under his skin, he is obviously very well insulated and would not lose heat very efficiently. He has, therefore, developed a very effective sweating system and has many millions of sweat glands in the skin. These enable him to secrete fluid which by evaporation allows loss of heat. He has retained hair only on his head. For an organism which is swimming, the protection of the head from the rays of the sun could conceivably explain why the hair has remained on the heads of most of us. When man cries, the glands in the eye produce tears which secrete salt. Man actually cries salt tears. Salt is also secreted in the sweat. Marine animals usually have some kind of a special mechanism for excreting salt; man's tears and sweat fulfill that function.

The large number of sweat glands enabled man to withstand the tropical climate and also to keep the layer of fat necessary for an aquatic life, so he was adapted for being both in and out of the water. In fact, Hardy says, "It seems to me likely that man learnt to stand erect first in the water and then, as his balance improved, he became better equipped for standing up on the shore when he came out, and indeed also for running. He would naturally have to return to the beach to sleep and to get water to drink. Actually I imagine him to have spent at least half his time on the land."

Hardy goes on to talk about the hand. I mentioned the human hand above, and the famous anatomist Wood Jones had this to say about it:

> In the first place, it seems to be perfectly clear that the human orthograde (walking upright) habit must have been established so early in the mammalian story that a hand of primitive vertebrate simplicity was preserved, with all its initial potentialities, by reason of its being emancipated from any office of mere bodily support. Perhaps the extreme structural primitiveness of the human hand is a thing that can only be appreciated fully by the comparative anatomist, but some reflection on the subject will convince anyone that its very perfections, which at first sight might appear to be

434

specializations, are all the outcome of its being a hand unaltered for any of the diverse uses to which the manus of most of the lower animals is put. Man's primitive hand must have been set free to perform the functions that it now subserves at a period very early indeed in the mammalian story.

Hardy points out that the hand of man is also a very sensitive instrument with the many touch organs just under the skin of the fingers that would make it ideal for exploring. Its fingers, moreover, could act like tentacles for feeling a sea bed for food, for grasping crabs, lobsters, and even fish, and for seeking out various shells in the sand such as oysters which could be broken open and eaten. He points out that the fish known as the gurnards have pectoral fins to which fingerlike projections are attached. They use these as sensitive feeling organs for hunting for food on the bottom of the sea and also for turning over stones to see what is underneath them.

He points out that there is a gap of 10,000,000 years or so between the fossil *Proconsul* and the first *Australopithecus* skull. He suggests that it may be during these years that man had his period in the sea which helped him establish his erect posture. We can perhaps interpret this theory then by saying that the animals which remained in the trees continued to remain in them and became apes. Those who came down out of the trees and began to forage in the water and spend a good deal of time in the sea were then being conditioned to develop the erect posture which enabled them then to inhabit the savannas and compete and live successfully on them. One of the arguments against this theory is that studies with humans subjected to long periods of experimental water immersion have been shown to result in cardiovascular deconditioning which makes the adoption of erect posture on land a difficult procedure. Thus marine life would have the reverse effect from that postulated by Hardy.

Anthropologists, up to date, have thought mainly about the anatomical factors involved in adopting the erect posture. There are some very important physiological factors as well. It is a well-known fact that when soldiers are on parade and have to stand at attention for a long period of time, occasionally one of them will collapse in a dead faint. Some of you may also have felt faint after standing erect for a long period of time. Sometimes, if you are bending down weeding in the garden and you stand up suddenly, you get a period of faintness. If you have spent a long period in bed, you find yourself very weak and dizzy when you first get up. All these factors are related to the physiological problems that are concerned with the erect posture.

435

The soldier who falls in a faint after a long period of standing has not done so because his muscles are tired. He has done so because of changes in his blood system. There are a number of deep veins in the legs. The blood which comes from the heart, down into the legs, and into the arteries has to be pumped back into the heart through these veins of the legs. The veins are elastic structures, and as more and more blood comes down, they tend to expand. If there are good strong muscles in the legs, they help support the walls of the veins so that the blood can continue to be pumped back to the heart. However, in some cases, the support the leg muscles are able to give to these veins is not adequate, especially if the legs cannot be moved, to help pump the blood back up to the heart. They will distend, and blood will tend "to pool" in the legs. In this way, the amount of circulating blood is reduced and there is an inadequate supply of blood pumped to the brain, so the individual faints.

When a person lies down, the necessity for the heart to work hard to pump blood through the legs is reduced, so the muscles in the legs do not have to support the walls of the veins in the same way in erect posture. This also happens to an astronaut when he is in a weightless condition. When the muscles do not have to be used in that way, they respond by losing water. There is a good deal of mobile water in the leg muscles, and if it is not needed, it is passed out into the bloodstream. When this happens, the amount of blood volume becomes too great, and a little nerve sensor in the vessels up near the heart causes the kidneys to excrete fluid from the blood so that the blood volume will drop back to a reasonable level. When the individual gets up or when, in the case of an astronaut, he returns to a 1 "g" environment, the blood will again start to pool in the legs and will be pooled much more effectively than before, because now the muscles do not have the additional water in them and they are flaccid and cannot support the veins as they would with their normal complement of water. The great veins of the legs become distended, and additional blood tends to accumulate there so that that blood volume is now less than that required for normal functioning. Again less blood goes to the brain, and the individual tends to feel faint. He comes back to normal after a few days when the muscle cells in the legs are being used again and start to take up water and become turgid so that they can support the veins once more. So here is a complex physiological mechanism which had to develop at the same time as the anatomical changes developed to enable man to adopt the erect posture.

At any rate, with all these drawbacks and deficiencies, man has

436

developed the erect posture long enough to enable him to use his hands for a variety of purposes. The freeing of the hands is one factor that has made him what he is; the other factor is the development of his brain.

The Brain

In general, the mammalian brain consists of two cerebral hemispheres, the surface of which is described as the neopallium or the neocortex. This cortex is the seat of consciousness and is the region where reflex actions are controlled, where voluntary movements are readjusted and subjected to adaptation, and onto which the reports from the sense organs are projected. The hind part of the brain, which is called the cerebellum, is composed of a central portion and two side portions called lateral lobes. Both the cerebral hemispheres and the cerebellum are attached to the brain stem which represents the ancient or reptilian part of the brain, the area of the brain which is the major part of the brain of frogs and reptiles. The cerebellum is a balancing organ and serves to coordinate voluntary movements. Behind the cerebellum the part of the brain called the medulla oblongata forms a sort of transition between the spinal cord and the rest of the brain.

Years ago, Professor F. Tilney worked out the relationship between the cerebral hemispheres and the whole brain for a number of animals, including different sorts of Primates. He came up with what was called a forebrain index. Animals, he said, whose front limbs were specialized to turn into wings or pedals or fins had a forebrain index of not more than 60 percent, usually much lower. Animals that had claws or hooves or paws had an index as high as 80 percent for the camel, the horse, and the dog. However, most of the animals in the classification are much below that.

In the Primates, the index was as follows: lemur, 81; *Tarsius,* 80; marmoset, 80; howler monkey, 81.6; baboon, 83; macaque, 84; gibbon, 81; orangutan, 83; chimpanzee, 83; gorilla, 84; and man, 86–89.

The cerebral hemispheres are divided into the frontal lobes; the parietal lobes, which extend from the top partly to the sides; the temporal lobes, which are on the side toward the back; and the occipital lobes, which are to the back and partly underneath the back of the hemispheres. As we will see in a while, certain functions are localized in those regions of the brain.

Anyone who has seen a photograph or observed firsthand a human brain or an ape brain will have noticed that it is a very complexly folded structure. The folds are known as the sulci, and the spaces between them are known as the gyri. The more primitive Primates have brains with very little folding; the more advanced Primates have brains with a great deal more folding.

The brain of the tree shrew, the tupaia, is very smooth. It also has well-defined olfactory bulbs, which are protuberances in front of the cerebral hemispheres. But the olfactory lobes in the tupaia are much smaller than those of other non-Primate animals of the same size, particularly the Insectivores. The neopallium, or cerebral cortex, is expanded toward the back and in the temporal region, which suggests an increase in the visual ability and in the hearing ability. These are increasing at the same time the smell brain is decreasing.

The occipital part, or the hind part, of the cerebral cortex extends so that it covers that part of the brain stem called the midbrain, which is situated between the cerebral hemispheres and cerebellum. This occipital projection not only covers the midbrain, but also covers the cerebellum. There are no sulci, or folds, in the tupaia brain with the exception of the occipital lobe, where there is a small sulcus called calcarine sulcus, marking the front boundary of the visual cortex.

If one cuts sections of the cerebral cortex in the tupaia and examines them under the microscope, this structure can be seen to be much more elaborate and complex than it is in the brains of Insectivores. It is, in fact, comparable to that of the small lemur *Microcebus*. *Microcebus* has better developed frontal and parietal regions, which are also called association areas. Messages from other parts of the brain and the different sense organs come together here and are integrated, and motor action is initiated as a result of this integration. The visual cortex of the tupaia is extremely well developed. The cerebellum, the balancing organ, is relatively simple and generalized, but it is also greatly folded to provide greater surface area. The folding of the cerebellum in the tupaia is more complicated than that of Insectivores of the same size and is actually more complex than the cerebellum of *Microcebus* and also of *Tarsius*. As compared with *Tarsius,* the tupaia is a much more arboreal animal, so balancing is more important to it than it would be to *Tarsius*.

I should mention that the reason for the folding of the surface of the brain is that there can be more cells connecting laterally to each other. The ability of the cerebral cortex to carry on complex activities is related to the number of cells it can contain, and there is only a limited number of cells that can be added from the point of view of depth. The number of cells can then most easily be increased in number by increasing the surface, so that more cells can be squeezed

438

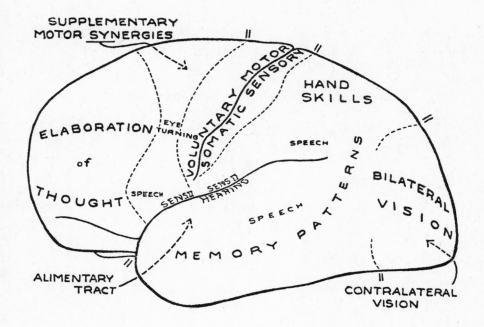

From Wilder Penfield, "Observations on cerebral localization of function," Proceedings of the Fourth International Congress on Neurology, 1949

Representation of the localization of certain functions in the cerebral hemisphere of man.

in per unit of area. The easiest way to squeeze the surface is to fold it; a fold of one inch deep provides two inches' surface area.

Among the lemurs the most primitive brain is found in *Microcebus*, the little mouse lemur from Madagascar. The olfactory lobes in *Microcebus* are smaller than those in the tupaia, but they are still better developed than in the other lemurs, which means that it is a more primitive animal and that smell plays a more important part in its life than it does in the other lemurs. Examination of sections of *Microcebus'* cortex under the microscope shows it to be simple, but there is an additional sulcus, or fold, so that two sulci can now be found in the cerebral cortex of this animal as compared with one in the tupaia. Thus, the complicated folded brain of higher Primates has begun.

The occipital lobes of the hind part of the cerebral cortex of *Microcebus* are large and extend backward at least to the extent that they do in the tupaia, but the frontal lobe is very poorly developed. In

439

the larger lemurs, the folding of the cerebral cortex is much more extensive, and the area of the cortex related to smell is progressively reduced. In lemurs the pattern of the sulci followed by all Primates is already set, but in some of them the pattern may be reduced in some areas. In the brain of *Daubentonia,* the aye-aye, the brain seems to be aberrant. It is different from the typical brain of the lemurs. In this animal the olfactory bulbs are relatively large compared with other lemurs, and the configuration of the sulci is more like that of Carnivores and small hoofed animals than it is like other lemurs or Primates in general.

It is possible to take the skulls of fossils and other Primates and, by pouring plaster or plastic fluid into the skull, obtain an endocranial cast. This shows the shape, size, and folding of the brain. One of the fossil skulls called *Archaeolemur,* which is certainly a Primate in its proportions, originally contained a globular brain which was richly convoluted and had cerebral hemispheres overlapping the cerebellum. The sulci are arranged in the same way as they are in monkeys, so that the brain of this animal has a definite monkeylike appearance. The greatest difference between it and the monkey brain is the poor development of the frontal lobes.

The *Tarsius* brain is a mixture of advanced and primitive characteristics. The cerebral hemispheres are large, and their proportions are rather like those of monkeys. The occipital lobes of the hemispheres project back substantially, so much so that they overlap the cerebellum almost as much as they do in the great apes. The olfactory lobes are correspondingly reduced. About half the cortex is made up of areas involved in the visual sensory processes. The visual cortex not only is very extensive, but also has a very complex structure. Despite the fact that the olfactory lobes are greatly reduced, the part of the cortex associated with smell is fairly extensive. This is a primitive factor not in keeping with the degree of development of other parts of the brain.

Another primitive characteristic in the brain of *Tarsius* is that there is no obvious surface folding. The brain, therefore, lacks the Sylvian fissure found in the brains of all living Primates; this is a surprisingly primitive factor.

The visual area of the galago differs very little from that of non-Primates; however, the beginnings of a primate pattern in the central area of the visual region of galagos can be seen. The galago has well-developed frontal and temporal lobes which compare favorably with *Tarsius,* where these lobes are poorly developed. Neither galago nor *Tarsius* possesses all the brain features that distinguish Primates from other mammals.

In the monkey brains the cerebral cortex is greatly advanced over that of the prosimians. The brains are also bigger. For example, in the case of the marmosets that have a body size similar to that of *Tarsius* and the galago, the brain may weigh about three times as much. On the other hand, the folding of the marmoset brain is probably the most primitive in the true Primates. However, the cerebral hemispheres are very large and have big frontal lobes which are not present in the lemurs, and the occipital lobes go so far back that they overlap not only the cerebellum, but also most of the medulla or the brain stem as well. In the squirrel monkey this particular feature is shown very well. The olfactory bulbs in all these animals are very greatly reduced, and so is the area of the cortex connected with smell.

The cortex is smooth in the marmoset, *Hapale jacchus,* except that there is a deep Sylvian fissure and in the visual cortex there is a calcarine sulcus. It is surprising that the marmoset brain is so smooth; this is undoubtedly a primitive characteristic since a number of lemurs have convoluted brains. The association areas in the cortex of the *Hapale* are, however, greatly expanded compared with the lemurs, and there is a well-developed visual cortex. The marmoset cerebellum is a fairly simple structure; but its lateral lobes are more expanded than the lemurs', and there are additional side lobes called the floccular lobes.

In the remaining New World monkeys and in the Old World monkeys, the cerebral cortex is very richly convoluted. It is surprising that the pattern of the fissures of the sulci in the New World monkeys is so similar to that of the Old World monkeys, considering the many millions of years ago that their ancestors were separated. This means that the common ancestor of these two types of monkeys must have already had a fairly well-developed brain, certainly no more primitive than that of the marmosets and possibly more advanced; and the pattern of the Primate brain had, in fact, already been laid down tens of millions of years ago.

The gibbon brain has a very complex folding, compared with the Old World monkeys, particularly in the parietal lobes where association areas are located. In the larger apes, the chimpanzee and especially in the gorilla, the pattern of convolution is very similar to that of the human brain, although a little more simplified. The parietal region enlarges so much in the human that the visual area is pushed very much to the back of the cortex and more toward the

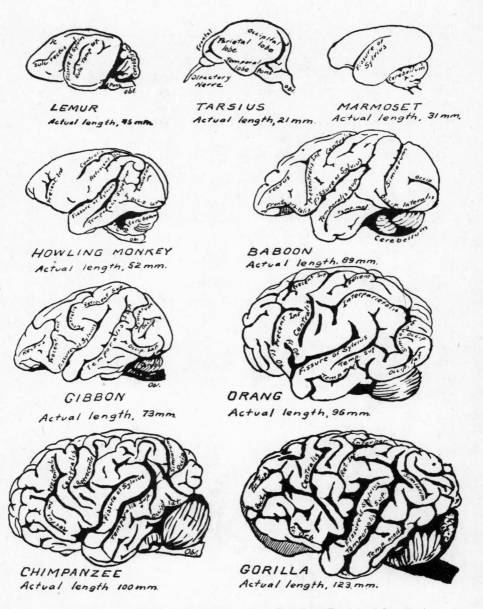

LEMUR
Actual length, 45 mm.

TARSIUS
Actual length, 21 mm.

MARMOSET
Actual length, 31 mm.

HOWLING MONKEY
Actual length, 52 mm.

BABOON
Actual length, 89 mm.

GIBBON
Actual length, 73 mm.

ORANG
Actual length, 96 mm.

CHIMPANZEE
Actual length 100 mm.

GORILLA
Actual length, 123 mm.

From E. Hooton, *Man's Poor Relations*, (New York, Doubleday Doran and Co., 1942)

Comparative sizes in increase of complexity of brain owing to folding, from lemur to gorilla.

center of the cerebral hemispheres. In this area in the monkeys there is a fold called a simian sulcus. Occasional specimens of apes may show this fissure, but in most of them it is a conspicuous difference between the monkey and the ape brain.

It is probable that the modern human brain, compared with that of apes, has a much greater complexity of folding because of the larger size of the brain in humans compared with the body weight.

The intracranial cast of *Proconsul* indicated that the folding pattern was still primitive as compared with that of modern apes. In fact, the pattern of convolutions was more monkeylike than apelike, and there was also evidence that it had a very well-defined floccular lobe on either side of the cerebellum, which again is characteristic of monkeys.

The size of the brain compared with the body weight provides a coefficient of encephalization. The brain of a mouse is smaller than that of a rat because the mouse's body is smaller than that of the rat. However if weight of the mouse's brain is divided by the weight of its body and the result is multiplied by 100 to make the product more than 1, you get a coefficient of encephalization which will in fact show little difference between the mouse and the rat (both around 2 or 3), showing that they are probably within similar areas of brain development. In the human the encephalization index is about three times what it is in the chimpanzee. In lepilemur the encephalization index is 2.4, whereas in man it is 28.8, twelve times higher. This indicates that man has a much greater brain in relation to his body than the lepilemur. The tupaia has an index of 3.0, not that much bigger than lepilemur; the indri has 3.3; and the lorisoids and galagos, about 4; *Tarsius,* 4.5; *Lemur fulva,* about 5.6; the aye-aye has a high index of about 7, which is misleading and probably due to a secondary reduction in body size. The talapoin also has a high index (11.5), higher, in fact, than the chimpanzee, but this is due to a secondary dwarfism in the former. The actual index for the chimpanzee is 11; the orang has one of nearly 9; the gibbon, 9; the spider monkey, 9.5; the cebus, 9.6; the baboon, about 9; and *Ceropithecus,* 8.4. Some of these high figures for these animals do not necessarily mean that they are more intelligent than the great apes; the latter have developed excessively heavy bodies which bring their index down. The significant factors of this index, therefore, are not differences of two or three degrees but the kind of difference one obtains between, say, man and lepilemur or man and chimpanzee or man and orang and so on. Estimates have also been made for the encephalization figures for some of the fossil men, in spite of the fact that it is difficult, of course, to estimate what they weighed, although this can be roughly

443

calculated from the size of the skeleton. *Homo erectus* and *Australopithecus* gave figures intermediate between those of the great apes and man. So encephalization figures do give some information about the evolutionary progress in the Primates.

Modern man's brain varies very extensively in size, from about 900 to 2,500 cubic centimeters. I have personally measured an Australian aboriginal skull which was adult and which had a cranial capacity of only 900 cubic centimeters, but I have also measured an aboriginal skull with a capacity of more than 2,300 cubic centimeters. Raymond Dart, the discoverer of *Australopithecus,* points out that skulls of Hottentots (a primitive people) had a cranial capacity of 2,000 cubic centimeters.

There seems to be no correlation between intellectual ability or genius and the size of the brain. Oliver Cromwell, leader of the revolutionaries during the English Civil War, had a brain of 2,231 cubic centimeters, and Lord Byron, the poet, 2,238 cubic centimeters. Jonathan Swift, who wrote *Gulliver's Travels*, Otto von Bismarck, the famous Prussian politician, and Ivan Turgenev, the Russian novelist, possessed skulls with a cranial capacity of about 2,000 cubic centimeters. On the other hand, equally brilliant individuals, such as the German anatomist Franz Gall, the famous French literateur Anatole France, and statesman Léon Gambetta, had brains with a cranial capacity of only about 1,100 cubic centimeters, a figure which has been quoted for the brains of the Pygmies of Africa. The average European has a cranial capacity around 1,200 to 1,300 cubic centimeters. Apparently even within the low capacity of 900 cubic centimeters, it is possible to contain most of the factors that are important for human intelligence. However, Dart points out that the smallest human brain known is 790 cubic centimeters, which is only 104 cubic centimeters more than the largest known gorilla brain (686 cubic centimeters). But we do not know the intelligence of the owner of this very small human skull.

The ape brain varies between 450 to 690 cubic centimeters, and the human brain is at least another 50 percent higher again. At that level the degree of complexity begins to be limited substantially by the space available in the skull. In the human brain the gray matter which is represented by the cerebral cortex has 50 percent more nerve cells than the chimpanzee cerebral cortex. This means that the fiber connections in the cortex of the human brain are much richer than they are in the chimpanzee.

However, the difference between human and ape brains appears to be primarily a quantitative difference rather than a qualitative. If we

444

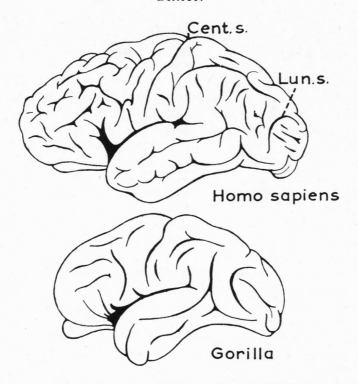

Brain of a gorilla compared with an exceptionally small and primitive brain of *Homo sapiens* (human). Notice greater complexity of folding in human which gives much greater total area of cortex of the cerebral hemispheres.

From W. E. Le Gros Clark, *The Antecedents of Man* (Chicago, Quadrangle Books, 1960)

look into the cranial capacities of some of the fossil skulls, we find that *Pithecanthropus (Homo erectus)* skulls were actually within the lower limit of the human range. In other words, they were somewhere around 1,000 cubic centimeters in cranial capacity, which meant that those creatures could have had human intelligence. Studies of *Homo erectus* in China show that they were clever hunters, that they were very good at the manufacture of stone implements, and that they knew how to make fire and how to use it for cooking. In *Australopithecus* the cranial capacity was not much bigger than a gorilla, but the body was small so that the encephalization coefficient was probably quite high. There is some evidence that the brain of this progenitor of man stayed fairly small until a million or so years ago and then began to expand very rapidly. So for several million years before that, the brain had not increased much beyond that of an ape. However, we should note the

445

large size of the brain (800 cubic centimeters—nearly twice that of an ape) which occupied the skull recently discovered in Africa by Dr. Leakey's son.

Development of the frontal lobes has obviously been important in the brain of Primates, and they are particularly well developed in humans. Up until the 1930's scientists were mystified by these large lobes in the front of the brain, for no particular characteristic had been attributed to them. However, between 1930 and 1937, Dr. Carlysle F. Jacobson and Dr. John F. Fulton, who was a professor of physiology at Yale, made a study of the lobes in two Yerkes' chimpanzees when the center was at Orange Park. The study was begun in 1933 by Dr. Jacobson on two chimpanzees with widely different temperaments. One of these, Lucy, was a six-year-old chimpanzee, very calm and even-tempered. The other, Becky, about four years old, was described as excitable and neurotic and fell into a rage whenever she made any error in a psychological test. Jacobson studied the behavior of these two chimpanzees in considerable detail, and then he and Fulton operated on the animals and cut out about half the frontal area of the brain of each of them. After this, testing was repeated, and the animals reacted pretty much the same as before. Their intelligence and their rapidity in solving intelligence tests did not seem to be affected. The experimenters then did another operation; they removed the remaining half of the frontal part of each animal's brain. In Becky this produced a most remarkable change in personality. Now instead of getting annoyed when she made an error in the psychological tests, she more or less shrugged her shoulders and showed no reaction to her failure or to not getting a reward. She became calm and relaxed.

In 1953, there was an international medical conference in England, and Dr. Jacobson and Dr. Fulton read a scientific paper on this work. It attracted considerable attention. In the audience was a Portuguese neurologist, Egas Moniz, who was director of a mental hospital in Portugal. Dr. Moniz thought that a number of men and women in his hospital who suffered from acute anxiety might be helped by this operation. Together with a surgeon in Lisbon, Dr. Moniz severed the nerve pathways between the prefrontal part of the brain and the rest of the brain in these patients, and in about half of them the intense fears and anxieties were relieved. Dr. Karl Lashley subsequently pointed out that the contribution of this single experiment at the Yerkes Laboratories to the knowledge of the functioning of the brain as well as its therapeutical value was so great that it was worth all the money that had been spent in the laboratory up to that time. So it

446

appeared that the frontal lobes were related to personality, and since that time it has been shown that they are primarily association areas. They process information from the visual, auditory, and somatosensory regions of the cerebral cortex and also have connections with the deeper parts of the brain, for example, the hypothalamus and the important brain structure called the limbic system. And in this way act as areas which integrate the thinking part of the brain with the emotional part. The product of this integration is the personality of the individual.

According to Ashley Montagu:

> The survival value of a higher development of mental capacities in man is obvious. Further, natural selection seemingly favors such a development everywhere. In the ordinary course of events in almost all societies those persons are most likely to be favored who show wisdom, maturity of judgment, and ability to get along with people. These are the qualities of the plastic personality, not a single trait but a general condition, and this is the condition which appears to have been at a premium in practically all human societies.

Primates in the Service of Man

The close relationship between man and his fellow Primates means that man has much to learn from the study of monkeys and apes.

The federal government is well aware of this importance and, through the National Institutes of Health, has established seven Primate Centers in different parts of the United States. One of these is the Yerkes Primate Research Center which is associated with Emory University in Atlanta, Georgia. The center and its scientific program was described in detail in my book *The Ape People.* Cutbacks in federal research funding may seriously hurt these Primate centers in the next few years and may delay for years the cure of many diseases which shorten our lives. It may be that public contributions will be needed to keep these important institutions going.

The principle behind the establishment of Primate centers was that it is much more economical to carry on research with Primates in several major, well-equipped centers than to set up small research facilities in many locations. Not only can the health and stability of the animal colonies be maintained more adequately, but the limited pool of skilled manpower, trained specifically for research with nonhuman Primates, can be utilized more effectively and the prohibitive costs of maintaining and duplicating facilities can be avoided.

447

The Primate centers have already made major contributions in many medical research areas, and the future programs of the centers will include studies of cancer-producing viruses, organ transplantation, nutrition, food additives, infectious disease, cancer, heart disease, geriatrics, mental health, population control, environmental pollution, child health and human development, effects of drugs on the developing embryo, and evaluation of pharmaceuticals. As the need for research in these areas becomes increasingly apparent, it underscores the key role the Primate centers will play in the years ahead.

Man's Future

The cerebral cortex is the latest development in animal evolution. In the lower vertebrates, such as the fish, it scarcely exists at all. But in Primates, as we have seen, it forms the massive convolutions covering the whole of the already-large cerebral hemisphere, and especially is this so in man. It is probable that many parts of the brain are related to the processes of consciousness, but it is also probable that the dominant role in consciousness is played by the cerebral cortex.

With his consciousness and with his expert ability to learn and to manipulate his hands, man has revolutionized the world in which he lives. In this century his technological developments have been immense. I remember my older son once saying to my younger son, "You know, Merfyn, when Dad was born, automobiles had just begun, but there were no airplanes, no movies, no television, no broadcasting." My younger son could scarcely believe it. So in the brief span of my own lifetime, I have seen the development, first, of ordinary aircraft and, second, of the jet plane; I have seen the splitting of the atom, the atom bomb, the development of nuclear power; I have seen the medical revolution brought about by the development of antibiotics.

Medicine itself is now developing into areas beyond antibiotics. Nothing has been more spectacular than heart transplantation. It is unfortunate that the mechanisms of rejection of foreign tissues were not sufficiently understood before it was done. With the surge of investigation in this area, it is, however, only a matter of time before the basic mechanism behind rejection will be understood and techniques for the reversal of this process developed.

We can then think not only in terms of transplantation of the heart, transplantation of the kidneys, and transplantation of the liver, which have already been done, but also in terms of transplantation of glands. Microsurgery of small blood vessels has now developed to the extent that it would be no problem to transplant an adrenal gland. Nor is there any real surgical problem in the transplantation of a complete testicle—only our inability to control the rejection mechanism of the body stops us from doing this.

Whether we will ever be able to transplant the brain is difficult to say. But it is my belief that this will eventually prove possible. One of the problems is that once you have cut through or damaged the spinal cord, as has been done many times accidentally, the unfortunate individual who has suffered this personal disaster (George Wallace is a recent and famous case) is then confined to a wheelchair. If the cord is cut through at higher levels, it can be fatal. The healing that follows the severing of the spinal cord (if the individual lives) is simply a deposition of fibrous tissue which gets in the way of any regrowth of the nervous tissue. In any case, there is very little regrowth of such tissue in the central nervous system. But there is already evidence of chemical substances which stimulate its growth, and it is only a matter of time before this type of material is developed to the extent that we will be able to use it in such cases. It has already been possible for many years to graft nerves and to join cut nerves together. The peripheral nerves are capable of regeneration if one provides a path that leads to the areas that they previously innervated and sprouts of the nerves will pass along those paths. It so happens that nerves contain a number of tubes and that when the nerve is cut, the tubes do not degenerate but serve as pathways for the sprouts which follow down to the muscle or the skin or whatever other part of the body the nerve was serving.

One of the things man is able to do is manipulate his environment. We know he can turn deserts into fertile land and fertile land into deserts; he can join seas by canals, alter the course of rivers, raze hills, flatten mountains, and perform many other marvels. We are already getting indications of possible control of weather. There is evidence that the United States played a part in the modification of weather in Vietnam by stimulating heavy rainfall when it was tactically useful.

Man has built himself extra corporeal brains in the form of computers. Because of the impetus of the space program, this has led to the development of computers to a level which would have been unthought of a few years ago and which certainly would have been unthought of if we had not a space program to do the stimulating. The computers, apart from anything else, form enormous memory banks, not only for individuals, but for society as a whole.

One of the things that man can look forward to in the future is the control of his behavior. At the Yerkes Center and in a number of other laboratories around the country and in other parts of the world, studies are being carried out by implanting electrodes in the brains of apes or monkeys, a procedure which is completely without pain to the animal. We know this because it does not cause pain when it is done, on some occasions, in humans. For example, some epileptics have had electrodes implanted in their brains and have been given a little switch which enables them to stimulate the appropriate part of the brain when they feel an epileptic seizure coming on. Some types of epilepsy respond to this interference with the neural activity which causes epilepsy. The behavior of monkeys and apes that have had these electrodes implanted in them is controlled according to where the electrode is placed. Stimulation of the electrode induces the behavior that part of the brain normally controls.

Dr. Adrian Perachio, at the Yerkes Center, has carried out studies with a male monkey which is dominant, a female, and a subdominant male which is submissive to the dominant animal. The social grouping of these three animals shows that the female associates with the dominant male and that the subdominant male is on the outskirts, is not allowed physical contact with the others, and often has difficulty in getting his share of food. However, if the subdominant male is implanted with an electrode which penetrates the aggressive part of his brain and if this is stimulated electronically, he immediately attacks the dominant male, who then becomes submissive. The female changes her allegiance and sides with the new dominant animal. If the experimenter takes his finger off the button, the process is reversed, and the temporarily dominant animal now becomes the submissive animal again. The female changes her allegiance back to her former mate.

Overaggressiveness in humans may be due to some overdevelopment or overactivity of the aggressive area; it could be coped with in a number of ways, either by surgery or by drugs. It would be easy to pass a little needle down into the part of the brain concerned with aggression and destroy that area wholly or in part, either by heat or by supercooling as in cryosurgery. Cryosurgery is already used to treat a disease known as "the shaking disease" (Parkinsonism). Such a possibility does, in fact, exist, and since we know the location of the various emotional behaviors in the brain, it would be possible to modify behavior by some form of surgery in those parts. This possibility was envisioned in the novel *A Clockwork Orange*, in which the young "hero" of the story permits the surgeon in the prison to operate on him and so reduce his aggressive tendencies but unfortunately finds

450

The author and Dr. Boris Lapin, director of the Russian Primate Center, talk about the future of primatology in the United States and Russia.

that he needs this aggressiveness to survive in the environment which is his normal everyday life.

It would also be possible to dim the aggressiveness of individuals by the use of drugs. Drugs are known which calm individuals and reduce their aggressive or violent tendencies, and these psychodynamic drugs have revolutionized the treatment of many types of mental illness. Earlier in this book I mentioned the studies at Yerkes in which Dr. Jacobson and Dr. Fulton removed the frontal cortex from the brain of a chimpanzee and showed that it completely changed its character so that it became a much more relaxed individual than formerly. This was then used by Dr. Moniz in Portugal to treat a large number of patients in his asylum. This operation, called prefrontal lobotomy, became widely used around the world as a treatment for some types of violence found in mental patients, not only violence to other people but also those who wished to do violence to themselves, even to the point of committing suicide. One of the problems in trying to treat an ordinary person with drugs is that you need to have his voluntary cooperation to take the drugs each day. Some drugs can be absorbed when implanted under the skin; if psychodynamic drugs could be designed to be administered the same way, a year's supply of the drug could be implanted under the skin at one time. One of the interesting things found by Dr. Adrian Perachio was that if the submissive animal was given the electrical treatment which stimulated the aggressive area of his brain often enough, his personality, after some months, began to change, and eventually he became naturally a much more aggressive animal than he ever was before. Gradually, by means of this technique, his personality was being altered. It is possible, therefore, that drugs used for a time or chronic stimulation of certain brain areas might reduce or add to the behavior which is being manipulated.

One of the important characteristics of man's brain is memory. Some animals have long memories; some have very short memories. Memories of places where food is buried are limited even in apes, although they can remember some things for quite an appreciable period of time, but rarely as long or as exactly as man can remember. One of the very important things about the development of the human civilization has been man's ability to memorize and to transmit what he has memorized and learned to other members of his community and to his children so that each child has a mass of information which has been accumulated by previous generations, has been stored, and can be transmitted to him. Man has developed a number of aids to memory, the most important among which are the computers; nevertheless, each individual of each generation still has to go through the labor of learning an enormous number of facts. It is impossible for him to acquire information any other way. There have been gimmicks in learning, such as the devices said to teach people while they are asleep or to pass information subliminally to them. Perhaps one of the most interesting developments which is still in the process of development is the possibility of being able to transmit memory by means of a chemical.

You will remember a little earlier I mentioned the billions of cells that make up the brain, some of them nerve cells and some of them supporting nutrient cells. The cells are the parts of the brain responsible for the phenomenon of memory. How do they store memory? Perhaps memory is not stored in the cells; perhaps it is stored as electric currents, but the cells are important factors in the generation of such currents. If you look up a telephone number, you can usually remember it long enough to dial it. In my case, the moment I have dialed it I have forgotten it. Usually if I have to dial it again, I have to look it up a second time, which is often very irritating.

However, if there is a number that I have to look up many times because I call it a lot, eventually, without conscious effort, it becomes a part of my memory, and when I want to call, I can dial that number without any trouble. There are many things I have said, many things I have read, many people I have met, many names I have heard that I have remembered for reasonably long periods of time. I may meet and be introduced to someone today and perhaps not meet him for a week or a few months later, at which time I will probably remember him. But if I meet him a couple of years later, I may not remember. Yet I have memories from my childhood, some of them dating back to the time when I was only eighteen months old, which is unusual; most people's memories do not go back that far. These early memories are more vivid than the things that happened yesterday.

This suggests that there are really two types of memory. I suggested this at a meeting of the Society for Experimental Biology in London many years ago. Perhaps in one type of memory there are electrical currents which circulate through a series of nerve cells. In other words, one can picture a current going continuously around and around through a series of cells. Isaac Asimov developed this idea in an article which he published recently in *Penthouse* magazine. He points out that it has been calculated that you can accumulate about 10 trillion bits of information a year and that it is possible that each of these bits of information is represented by a particular circuit. For example, if you have a group of cells, say, four cells, the current can pass through these cells in various orders. For example, it can go 1-2-3-4 or 1-3-2-4 or 3-4-1-2, etc. The number of paths that could be produced by ten cells is 3,628,800. If eighteen cells were involved in such a pathway, then there would be 6,400,000,000,000,000, which is a lot of circuits. Perhaps each one of these pathways could represent a single bit of information so that there would be many more pathways, in fact, than would be necessary.

It is, of course, unlikely that most memory that is held for any length of time would simply be restricted to eighteen cells because eighteen cells could be damaged and one could lose one's whole memory in a flash. In actual fact, it is almost certain—and experiments by Dr. Lashley have demonstrated this—that memory is widely distributed over different parts of the brain and there is probably an enormous amount of redundancy in the storage of memory.

A very fleeting memory, however, such as the looking up of a telephone number and immediately forgetting it, may, in fact, by simply a local electrical circuit developing in relatively few cells. Most of these short-term memories, unless they are continuously reinforced, eventually fade away. It would seem that most short-term memories are probably something of an electrical nature. Long-term memories, however, such as those that are strongly fixed in one's lifetime and date back to an early period in life, seem more likely to have a structural basis.

There is some evidence now that the nucleic acid, RNA, of nerve cells can carry memory. The RNA present in the cells controls the synthesis of proteins in the cytoplasm of the cells. Whether memory is actually contained in the pattern of the RNA molecules or is contained in a polypeptide compound related to protein or to a protein itself which is produced by the RNA, one cannot say for certain at the moment. But whether it is a protein or RNA, it has to be associated with a type of mechanism which can result in the liberation of an electrical impulse when required, because the actual retrieval of memory is undoubtedly an electrical process.

The first evidence of the relationship of RNA to memory was reported in a series of articles in a journal known as the *Worm Runner's Digest*. They dealt with a series of experiments carried out on a flatworm called a planarian. This is a flatworm looking a bit like a piece of liver. This animal has a number of characteristics which make it suitable for this type of experimentation. If a worm is cut in two, each half regenerates so that you have two planarians instead of one. Even pieces of a planarian will often develop into new planarians. The worm can also be trained. If you put it in a trough filled with water and flash a light at it, then give it an electric shock, it will contract on the electric shock. It does not take it very long before it will contract whenever the light is flashed without even having the electric shock. It has now been conditioned to the response of the light. The worm can also be taught how to find its way out of a simple maze.

When you cut such a trained worm in two and let each half regenerate, they both have to be retrained again to perform what they were doing before, but they are able to do it in about half the time that an untrained planarian takes. Also, if you take an untrained planarian and chop up a trained planarian and let the untrained animal eat it, he learns to respond to the light much faster than animals that do not eat trained planarians.

The next step was to extract RNA, ribonucleic acid, from planarians that had been trained. It was then injected into untrained worms, and it was found that these, too, were able to learn in half the time it took uninjected planarians to learn.

Subsequently a group of investigators at the University of California published in *Science* the results of research in which they obtained similar findings in rats. They had a group of rats in a cage. Every time a click sounded, food was placed in a food cup. Eventually the rats would go to the food cup every time the click was sounded, irrespective of whether food was put into it or not. So they were conditioned to respond to the click by going to the food cup. A portion of their brains was minced up, and the RNA was extracted from it and was injected into the body cavity of untrained animals. Similarly RNA from the brain of untrained rats was injected into a series of control rats. Both lots had RNA; but in one case the RNA had come from trained animals, and in the other case it had come from untrained animals. The animals that had had the RNA from the trained animals demonstrated that they were able to learn to respond to the click more rapidly than those that had the RNA from the untrained animals.

It is not outside the bounds of possibility that one day it will be possible to take an RNA molecule and actually synthesize onto it the various units for the particular types of memory which are to be transferred.

In the 1950's Holger Hyden, who was working in the University of Gothenberg, stimulated this type of work. In one of his experiments he trained rats to balance on a wire for sixty seconds each day, and then he found that the amount of RNA in brain cells of such trained rats had increased by about 12 percent and was different in composition from those in untrained rats.

Other studies of the transfer of memory with RNA were carried out on goldfish. Then some scientists found that by giving animals a drug which increased the rate at which RNA was produced in brain cells by 30 to 40 percent, it was possible for them to learn faster. There were some studies carried out on this particular drug with humans, but I do not believe that they were very successful. At any rate, it seems possible that in the future man is going to develop some kind of chemical substance which will stimulate his memory. Man is not now waiting for nature to develop his brain further. Man is himself stepping in and altering his own destiny.

Perhaps another area in which this destiny may also be altered, again by drugs, is in the phenomenon of aging. The English biologist Alex Comfort said in a recent article: "However successfully we dodge the misfortunes of life, however cautious, heroic or lucky we may be, the mere passage of time kills us. And before doing so, it impairs us which is worse."

The human life-span has probably not changed much since human history began. In developed countries, more people are given the opportunity now of reaching this predetermined life-span, so that they now die of degenerative diseases such as cardiovascular disease and of the diseases of older people, such as cancer, more than do people in undeveloped countries where fewer get the opportunity to reach old age.

There are a number of environmental factors which age people before their time. Smoking is probably one of them, overeating is another, but all of us have built into us an aging time clock which is genetic in nature and is carried by the DNA (desoxyribonucleic acid) contained in our nuclei. We do not know exactly how this aging factor functions. There have been many views of its nature, but there is not space to discuss them in this final chapter.

Attempts are certainly being made to circumvent this factor or factors. What actually happens to the body when it ages? We all know what the symptoms are—the skin becomes less elastic and more wrinkled; fibrous tissue accumulates in active organs, such as muscles; our reproductive capacities decline or disappear; and many other familiar things happen. It is well known that cells die out progressively in the brain over the years so that the brain of a man of ninety has a significantly smaller number of cells than the brain of a young man of twenty. We still do not know what it is that starts the various cells of the body into the aging process. Attempts to circumvent these changes have been made.

In fact, life can be prolonged now in animals such as rats and mice. Dr. Clive McCay, some years ago, did some fascinating experiments using rats in which he gave young growing rats a diet low in calories but adequate in minerals and vitamins. The number of calories was smaller than the number necessary to permit the young rats to grow. They remained in youthful condition, and Dr. McCay kept them in this condition for a normal lifetime of a rat, which is about three years. At that time they still looked and behaved very youthfully. He then gave them the right number of calories, and the animals immediately became sexually mature because they had not matured sexually up to that time and then went ahead and lived again for the normal lifetime of a rat. So he had doubled their lifetime, but had kept the first half of it in a juvenile condition. This was a spectacular experiment, and I feel sure that there would be a good possibility that we could prolong the lifetime of human beings the same way, at least to some extent, if it were possible to carry out this type of experiment. We plan to do an experiment at the Yerkes Center to try to double the lives of chimpanzees in this way. I feel sure that it will work with them and that it would also work with humans.

A few days after the Japanese surrender in World War II, when I

456

was in Singapore, I was charged with the duty of investigating the degree of malnutrition in the civilian population of Malaya and remedying it. It soon became obvious that malnutrition was widespread, especially among children, and I began to exert pressure on the British military administration to do something about it. However, I had problems from one or two generals who claimed that in driving through Singapore the children they saw in the street looked quite healthy. Our investigation showed, however, that these children were greatly retarded in growth; what looked like a twelve-year-old boy or girl was actually a sixteen-, seventeen-, or eighteen-year-old. They had been stunted by the diet they had received but otherwise looked reasonably healthy despite the reduction in the quality as well as the quantity of food. How long would this reduced diet have kept them in this undeveloped condition and would it have prolonged their lives? In this case it probably would not, because the diet was deficient in vitamins and minerals, as well as calories.

One of the areas which seems to have possibilities is in the use of various drugs to prolong life. One drug called centrophenoxine has been used to reduce the amount of a pigment, called lipofuscin, that accumulates in the brain with age. Lipofuscin is an inactive material, brownish-yellow in color, which accumulates in nerve cells and in heart muscle cells and which probably interferes with their normal functioning. The accumulation of this pigment is one of the cellular symptoms of aging. Centrophenoxine was discovered some years ago. Ten years ago, when I was chairman of anatomy at Emory University, I had an Indian doctor, Dr. Kalidas Nandy, working with me. I asked him to investigate the effect of the drug centrophenoxine on the brains of old guinea pigs. The reason I suggested this was that I was aware of the fact that this drug was being used in Europe to treat humans suffering from a type of mental senility in which their mental processes were affected and they were not able to think clearly. The drug was said to improve this condition.

Dr. Nandy injected the drug in old guinea pigs, and we found that it eliminated the fatty pigment lipofuscin from the nerve cells. More recently, my wife and I have found that the same drug, when fed to a squirrel monkey, will remove the lipofuscin from its brain cells also, but that it does not affect the lipofuscin present in the heart muscles. It looks as though this drug is capable of removing one of the causes of aging, at least in nerve cells.

More recently some studies have been made by Dr. Denham Harman and others in Omaha concerning the role in aging of a process that is known as peroxidation. This particular type of oxidation in cells is said to occur as a result of the presence of what are known as free radicals and to be responsible for the aging process. A free radical is a chemical unit which, according to Dr. Alex Comfort, has been likened to a convention delegate away from his wife: "It is a highly reactive chemical agent that will combine with anything suitable that is around." It is believed that the free radicals cause peroxidation in the tissues and that it is this peroxidation which causes the aging of the chemical molecules in the cells.

There are certain reducing substances in the food. One of them is vitamin E. Dr. Harman has shown that, at least in mice, the use of vitamin E and vitamin E-like chemicals has been able to prolong the life of mice. Whether this would work in human beings, there is no way of saying. We have no idea of the level of vitamin E that would be required for this purpose.

Despite the fancy claims one reads in the papers from time to time of the "important" doctors of clinics who claim they can rejuvenate humans, particularly wealthy humans, we should realize that there is no such person or clinic in existence at this moment that can genuinely rejuvenate or prolong life significantly in human beings. You can retard the aging process and possibly death for a certain period of time by eating a sensible diet and carrying out other sensible health precautions, such as regular exercise, but the genetic programming in your cells moves you inexorably toward the end of your allotted span.

Dr. Comfort has given what he describes as a position paper on the practical side of age slowing in animals. In 1971, he said that it would run roughly like this:

> 1) We now have perhaps half a dozen ways of slowing down aging or lengthening life or both in rats and mice. 2) The exact way these methods fit together and the nature of the aging clock and whether there is one clock or more are unanswered questions, but we should be close to an answer within five years. 3) It is not certain whether any of the known age slowing methods would work in man. 4) Whether they would and whether they would work in adult life can be found only by trying them. 5) If they don't, then it's likely that similar and equally effective methods will. 6) Human experiments will be started within the next three to five years, probably at more than one center.

So man has indeed reached the point of being very close to prolonging healthy life for a significant period of time. It is difficult to say just when this can occur. It could happen quite suddenly if the right drug were discovered, but there is no doubt that this brain, this product of millions of years of evolution, has reached the stage when it is probably going to be able to affect or to control its own life-span.

There has been a recent fad of deepfreezing the bodies of people who have died, some of them from incurable diseases, with the claim that they can be unfrozen many years later, brought back to life, and their disease cured. It is a pious hope. Cells and tissues of animals can be deepfrozen and kept for long periods of time, be thawed, and come back to life again. Sperm banks are now well known. Even very small whole animals or embryos can be frozen and subsequently brought back to life. Regularly, in our laboratories at the Yerkes Center, we deepfreeze pieces of brain and other tissues at the temperature of solid carbon dioxide and keep them for long periods of time at a temperature of $-120°$ C. The enzymes in these tissues, which make the life of the cell possible, remain intact and active, and when the tissue is thawed, the enzymes can be made to work as they did when the tissue was freshly removed from the animal. However, if the pieces of tissue which are frozen are not very small, the tissue freezes too slowly, and ice crystals form in the cells and destroy them. How then is it possible to freeze such a large structure as the human body quickly enough to prevent this happening, even with the use of fluids which reduce ice crystal formation? This is the real hurdle which makes revival after freezing little more than wishful thinking. This is not to say that it will not be possible in the future, but it is a future that is still a long way off.

Man has already reached the stage of being able to control his own reproduction by the means of drugs. The development of the contraceptive pill has revolutionized contraception and at the same time revolutionized our whole attitude toward sex, particularly the attitude of women toward sex. This is producing a social revolution of the first magnitude.

If the contraceptive pill could be applied to the control of populations throughout the world, it would be of even greater importance. At the moment man, for all his big brain, is headed on a downward path because of his inability to control the rate of increase in population. It is being controlled in some of the more developed countries and can be controlled in others if the right people want to do it. It has been controlled very effectively in Japan, where the population level has now remained stationary for some years. Those individuals who, for religious or other reasons, refuse to take this threat of overpopulation seriously and to take the necessary action now are a menace to the world and to the future of mankind.

Pollution and the destruction of the ecology are due partly to the increased population which is in itself a pollution and partly to technologies that have developed without any thought or consideration of their effect on the environment around them. Man has the brain to solve problems like the destruction of the environment and overpopulation, has the brain to solve the fuel crisis. All he lacks, apparently, is the will do to it. When I say that man has the ability to do it, I mean certain members of the community have the will to do it; there are others who do not have the ability to understand the direction in which we are going, others who do not care, and others who do not want to. With his millions of years in evolution, man can solve pretty well all the problems in his environment if he wishes to.

Will man's brain develop further? The time that is required for significant evolutionary change to develop is so great that this is not something that we can look forward to in the near future. Yet any country which sets up a policy of seeking out its most intelligent children, pushing them to their capacity in learning, and seeing that they subsequently get full opportunity to use their brains and their education would reproduce their kind and, in a generation or two, take over the world.

One area in which we may perhaps see a further development is that of extrasensory perception. Does it exist? Some experiments have shown that from the statistical point of view, extrasensory perception can occur. There are occasions on which I feel that I myself have had extrasensory perception, and yet in all cases I feel that the same perception could have been explained in other ways.

In the Soviet Union there seems to be a revival of interest in ESP. The Russians have a history of being interested in strange rays or impulses. For example, in the early 1930's a book was published in the Soviet Union on *mitogenetic rays*. These were rays supposed to be emitted by dividing cells which would stimulate nearby cells to divide; for example, if a number of seeds were placed close together, the seedlings would all grow faster than if the seeds were planted some distance apart. It was not long before other odd rays were "discovered." The fingertips were supposed to produce rays which would stimulate the growth of a culture of yeast cells if the fingers were held near their container. Washings from the skin of menstruating women also gave off radiation, and even bodies, when they died, gave off a burst of radiation—the so-called necrobiotic rays—perhaps this was the "soul" leaving the body. No one, however, was able, using the most sophisticated physical equipment for detecting various sorts of rays, to find any signs of any of them, so if they exist, they cannot be

460

detected by methods known to physical science. The parallel with ESP is close. It is impossible to conceive from known physical and biological laws a mechanism which would permit ESP to take place. One Soviet scientist has suggested that ESP may be transmitted by some time-energy system rather than by electromagnetic waves such as radio.

Extrasensory perception is also called psi or parapsychology, and it has four different branches: (1) telepathy, or "mind reading"; (2) clairvoyance, the ability to know that an event has taken place or what is in a locked box or a locked room, what cards your opponent in a card game holds, and so on; (3) precognition, the ability to forecast future events, and (4) psychokinesis, the ability to control or move objects by "willing" them to—making the little ball fall into the right slot in a roulette table is an example.

Much of the existing evidence for ESP is not evidence at all; it is hearsay, coincidence, imagination, or fraud. Among all the confusion produced, however, there exist some experiments which indicate that some degree of extrasensory perception may occur, and the Parapsychological Society, founded in 1957 to study ESP phenomena, was admitted to the American Association for the Advancement of Science in 1969. Among the experimenters who helped make this possible were Douglas Dean, who used a finger plethysmograph (which measures volume changes in a finger caused by changes in blood volume in the finger). Emotions can cause a change in the blood volume of a finger (and other organs, of course), and investigators have been able to cause such a change in one person when somebody else in another room or some distance away thought of something emotionally important to him. Probably one of the most famous names in the realm of parapsychology is Dr. J. B. Rhine, until recently director of the laboratory for parapsychology at Duke University. He used a series of cards with various symbols on them, and individuals had to guess what symbol would be on the card that had been turned up even though they could not see the card or the person turning it. Some individuals were found to anticipate this at a significantly higher level than chance. However, I remember some years ago attending a lecture by Professor Rhine in which he said the most important thing that came out of his experiments was the discovery that some people, while their guessing was consistently incorrect for the symbol on the card that had been turned up, were correct for the next card that was to be turned up, so they were actually predicting a future event (precognition).

The U.S. Air Force, using a mechanical device, was, however, unable to produce any evidence for ESP in its subjects; but other workers have indicated that some people are very much more likely to have this ability than others, and one might suspect that the subjects of the ESP studies used by the Air Force would not be the likeliest candidates for the possession of this ability. One of the Apollo astronauts tried to send ESP messages back to earth while he was on the moon. I understand that he was unsuccessful.

My own feeling about ESP is that it may occur, but I have not myself seen enough evidence yet to convince me that it exists. The fact that modern physics does not throw any light upon the mechanism of transmission of ESP need not inhibit us in considering the possibility of its existence. Twenty years ago I would not have believed that a ray of ruby-colored light could be used to burn a hole through a half dollar as a laser beam can. Some future discovery may shed light on this elusive subject, and it is something that may be a factor in man's further evolution.

There is no doubt that the most spectacular and striking advance that man has made, which really shows the degree of technical proficiency to which his brain has brought him, is the Apollo space program, recently brought to such a magnificent and spectacular end by NASA. Very few achievements of man are worthy of greater admiration than this particular feat. If I may predict the future, I say that this is probably the area in which man is most likely to develop. He will develop his ability to move around the universe, and, with luck, he will develop it in time to enable nuclei of humans to colonize other planets before they completely eliminate themselves from the earth. I do not feel optimistic that they will not ultimately do this despite the fact that man's brain can be very logical and constructive at times. It has become essential to start our move to other planets, but there are many among us whose feet are still immersed in the primordial slime from which we all emerged and from which they themselves will never be able to climb. It is our misfortune that they will try, and in fact are trying now, to drag back into the slime those of us who have seen the stars and want to go there, for that is where the future of the Primate odyssey lies.

462

Bibliography

THE FOLLOWING books and articles have been referred to or quoted from in the compilation of this book.

ADLER, IRVING, *The Hoaxes of Science*. New York, Collier Books, 1962.

ALLEN, J. A., "Primates of the Congo." *Bulletin of the American Museum of Natural History,* Vol. 47 (1925), p. 283.

Animal Behavior. Washington, D.C., National Geographic Society, 1972.

ARDREY, ROBERT, *African Genesis*. New York, Dell Publishing Company, 1961.

AUGSPURG, GUS, and AUGSPURG, CASEY, *Monkey Business*. Jersey City, T. F. H. Publications, 1964.

BENCHLEY, BELLE, *My Life in a Man-Made Jungle*. Boston, Little Brown and Company, 1940.

BERGWIN, C. R., and Coleman, W. T., *Animal Astronauts*. Englewood Cliffs, New Jersey, Prentice-Hall, Inc., 1963.

BERTRAND, M., "The Behavioral Repertoire of the Stumptail Macaque." *Bibliotheca Primatologica*, No. 11. Basel, Switzerland, S. Karger, 1969.

BOLWIG, N., "Facial Expression in Primates," in *Behavior*, Vol. 22 (1962).

BOULENGER, E. G., *Apes and Monkeys*. New York, Robert M. McBride and Company, 1930.

BOURNE, G. H., *The Ape People*. New York, G. P. Putnam, 1971.

———, *The Chimpanzee*. Basel, Switzerland, S. Karger, 1971. 6 vols.

BUETTNER-JANUSCH, JOHN, *Evolutionary and Genetic Biology of Primates*, vols. 1 and 2. New York and London, Academic Press, 1963.

———, and HILL, ROBERT L., "Molecules and Monkeys." *Science*, Vol. 147 (1965), p. 836.

BURBRIDGE, BEN, *Gorilla*. New York, The Century Company, 1928.

BURT, OLIVE, *Space Monkey*. New York, John Day Company, 1960.

BURTON, MAURICE, "The World of Science. Gaily Dressed Monkeys." *Illustrated London News* (June 9, 1962), p. 940.

BUTLER, ROBERT A., "Curiosity in Monkeys." *Scientific American* (February, 1954), p. 3.

CARPENTER, C. R., *The Naturalistic Behavior of Non-Human Primates*. University Park, Pennsylvania, Pennsylvania State University Press, 1964.

CHARNEY, MICHAEL, and McCRACKEN, R. D., "Intestinal Lactose Deficiency in Adult Nonhuman Primates." *Social Biology*, Vol. 18 (1972), p. 416.

CHIARELLI, A. B., *Comparative Genetics in Monkeys, Apes and Man*. New York and London, Academic Press, 1971.

———, *Taxonomic Atlas of Living Primates*. New York and London, Academic Press, 1972.

———, *Taxonomy and Phylogeny of Old World Primates*. Torino, Italy, Rosenberg and Sellier, 1968.

CICALA, GEORGE A., *Animal Drives*. Princeton, New Jersey, Van Nostrand, 1965.

CLARK, HOWELL F., *Early Man*. New York, Time-Life Books, 1968.

CLARK, W. E. LE GROS, *The Antecedents of Man*. Chicago, Quadrangle Books, 1960.

———, *History of the Primates*. Chicago, University of Chicago Press, 1961.

———, *Man-Apes or Ape Men*. New York, Holt, Rinehart and Winston, 1967.

COMFORT, ALEX, The New Scientist 1972. *Playboy* (1972).

CONROY, GLENN C., and FLEAGLE, J. G., "Locomotor Behavior in Living and Fossil Pongids." *Nature*, Vol. 137 (1972), p. 103.

CRAIG, HUGH, *The Animal Kingdom*. New York, 1897.

CURTIS, R. F.; BALLANTYNE, J. A.; KEVERNE, E. B.; BONSALL, R. W.; and MICHAEL, RICHARD P., "Identification of Primate Sexual Pheromones and the Properties of Synthetic Attractants." *Nature*, Vol. 232 (1971), p. 396.

DAY, MICHAEL H., *Fossil Man*. New York, Grosset and Dunlap, Inc., 1970.

DELGADO, J. M. R., *Physical Control of the Mind*. New York, Harper and Row, 1969.

464

DENENBERG, VICTOR H., *Readings in the Development of Behavior*. Stamford, Connecticut, Sinauer Associates, 1972.

DE VORE, I., *Primate Behavior*. New York, Holt, Rinehart and Winston, 1965.

DU CHAILLU, PAUL, *Africa*. London, John Murray, 1861.

———, *Stories of the Gorilla Country*. New York, Harper and Brothers, 1868.

DURO, EDWARD H., "Notes on a Lemur Catta with Epileptic Seizures." *Primate News*, Vol. 9 (1971), p. 4.

EIBL-EIBESFELDT, I., *Love and Hate*. New York, Holt, Rinehart and Winston, 1972.

EIMERL, SAREL, and DE VORE, IRVEN, *The Primates*. New York, Time-Life Books, 1965.

EISENBERG, J. F., and DILLON, WILTON, S., *Man and Beast, Comparative Social Behavior*. Washington, D.C., Smithsonian Institution Press, 1971.

FAST, JULIUS, *Body Language*. New York, M. Evans and Company, 1970.

FELCE, WINIFRED, *Apes*. London, Chapman and Hall, 1948.

FERSTER, CHARLES B., "Arithmetic Behavior in Chimpanzees." *Scientific American* (May, 1964), p. 2.

FITZSIMMONS, F. W., *Natural History of South Africa*. London, Longmans Green, 1919.

FOODEN, JACK, "Color and Sex in Gibbons." *Bulletin of the Field Museum of Natural History*, Vol. 42 (1971), p. 2.

FORBES, H. O., *Monkeys*, Vols. 1 and 2. London, Edward Lloyd, 1896.

FOSSEY, DIAN, and CAMPBELL, ROBERT M., "Making Friends with Mountain Gorillas." *National Geographic*, Vol. 137 (1970), p. 48.

———, "More Years with the Mountain Gorillas." Vol. 140 (1971), p. 574.

FUCHS, L. H., *Family Matters*. New York, Random House, 1972.

GALLOP, GORDON G., "Chimpanzees: Self-Recognition." *Science*, Vol. 167 (1970), p. 86.

———, "Minds and Mirrors." *New Society* (November, 1971), p. 976.

GARDNER, BEATRICE T., and GARDNER, R. ALLEN, "Two Way Communication with an Infant Chimpanzee," in *Behavior of Non-Human Primates*, Vol. 4, A. Schrier and F. Stollnitz, eds. New York, Academic Press, 1971.

GARNER, R. L., *Apes and Monkeys*. New York, Atheneum Press, 1900.

GARTLAN, J. S., and BRAIN, C. K., "Ecology and Social Variability in Cercopithecus aethiops and C. nietis," *Primates*. Phyllis C. Jay, ed. New York, Holt, Rinehart and Winston, 1968.

GATTI, ATTILIO, *The King of the Gorillas*. New York, Doubleday, Doran and Company, 1932.

GAZANIGA, MICHAEL S., "One Brain, Two Minds." *American Scientist*, Vol. 60 (1972), p. 331.

GROVES, COLIN P., *Gorillas*. New York, Arco Publishing Company, 1970.

HADDON, A. C., *Races of Man*. London, Milner and Company, 1925.

HAHN, EMILY., *On the Side of the Apes*. New York, Thomas Y. Crowell, 1971.

465

HAINES, D. E., and SWINDLER, D. R., "Comparative Neuroanatomical Evidence and the Taxonomy of the Tree Shrews." *Journal of Human Evolution*, Vol. 1 (1972), p. 407.

HALL, EDWARD, and HALL, MILDRED, "The Sounds of Silence." *Playboy* (1973), p. 138.

HALL, K. R. L., "Behavior and Ecology of the Wild Patas Monkey." *Primates*, Phyllis C. Jay, ed. New York, Holt, Rinehart and Winston, 1968.

HARDY, ALISTER, "Was Man More Aquatic in the Past?" *New Scientist* (March 17, 1960), p. 642.

HARRISSON, BARBARA, "Getting to Know About Tarsius." *Malayan Nature Journal*, Vol. 16 (1962), p. 198.

———, *Orang Utan*. New York, Doubleday and Company, 1963.

HARTMANN, R., *Anthropoid Apes*. London, Kegan Paul, French and Company, 1885.

HASS, HANS, *The Human Animal*. New York, G. P. Putnam's, 1970.

HEDIGER, H., *Man and Animal in the Zoo*. New York, Delacorte Press, 1969.

HILL, CLYDE A., "The Last of the Golden Marmosets." *Zoonooz*, Vol. 43 (1970), p. 12.

HILL, W. C. OSMAN, *Primates*, Vols. I, II, III, VIII. Edinburgh, Scotland, Edinburgh University Press, 1957.

HOOTON, ERNEST, *Man's Poor Relations*. New York, Doubleday, Doran and Company, 1942.

———, *Up from the Apes*. New York, Macmillan Company, 1947.

———, *Why Men Behave Like Apes and Vice Versa*. Princeton, New Jersey, Princeton University Press, 1940.

HORNADAY, W. T., *Two Years in the Jungle*. New York, Charles Scribner's Sons, 1929.

HORR, DAVID A., "The Borneo Orang Utan." Paper presented at the meetings of the American Association of Physical Anthropologists, April, 1972.

HOWARD, C. F., "Spontaneous Diabetes in the Colony of Celebes Apes." *Primate News*, Vol. 10 (1972), p. 2.

HOWELL, F. CLARK, and BOURLIERE, F., *African Ecology and Human Evolution*. Chicago, Aldine Publishing Company, 1963.

IRVINE, WILLIAM, *Apes, Angels and Victorians*. London, Weidenfeld and Nicolson, 1956.

JANSON, H. W., *Apes and Ape Lore*. London, The Warburg Institute, 1952.

JOLLY, ALLISON, *Lemur Behavior*. Chicago, University of Chicago Press, 1966.

KAFKA, FRANZ, "Ein Bericht für eine Akademie," in *Gesammelte Schriften*, Max Brod, ed. Berlin, 1935.

KAY, HELEN, *Apes*. New York, Macmillan Company, 1970.

KEARTON, CHERRY, *The Happy Chimpanzee*. New York, Dodd, Mead and Company, 1928.

———, *My Friend, Toto*. London, Arrowsmith, 1925.

KENT, LIONEL, "The Animal Torturers." *Male*, Vol. 22 (1972), p. 36.

KINGDON, J., *East African Mammals,* Vol. 1. London and New York, Academic Press, 1971.

KLOPFER, PETER H., "Mother Love: What Turns It On." *American Scientist,* Vol. 59 (1971), p. 404.

KORTLANDT, Adrian, *New Perspectives on Ape and Human Evolution.* Amsterdam, Stichting voor Psychobiologie, 1972.

KRAUS, R. F., "Implications of Recent Developments in Primate Research for Psychiatry." *Comprehensive Psychiatry,* Vol. 11 (1970), p. 328.

KURTH, G., *Evolution and Hominisation.* Stuttgart, Germany, Gustav Fischer, 1968.

LANG, ERNST M. "The Care and Breeding of Anthropoids," in *Some Recent Developments in Comparative Medicine.* New York, Academic Press, 1966.

————, *Goma, the Gorilla Baby.* New York, Doubleday and Company, 1963.

————, "Jambo, the Second Gorilla Born at Basel Zoo." *International Zoo Yearbook,* Vol. 3 (1962), p. 84.

————; SCHENKEL, R.; and SIEGRIST, E., *Gorilla Mutter und Kind.* Basel, Switzerland, Basilius Press, 1965.

LANG, K., *Die Grussitten,* Vienna, Volkerlade, 1926.

LAPIN, BORIS, and FREEDMAN, E., *Monkeys for Science.* Moscow, Novasti Press Agency, 1969.

LASKER, G. W., *The Evolution of Man.* New York, Holt, Rinehart and Winston, 1961.

LEWIS, O. J., "Brachiation and the Early Evolution of the Hominoidea." *Nature,* Vol. 130 (1971), p. 577.

LIEBERMAN, P.; HARRIS, K. S.; WOLF, P.; and RUSSELL, L. H., "Newborn Infant Cry and Non Human Primate Vocalization." *Journal of Speech and Hearing Research,* Vol. 14 (1971), p. 718.

————, and Crelin, Edward S., "On Speech of Neanderthal Man." *Linguistic Inquiry,* Vol. 2 (1971), p. 203.

————; CRELIN, EDWARD S.; and KLATT, D. H., "Phonetic Ability and Related Anatomy of the Newborn and Adult Human, Neanderthal Man and the Chimpanzee." *American Anthropologist,* Vol. 74 (1972), p. 287.

————; KLATT, DENNIS H.; and WILSON, WILLIAM H., "Vocal Tract Limitations on the Vowel Repertoires of Rhesus Monkey and Other Non-Human Primates." *Science,* Vol. 164 (1969), p. 1185.

McDERMOTT, W. C., *The Ape in Antiquity.* Baltimore, Johns Hopkins Press, 1938.

MacDONALD, JULIE, *Almost Human.* Philadelphia, Chilton Books, 1965.

MacROBERTS, BARBARA R., and MacROBERTS, MICHAEL H., "The Apes of Gibraltar." *Natural History,* Vol. 80 (1971), p. 38.

MASON, WILLIAM A., "Scope and Potential of Primate Research," in *Science and Psychoanalysis,* Vol. 12, Jules H. Masserman, ed. New York, Grune & Stratton, 1968.

MENZEL, E. W., "Communication About the Environment in a Group of Young Chimpanzees," *Folia Primatologia*, Vol. 17 (1972), p. 220.

———, "Spontaneous Invention of Ladders in a Group of Young Chimpanzees." *Folia Primatologia*, Vol. 17 (1972), p. 87.

MONTAGU, M. F. ASHLEY, "Meet Tarsius." *The Technology Review*, Vol. 46 (1944), p. 1.

MORGAN, ELAINE, *The Descent of Woman*. New York, Stein and Day, 1972.

MORRIS, DESMOND, *Intimate Behavior*. New York, Random House, 1971.

———, *The Naked Ape*. New York, McGraw-Hill, 1967.

MORRIS, R., and MORRIS, D., *Man and Apes*. New York, McGraw-Hill, 1966.

NAUTA, W. J. H., "The Problem of the Frontal Lobe." *Journal of Psychiatric Research*, Vol. 8 (1967), p. 197.

NAPIER, J. R., and NAPIER, P. H., *A Handbook of Living Primates*. London and New York, Academic Press, 1967.

———, *Old World Monkeys*. New York and London, Academic Press, 1970.

NAPIER, JOHN, *The Roots of Mankind*. Washington, D.C., Smithsonian Institution Press, 1970.

NESTURKH, M., *The Origin of Man*. Academy of Sciences of the USSR, 1958.

OGLE, NATHANIEL, *Memoirs of Monkeys*. London, George B. Whittaker, 1825.

PENFIELD, W., "Observations on Cerebral Localization of Function." *Fourth International Congress of Neurology*, Vol. III, p. 27.

Penthouse (August, 1973).

PERRY, JOHN, "In Danger, Golden Lion Marmosets." *Smithsonian* (December, 1972), p. 49.

PIETSCH, PAUL, "Shuffle Brain." *Harper's Magazine* (May, 1972), p. 48.

PILBEAM, D., *The Ascent of Man*. New York, Macmillan Company, 1972.

POIRIER, F. E., *Primate Socialization*. New York, Random House, 1972.

PREMACK, ANN J., and PREMACK, D., "Teaching Language to an Ape." *Scientific American*, Vol. 227 (1972), p. 92.

PREMACK, DAVID, "Language in Chimpanzee." *Science*, Vol. 172 (1971), p. 808.

PREUSCHOFT, H., "Body Posture and Mode of Locomotion in Early Pleistocene Hominids." *Folia Primatologia*, Vol. 14 (1971), p. 209.

RAHAMAN, H., and PARTHASARARTHY, M. D., "Studies on the Social Behavior of Monkeys." *Primates*, Vol. 10 (1969), p. 149. Published by the Japan Monkey Center.

REYNOLDS, VERNON, *The Apes*. New York, E. P. Dutton and Company, 1967.

ROBINSON, J. T., *Early Hominid Posture to Locomotion*. Chicago, University of Chicago Press, 1972.

———; FREEDMAN, L.; and SIGMON, B. A., "Some Aspects of Pongid and Hominid Bipedality." *Journal of Human Evolution*, Vol. 1 (1972), p. 361.

ROLAND, CHARLES E., "Tyson's Anatomy of a Pygmie, 1699." *Mayo Clinic Proceedings*, Vol. 46 (1971), p. 561.

ROSENBLUM, L. A., *Primate Behavior*, Vols. 1 and 2. New York, Academic Press, Inc., 1971.

RUMBAUGH, D., *Gibbon and Siamang*, Vol. 1. Basel, Switzerland, S. Karger, 1972.

SACKETT, GENE P., Exploratory behavior of rhesus monkeys as a function of rearing experience and sex. *Developmental Psychology*, Vol. 6, p. 260, 1972.

SANDERSON, IVAN, *The Monkey Kingdom*. New York, Hanover House, 1957.

SASSELLA, DANIELA, "Some Aspects of the Regulation of the Circulating Blood Volume." *Rassegna Medica*, Vol. 49 (1972), p. 34.

SCHALLER, GEORGE B., *The Mountain Gorilla*. Chicago, University of Chicago Press, 1965.

SCHRIER, ALLAN M., and STOLLNITZ, FRED, *Behavior of Non Human Primates*, Vols. 1–4. New York, Academic Press, Inc., 1971.

SCHULTZ, ADOLPH H., *The Life of Primates*. London, Weidenfeld and Nicolson, 1969.

SEBEOK, T. A., "Animal Communication." *Science*, Vol. 147 (1965), p. 1006.

SIEGEL, JULES, "Sixth Sense." *Playboy* (1972).

SIGMON, B. A., "Bipedal Behavior and the Emergence of Erect Posture in Man." *American Journal of Physical Anthropology*, Vol. 34 (1971), p. 55.

SIMON, N., and GEROUDET, P., *Last Survivors*. New York, World Publishing Company, 1970.

SKINNER, B. F., *Beyond Freedom and Dignity*. New York, Alfred A. Knopf, 1972.

SMITH, C. U. M., *The Brain*. New York, G. P. Putnam's Sons, 1970.

SOUTHWICK, O. H., and CADIGAN, E. C., "Population Studies of Malaysian Primates." *Primates*, Vol. 13 (1972), p. 1.

STRUGHOLD, H., *Your Body Clock*. New York, Charles Scribner's Sons, 1971.

STRUHSAKER, T. T., and HUNKELER, P., *Folia Primatologia*, Vol. 15 (1971), p. 212.

SUOMI, S., and HARLOW, H. F., "Social Rehabilitation of Isolate Reared Monkeys." *Developmental Psychology*, Vol. 6 (1972), p. 487.

———; ———; and KIMBALL, S. D., "Behavioral Effects of Prolonged Partial Social Isolation in the Rhesus Monkey." *Psychological Reports*, Vol. 29 (1971), p. 1171.

———; ———; and McKINNEY, W. T., "Monkey Psychiatrists." *American Journal of Psychiatry*, Vol. 128 (1972), p. 927.

SUZUKI, A., "Carnivority and Cannibalism Observed Among Forest Living Chimpanzees." *Journal of the Anthropological Society of Nippon*, Vol. 79 (1971), p. 30.

TELEKI, Geza, "The Omnivorous Chimpanzee." *Scientific American*, Vol. 220 (1973), p. 31.

THORINGTON, R. W., "Importation, Breeding, and Mortality of New World Primates in the United States." *International Zoo Yearbook*, Vol. 12, 1971.

TRUEPENNY, CHARLOTTE, *Dear Monkey*. London, Victor Gollancz Ltd., 1963.

TUTTLE, R., *The Functional and Evolutionary Biology of Primates*. Chicago, New York, Aldine/Atherton, 1972.

VAN HOOFF, J. A. R. A. M., "A Comparative Approach to the Phylogeny of Laughing and Smiling, in Non-Verbal Communication, R. A. Hinde, ed. Cambridge University Press, 1971.

VAN LAWICK-GOODALL, J., *In the Shadow of Man*. Boston, Houghton Mifflin Company, 1971.

————, *My Friends, the Wild Chimpanzees*. Washington, D.C., National Geographic Society, 1969.

VON GULICK, R. H., *The Gibbon in China*. Leyden, Holland, E. J. Brill, 1967.

WALKER, ERNEST P., *Mammals of the World*, Vol. 7. Baltimore, Johns Hopkins Press, 1966.

WASHBURN, S., and DE VORE, I., "The Social Life of Baboons." *Scientific American*, Vol. 204 (1961), p. 62.

————, *The Monkey Book*. New York, Macmillan Company, 1954.

WATSON, PETER, "In Search of Memory." *Illustrated London News* (November, 1971).

WELLS, L. H., "Africa and the Ancestry of Man." *Die Suid-Afrikanse Tydskrif von Wetenshap* (April, 1971), p. 278.

WILLIAMS, LEONARD, *The Dancing Chimpanzee*. London, Andre Deutsch, 1967.

————, *Man and Monkey*. Philadelphia, J. B. Lippincott, 1968.

WILSON, J. R., *The Mind*. New York, Time-Life Books, 1964.

WINDLE, B. C. A., *Tyson's Pygmies of the Ancients*. London, David Nutt, 1894.

WOOD JONES, F., *Arboreal Man*. London, Edward Arnold, 1926.

————, *Man's Place Among the Mammals*. London, Edward Arnold, 1929.

————, and PORTEUS, S. D., *The Matrix of the Mind*. London, Edward Arnold, 1929.

YERKES, R. M., *The Mind of a Gorilla*. (Genetic Psychology Monographs.) Worcester, Massachusetts, Clark University, 1927.

————, and LEARNED, B. M., *Chimpanzee Intelligence*. Baltimore, Williams and Wilkins, 1925.

————, and YERKES, A. W., *The Great Apes*. New Haven, Connecticut, Yale University Press, 1929.

ZUCKERMAN, S., *The Social Life of Monkeys and Apes*. London, Routledge and Kegan Paul, 1932.

————, *Functional Affinities of Man, Monkeys and Apes*. New York, Harcourt, Brace and Company, 1933.

Index

Abominable Snowman, 354 n.
Aelian, 164
Aepyornis, 27
Africanus, Leo, 203
After Many a Summer Dies the Swan (Huxley), 175
Aggressive behavior, 70, 71; chimpanzees, 305; gibbons, 268, 276; Japanese macaques, 139–40; lemurs, 34; lion-tailed macaques, 162; man, 450–51; pygmy chimpanzees, 344; rhesus monkeys, 110, 127; spider monkeys, 101–05; tree shrews, 19
Agile gibbons (*Hylobates agilis*), 258, 262
Agile mangabeys, 167
Aging of man, phenomenon of, 455–59
Air Force, U.S., 462
Akeley, Carl, 398, 400
Alabama Space and Rocket Center, 86
Albert National Park, 400
Allen's swamp monkeys (*Allenopithecus nigroviridis*), 107–08, 166, 200, 209
Almost Human (MacDonald), 178
Amenhotep II, 202
American Association for the Advancement of Science, 461
American Museum of Natural History, 398, 400
Amino acids, 253
Anagale, 18
Angwantibos, 49, 58
Animal Kingdom, The (Craig), 53
Ape and Essence (Huxley), 176
Ape People, The (Bourne), 30, 87, 130, 141, 143, 151, 300, 346, 376, 378, 384, 413, 447
Apes, 28, 29–30, 249–56; brain of, 444; Celebese black, 107, 152–56; chimpanzees, 13–15, 114, 118, 151, 166, 223, 272, 278–341, 387, 428–29, 437, 443; Dryopithecone, 278; gibbons, 101, 257–77, 428, 431, 441; *Gigantopithecus*, 354; gorillas, 114, 150, 251, 279, 280, 346, 387–426, 428–29; great, 278–426; lesser, 257–77; orangutans, 114, 181, 279, 280, 281, 345–87, 428, 437; pygmy chimpanzees, 279, 283, 289, 341–45; siamangs, 257, 258, 262, 265
Apes and Monkeys (Garner), 396
Apollo space program, 462
Apology Addressed to the Travellers' Club, or Anecdotes of Monkeys, 284–85
Arashiyama West Primate Research Ranch, 140
Archaeolemur, 440
Aristotle, 197, 202
Ashurbanipal II, 202
Asimov, Isaac, 453

Assam macaques (*macaca assamensis*), 141–43
Atlanta *Journal*, 111
Atlanta Zoo, 375
Aurignacian period, 321
Australia, 27
Australopithecus, 253–54, 320, 321, 344, 355, 432–33, 435, 444, 445
Aye-aye lemur (*Daubentonia*), 34–35, 440

B virus, 182
Baboons (*Papio*), 107, 108, 168–93, 223, 437; chacma, 173, 177; gelada, 175, 177; Guinea, 177; olive, 177; sacred, 173, 177, 178, 182; yellow, 177
Bako National Park, Borneo, 376
Bald-headed tamarins, 71
Banded-leaf monkeys, 244
Barbary apes (*Macaca sylvana*), 107, 109, 124, 156–62, 202, 203
Barcelona Zoo, 413
Bartholin, Caspar Thomèson, 197
Basel Zoo, 416, 420, 422
Battell, Andrew, 280, 388
Baumgartel, Walter, 401
Bear macaques, *See* Red stump-tail macaques
Bearded sakis monkeys, 75, 80
Beeckman, Daniel, 348
Behavior patterns, chimpanzees, 297–300; gorillas, 404–12; red stump-tail macaques, 147–51; rhesus monkeys, 125; tree shrews, 20; woolly monkeys, 98; *See also* Aggressive behavior; Courtship and mating behavior; curiosity behavior; Grooming behavior; Hierarchy, dominance; Jealous behavior; Social relationship; Washing behavior
Beischer, Dietrich, 83
Belgium Zoological Garden, 391
Benchley, Belle, 268
"Berich für eine Akademie, Ein" (Kafka), 285–86
Beringe, Oscar von, 399
Berlin Aquarium, 287, 390
Bernstein, Irwin, 126, 127, 130–33, 143, 196
Berosini, Mr., 319
Bertrand, Mireille, 148–49, 150, 151
Bingham, Harold C., 400
Bipedalism, 428–31
Birdwhistell, R. L., 317
Bismarck, Otto von, 444
Black lemur (*Lemur macaco*), 31
Black mangabeys, 167
Black-faced tamarin, 71
Blair, W. Reid, 391
Bland-Sutton, John, 124–25
Body Language (Fast), 316

Body language, chimpanzees, 316–17
Bolwig, N., 312
Bonnet macaques (*Macaca radiata*), 109, 144–45
Bontius, Jacobus, 283, 345
Bosman's potto, 56
Boston Journal of Natural History, 389
Bourne, Merfyn, 93
Bowdich, Thomas, 388–89
Brachiation, 18, 101, 265–66, 428, 431
Brain, C. K., 220, 221, 223
Brain, the, 21–23, 437–47; capuchin monkeys, 87; gibbons, 265; guenon monkeys, 204; lemurs, 28–29; man, 452, 459, 460–62; orangutans, 358; porpoises, 23; siamangs, 265; squirrel monkeys, 81; tarsiers, 65
Brain stem, 437
Bretonne, Restif de la, 346
Breughel, Pieter, 175
British Gynecological Journal, 124
British Museum, 21, 396
Brontë, Charlotte, 284
Brooke, Sir James, 351
Brown, James, 341
Brown, Sergeant, 158
Brown lemurs (*Lemur fulvus*), 31
Buffon Georges de, 60, 198, 203, 260, 286, 346
Bulletin of the Field Museum of Natural History, 260
Burbridge, Ben, 391
Burt, Olive, 86
Burton, Sir Richard, 170
Bush babies (galogos), 28, 31, 49–53
Byron, George Lord, 349, 444

Calabar potto, 57–58
Camelli, Father, 60
Capped langurs, 244
Capuchin monkeys, 71, 87–93, 149, 430
Cardano, Girolomo, 280
Carpenter, C. R., 101, 105, 124, 149, 265, 266, 267, 272, 273, 276, 277
Carrington, W., 157
Ceboids, 87–106
Cebus monkeys, 66, 71, 275
Celebese black apes (*Macaca nigra*), 107, 152–56
Celebese islands, 156
Centrophenoxine, 457
Cercle Zoologique Congolais, 341
Cercopithecoid monkeys, 200–48
Cerebellum, 437
Cerebral cortex, 21, 22, 23, 65, 437
Chacma baboons (*Papio ursinus*), 173, 177
Chapin, James P., 343
Cheek pouches, 110, 139, 166, 205
Chest rubbing, ritual of, 97–98
Chester Zoo, England, 270
Chimpanzees (*Pan troglodytes*), 13–15, 114, 118, 151, 166, 178, 223, 272, 278–341, 387, 428–29, 437, 443; common or masked, 288; eastern or long-haired, 288; pygmy, *See* Pygmy chimpanzees
Churchill, Winston S., 159

Cicero, 174
Clark, M. C., 253
Clark, Sir Wilfrid Le Gros, 60, 250, 355
Clavicle, 26
Clayton, Gil, 407, 408–09
Clockwork Orange, A, 450
Coilus pigtail monkeys, 133
Coimbra-Filho, A. F., 72
Collarbone, 26
Coloboid monkeys, 200
Colobus or guereza monkeys, 108, 115, 232–38
Color, reactions to, 92, 306
Columbus Zoo, 420
Comfort, Alex, 128, 129, 455, 458
Communication, baboons, 186; chimpanzees, 291–93, 294, 308–33; gibbons, 264, 267–68; gorillas, 411; green monkeys, 222, 223; macaques, 108; mangabey monkeys, 166; monkeys, 91–92; orangutans, 360–63, 378; patas monkeys, 228; pygmy chimpanzees, 343–44; red stump-tail macaques, 151–52; rhesus monkeys, 126; woolly monkeys, 98–100
Comoro islands, 34
Concentration, powers of, 88
Cook, Captain James, 21, 346
Coolidge, Harold J., 276, 343, 344
Copulation, *See* Courtship and mating behavior
Cottontop marmosets, 75
Courtship and mating behavior, baboons, 189; gibbons, 272; gorillas, 402–04; green monkeys, 222–23; lemurs, 40–41; orangutans, 372–73; patas monkeys, 230–32; rhesus monkeys, 113–14; woolly monkeys, 100
Crab-eating monkeys, *See* Java monkeys
Craig, Hugh, 53
Cretaceous period, 16, 18
Cro-Magnon man, 321
Cryosurgery, 450
Cuma, 149
Curiosity behavior, 20, 112, 219
Cuvier, Baron, 193
Cuyters, Gerard, 308
Cynocephalidae, 177
Cynopithecoids, 165, 200

Dance carnivals, chimpanzees, 294–97
Dancing Chimpanzee, The (Williams), 296
Dart, Raymond, 173, 255, 432, 444
Darwin, Charles, 254, 281, 282, 312, 351
Davenport, Richard, 118, 119, 333, 366, 367, 371
Dawson, Charles, 354–55, 358
Dean, Douglas, 461
De Brazza monkeys (*Cercopithecus neglectus*), 209–13
Deepfreeze, 459
Delta Primate Research Center, 291, 294
De Meyer, Henri, 425
Demidoff's bush baby (*galagoides demidovii*), 51
Descent of Man, The (Darwin), 254, 281
Descent of Woman, The (Morgan), 256

Desoxyribonucleic acid (DNA), 456
DeVore, Irvin, 150, 191
Diabetes, Celebese black apes, 153
Diana monkeys, 206–08
Dog monkeys, 165
Donisthorpe, Jewel, 401
Douc langurs, 244
Drawing ability, chimpanzees, 302
Drills (*Mandrillus leucophaeus*), 107, 177, 193–99
Drugs, use of, 451–52, 455, 457, 459
Dryopithecine apes, 278
Dryopithecus, 251, 351–54
Dubois, Eugene, 354
Du Chaillu, Paul Belloni, 392–96
DuMond, Frank, 73–74
Duro, Edward H., 47–48

East Africa, 25
East African Mammals (Kingdon), 211
Eastern chimpanzees (*troglodyte schwein-furthii*), 288
Ecology, pollution and destruction of, 460
Ellefson, John, 277
Ellis, William, 21
Emblen, John T., 140
Emotions de Polydore Marasquin, Les (Gozlan), 285
Emory University, 111, 112, 447, 457
Eocene epoch, 19, 24, 60
Erect posture, 427–36
Erliker, James, 341
Essay of the Learned Martinus Scriblerus, Concerning the Origin of Sciences, 283–84
Estrogen, 129
Evolution, 10, 19, 22, 69, 284, 448
Evolution of Primate Behavior, The (Jolly), 275
Extrasensory perception (ESP), 460–62

Facial expressions and emotions, chimpanzees, 310–15
Falkenstein, Dr., 390
Fast, Julius, 316
Fayum depression, Egypt, 249–51
Felce, Winifred, 386, 387
Female sex hormones, 128, 129
Field Museum of Natural History, 260
Fitzsimmons, F. W., 214
Flacourt, Étienne de, 27
Flaubert, Gustav, 346
Fluorine tests, 356
Fooden, Jack, 260
Foramen magnum, 430
Forbes, H. O., 262–63, 264
Ford, Henry A., 390
Forebrain index, 437
Formosan rock macaques (*Macaca cyclopis*), 109, 110, 143
Fossey, Dian, 409–12
France, Anatole, 444
Frankfurt Zoo, 422
Frazer, George, 407, 408
Freedman, L., 432

Frontal lobes, 437, 446
Fulton, John F., 446, 451
Future of man, 448–62

Galen of Pergamon, 124, 161, 203
Gall, Franz, 444
Gallico, Paul, 160
Gallop, Gordon, 306–07
Gambetta, Léon, 444
Gardener, Edward, 140
Gardner, Allen and Beatrice, 317, 321, 322, 323
Garner, Robert L., 91, 92, 294, 365, 391, 396–98
Gartman, J. S., 220, 221, 223
Gelada baboons (*Papio theropithecus*), 175, 177, 188–90
Geoffroy's marmoset, 75
Georgia Mental Health Institute, 112
Georgia State Crime Laboratory, 111
Gesner, Conrad, 197, 280
Gestures, communication by, 45, 321–22
Gibbons, 101, 257–77, 428, 431, 437, 441; agile, 258, 262; hoolock, 258; wau-wau, 258, 264, 268–69
Gibraltar, 156–62
Gigantopithecus, 354
Giraudoux, Jean, 176
Glaserfeld, Ernst von, 330
Gluteal muscles, 430
Goeldi's marmoset (*Callimico goeldii*), 69, 75, 105
Golden lion tamarin (*Leontocebus*), 71–75
Golden marmoset, 105
Goma, the Gorilla Baby (Lang), 422
Goodall, Jane, 178, 290, 297, 298, 409, 428, 429
Gorillas, 114, 150, 251, 279, 280, 346, 387–426, 428, 437
Gozlan, Léon, 285
Grandidier, A., 29
Grant, James Augustus, 399
Great apes, the, 278–426; chimpanzees, 278–341, 387; gorillas, 114, 150, 251, 279, 280, 346, 387–426, 428, 437; orangutans, 114, 181, 279, 280, 281, 345–87, 428, 437; pygmy chimpanzees, 279, 283, 289, 341–45
Great cheeked mangabeys, 167
Green monkeys (*Cercopithecus sabacus*), 201, 205, 213–23
Grey, John Edward, 396
Griswold, Augustus, 276
Grivet monkeys (*Cercopithecus aethiops*), 201–02, 213–14
Grooming behavior, apes, 29–30; baboons, 184, 189; gorillas, 406; lemurs, 39, 40, 47; lorises, 51; monkeys, 29–30; marmosets, 70–71; tree shrews, 19–20
Ground monkeys, 227
Ground shrews, 17–18
Groves, Colin, 412
Guenon monkeys (*Cercopithecus sabaeus*), 107, 200, 201–27; Allen's swamp, 107–08, 166, 200, 209; De Brazza, 209–13; Diana, 206–08; green, 201, 205, 213–15;

473

grivet, 201–02, 213–14; mona, 205–08; white-throated, 223–27
Guereza monkeys, See Colobus monkeys
Guinea baboons (Papio papio), 177
Gulliver's Travels (Swift), 175, 284

Hagenbeck, Karl, 179
Hahn, Emily, 268–69
Hair, loss of, 433–34
Hall, Edward and Mildred, 314, 315
Hall, K. R. L., 227, 228, 230–32
Hamadryad baboon, See Sacred baboons
Hand, use of, 431–32, 434–35, 437, 448
Hanno, 387–88
Hanuman langurs, 239–43
Hapalines, marmosets, 66–71; tamarins, 71–75
Hardy, Sir Alister, 256, 433, 434, 435
Harlow, Harry F., 115–18, 119, 121, 145
Harman, Denham, 457, 458
Harrison, Barbara, 351, 362, 364, 376, 378, 382, 384
Harrison, Tom, 345, 346, 349, 351, 354
Hartman, Julia, 123–24
Hartman, Robert, 286–87
Hass, Hans, 314
Hatshepsut, Queen, 172
Hayes, Dr. and Mrs., 319, 321
Hearing, sense of, 23
Hemoglobin, 253
Hendrick, Grant H., 130
Hermes, Dr., 288
Heuvelmans, Bernard, 171
Hierarchy, dominance, Barbary apes, 161; bonnet macaques, 144; chimpanzees, 298; rhesus monkeys, 121–24
Hill, Osman, 143, 189, 198
Historia animalium (Gesner), 280
Homo erectus, 253, 254, 255, 354, 355, 444, 445
Homo habilis, 255
Homo sapiens, 283
Homo sylvestris, 284
Homo troglodytes, 283
Honey bear, 55
Hong, S., 382
Hoolock gibbons, 258
Hooton, Ernest A., 36–37
Hoppius, 345
Hornaday, William Temple, 349–50, 359, 366, 367
Horr, David, 362, 363, 369, 371
Howler monkeys, 66, 67, 87, 105–06, 149, 437
Huai-nan-tzu, 260
Human Animal, The (Hass), 314
Hunkeler, P., 290
Hunter, John, 198
Hussar monkeys (Erythrocebus), 108, 227–32
Huxley, Aldous, 175–76
Huxley, Thomas Henry, 29, 58
Hyden, Holger, 455

Ice Age, 355
Ikhnation, 172

Indian langurs, 239–43
Indri lemur, 35–36
Insectivores, 18
Intelligence, primate, 21–22; capuchin monkeys, 87; chimpanzees, 304–05; gibbons, 274–75; Japanese macaques, 138–39; lion-tailed macaques, 162; marmosets, 71; orangutans, 346; pygmy chimpanzees, 343; rhesus monkeys, 110–11; woolly monkeys, 98
International Congress of Primatology (1968), 312
International Union for the Conservation of Nature, 72
Ischial callosites, 108, 166
Itani, J., 139, 148
Iwata, Sonosuka, 140
Iwatayama Monkey Sanctuary, 140

Jacobson, Carlysle F., 446, 451
Janson, H. W., 280, 282, 285, 388
Japan Monkey Center, 134
Japanese macaques (Macaca fuscata), 134–40, 149, 150, 152, 430, 433
Java (crab-eating) macaques (Macaca fascicularis), 109, 132, 146, 149
Java man (Pithecanthropus erectus), 354
Jealousy behavior, gibbons, 263; green monkeys, 220
Jerboa mice, 20
Johns, John, 114–15
Johns, June, 270
Jolly, Allison, 39, 40, 41, 44, 45, 115, 145, 275
Jones, Frederick Wood, 59–60, 62, 250, 434
Jones, Herman, 111
Journal of Reproduction and Fertility, 127

Kafka, Franz, 285–86
Kawamura, S., 139
Keeling, Dr., 339
Keith, Sir Arthur, 355
Keller, Helen, 92–93
Kellogg, Dr., 310
Kent, Lionel, 181, 182
Kenyapithecus, 256
Kinesics, science of, 316–17
Kingdon, Jonathan, 211, 290, 298, 299
Kinkajou (Potos), 51, 55
Kiowa, USS, 86
Klüver, Heinrich, 87, 88
Knattnerus-Nerer, Theodore, 169, 170
Kohts, Ladygin, 310
Kortland, Adrian, 290, 298–99, 428
Krupa, Gene, 296
Kyoto Zoo, 422

La Brosse, N. de, 283
Landsteiner, Dr., 125
Lang, Ernst M., 418, 419, 420, 422
Lang, K., 133
Language communication, chimpanzees, 317–33
Langurs, 108, 288–44; banded leaf monkeys, 244; capped, 244; douc, 244; In-

474

dian or Hanuman, 239–43; purple-faced, 243
Lapin, Boris, 179
Lashley, Karl, 446, 453
Lateral lobes, 437
La Tour, Leschenault de, 21
Lawick-Goodall, Jane Van, *See* Goodall, Jane
Leaf-eating monkeys, 232–38; colobus or guereza, 232–38; snub-nosed, 232
Leakey, Louis S. B., 60, 251, 254–56, 409
Leakey, Mary (Mrs. L. S. B.), 254, 278
Leakey, Richard E., 253, 254, 446
Le Comte, Jacob, 345
Le Comte, Louis, 262
Lemurs, 18, 19, 24–48, 437, 438, 439–40; aye-aye, 34–35, 440; black, 31; brown, 31; indri, 35–36; lepilemurs, 31; mongoose, 31, 34; mouse, 30–31; red-bellied, 31; ring-tailed, 29, 37, 41–45, 47; ruffed, 31; sifaka, 35, 37–41; woolly, 35
Leong Khee Meng, 372
Lepilemurs, 31
Lesser apes, gibbons, 257–77; siamangs, 257, 258, 262, 265
Lewis, O. J., 428
Lieberman, Philip H., 317, 318, 319, 320–21
Life magazine, 384
Limnopithecus, 251
Lincoln Park Zoo, 422
Linnaeus, Carl, 29, 283, 348–49
Lion Country Safari, Atlanta, 426
Lion-tailed macaques (*Macaca silenus*), 109, 149, 162–64
Lipofuscin, 457
Livingstone, David, 399
London Magazine, 283
London Zoological Garden, 158, 189, 287
Long, R. L., 111
Long-haired chimpanzees (*Troglodyte schweinfurthii*), 288
Lorises, 49, 53–58; angwantibos, 49, 58; kinkapu, 51, 55; potto, 49, 55–58; slender, 49, 53–54; slow, 49, 55
Lorisoids, 26, 49–58; bush babies, 28, 31, 49–53; lorises, 49, 53–58

Macaques (Cercopithecidae), 88, 107–164, 165, 429–30, 437; Assam, 109, 141–43; Barbary apes, 107, 109, 124, 156–62, 202, 203; bonnet, 109, 144–45; celebese black apes, 107, 152–56; formosan rock, 109, 110, 143; Japanese, 134–40, 149, 150, 152, 430, 433; Java (crab-eating), 109, 132, 146, 149; lion-tailed, 109, 149, 162–64; Moor, 109, 152–56; pigtail, 109, 130–34, 143, 145; red stump-tail, 109, 146–52; rhesus, 108–09, 110–30, 140, 149, 150, 151; stump-tail, 429–30; toque, 109, 145–46
MacDonald, Julie, 178–79, 182, 183, 185, 186
MacRoberts, Barbara and Michael, 156–57, 161–62
Madagascar, 25, 27, 30, 34, 41

Malayan Nature Journal, 378
Male magazine, 181
Male sex hormones, 127, 129
Malnutrition, 457
Mammals, 18–19
Man (*Homo erectus*), 355; aggressiveness of, 450–51; aging of, 455–59; bipedalism, 429–31; brain of, 444, 452, 459, 460–62; Cro-Magnon, 321; drugs, use of, 451–52, 455, 457, 459; ecology, pollution and destruction of, 460; erect posture of, 427–36; extrasensory perception and, 460–62; forebrain index, 437; future of, 448–62; hair, loss of, 433; hand, use of, 431–32, 434–35, 437; Java, 354; malnutrition, 457; memory of, 452–55; Neanderthal, 318, 319, 320, 321, 355; origin of, 250, 251, 253–54, 256, 344; Piltdown, 355–58; primates in service of, 447–48; reproduction, control of, 459
Man and Monkey (Williams), 96
Mandrills (*Mandrillus sphinx*), 107, 177, 193–99
Maned tamarins (*Leontocebus*), 71–75
Mangabeys (*Cercoebus*), 107, 166–68; agile, 167; black, 167; great cheeked, 167; sooty, 167; white-collared, 167
Mannix, Daniel P., 170
Marceaux, Marcel, 316
Marmosets, 66–71, 437, 441; common (*Hapale jacchus*), 70; cottontop, 75; Geoffroy's, 75; Goeldi's, 69, 75; golden, 105; lion, 71–75; pygmy, 69–70; tamarin, 66, 69, 71–75; titi monkeys, 66, 75; white-handed, 75
Marsupials, 20, 27
Masked chimpanzees (*Troglodytes verus*), 288
Mating behavior, *See* Courtship and mating behavior
McCann, S., 150
McCay, Clive, 456
McClintock, M. K., 129
McKinney, W. T., 118
Mechanical ability, capuchin monkeys, 87–90
Medical World News, 225
Medulla oblongata, 437
Megalodapis, 27
Melincourt (Peacock), 284, 285
Memory, capuchin monkeys, 93; man, 452–55
Mentawai leaf monkeys (*Simias cocolor*), 244–46
Menzel, Emil, 291–92, 293–94, 314
Mesozoic era, 18
Michael, Richard, 127, 128
Michaelangelo, 175
Midbrain, 438
Military or red monkeys, 227–32; ground, 227; Hussar, 108, 227–32; nisnas, 227, 232; patas, 227–32
Milne-Edwards, Alphonse, 29, 390
Ming Dynasty, 197, 261
Miocene epoch, 25, 251, 279, 428
Mirrors, reaction to, 151, 218, 306–08

Mission from Cape Coast Castle to Ashanti (Bowdich), 388
Mitogentic rays, 460
Mona monkeys, 205–06
Monboddo, Lord, 284, 388
Mongoose lemurs (*Lemur mongoz*), 31, 34
Moniz, Egas, 446, 451
Monkey lemurs, 35, 37–41
Monkey psychiatrists, 120
Monkeys, 25, 28, 29–30, 38, 87; Allen's swamp, 107–08, 166, 209; Assam macaque, 109, 141–43; aye-aye, 34–35, 440; baboons, 107, 108, 168–93, 223, 437; Barbary apes, 107, 109, 124, 156–62, 202, 203; bonnet, 109, 144–45; capuchin, 71, 87–93, 149, 430; ceboids, 87–106; cebus, 66, 71, 87–106, 275; Célebese black ape, 107, 152–56; Cercopithecoid, 200–48; coloboid, 200; colobus or guereza, 108, 115, 232–38; Cynopithecoid, 165, 200; De Brazza, 209–13; Diana, 206–08; dog, 165; drills, 107, 177, 193–99; Formosan rock macaque, 109, 110, 143; green (vervet), 201, 205, 213–23; grivet, 201–02, 213–14; ground, 227; guenon, 107, 200, 201–27; hapalines, 66–75; howler, 66, 87, 105–06, 149, 437; Hussar (Patas), 108, 227–32; Japanese macaque, 134–40, 149, 150, 152, 430, 433; Java (crab-eating), 109, 132, 146, 149; langurs, 108, 238–44; leaf eaters, 232–38; lion-tailed macaque, 109, 149, 162–64; macaques, 88, 107–64, 165, 430, 433, 437; mandrills, 107, 177, 193–99; mangabeys, 107, 166–68; marmoset, 66–71, 437, 441; Mentawai leaf, 244–46; military or red, 227–32; mona, 205–08; Moor macaque, 109, 152–56; mustached, 208; New World, 66–106, 249–50, 441; nisnas, 227, 232; Old World, 107–248, 249–50, 441; owl, 66, 75–77; pigtain, 109, 130–34, 143, 145, 149; pinches, 69, 75; proboscis, 108, 149, 232, 248; purple-faced, 243; red stump-tail macaque, 109, 146–52; rhesus, 108–09, 110–30, 140, 149, 150, 176, 239, 275, 430; sakis, 75, 80, 105; sakiwinkis, 75, 77–80; snub-nosed, 232, 244–48; spider, 66, 87, 101–05, 108; squirrel, 66, 75, 81–82, 105–06; swimming, 137, 148–49, 365–66, 433–34; talapoin, 149, 166, 200, 204; tamarin, 66, 69, 71–75; titi, 66, 69, 75; toque, 109, 145–46; uakaris, 75, 80, 105; white-throated guenon, 223–27; woolly, 87, 93–101, 108
Monroe, Marilyn, 316
Monsieur Hulot's Holiday (film), 316
Montagu, N. F. Ashley, 64, 114, 282, 447
Moor macaques (*Macaca maura*), 109, 152–56
Morgan, Elaine, 256
Morris, Desmond, 188
Mouse lemurs, 30–31
Mundy, Rodney, 351
Munich Animal Park, 386
Mustached monkeys, 208
Mustached tamarins, 71

My Friends, the Wild Chimpanzees (Goodall), 178

Nadler, Ronald, 415
Nandy, Kalidas, 457
Napier, John, 279
Narmer, King, 172
National Academy of Sciences, 344
National Geographic Society, 409
National Institutes of Health, 447
National Zoo, 351, 392, 422
Natura animalium, De (Aelian), 164
Natural History of South Africa (Fitzsimmons), 214
Nature magazine, 29, 129
Naval Aerospace Medical Center, 83
Naval Institute of Aerospace Medicine, 86
Neanderthal man, 318, 319, 320, 321, 355
Necrobiotic rays, 460
Nesturkh, M., 250
New Scientists magazine, 128
New World monkeys, 66–106, 249–50, 441; capuchin, 71, 87–93, 149, 430; ceboids, 87–106; cebus, 66, 71, 87–93, 275; hapalines, 66–75; howler, 66, 67, 87, 105–06, 149, 437; marmoset, 66–71, 437, 441; owl, 75–77; pinches, 69, 75; sakiwinkis, 75, 77–80; spider, 66, 87, 101–05, 108; squirrel, 66, 75, 81–86, 105–06; tamatins, 66, 69, 71–75; titi, 66, 69, 75; woolly, 87, 93–101, 108
Nisnas monkeys, 227, 232
Nissen, Henry, 400, 429
Nucleic acid (RNA), 454–55

Oakley, Kenneth, 355
Occipital lobes, 437
Olduvai Gorge, Tanzania, 254
Old World monkeys, 107–248, 249–50, 441; baboons, 107, 108, 168–93, 223, 437; Barbary apes, 107, 109, 124, 156–62, 202, 203; capuchin, 71, 87–93, 149, 430; drills, 107, 177, 193–99; guenon, 107, 200, 201–27; langurs, 108, 238–44; leaf-eaters, 232–38; mandrills, 107, 177, 193–99; mangabeys, 107, 166–68; military or red, 227–32; proboscis, 108, 149, 232, 248; rhesus, 108–09, 110–30, 140, 149, 150, 151, 176, 239, 275, 430; snub-nosed, 232, 244–48
Oligocene period, 18, 25, 249
Oligopithecus, 249
Olive baboons (*Papio anubis*), 177
Orang-outang, sive Homo Sylvestris (Tyson), 282
Orangutans, 114, 181, 279, 280, 281, 345–87, 428, 437
Oregon Primate Research Center, 47, 139, 153
Osborne, Rosalie, 401
Owen, Sir Richard, 389
Owl monkeys (*Douroucoulis*), 66, 75–77

Paget, Sir Richard, 321
Paine, Arnold, 111
Paleocene period, 16

Paranthropas, 432–33
Parapithecus, 250
Parapsychological Society, 461
Parapsychology, 461
Parietal lobes, 437
Paris Natural History Museum, 21, 396
Parkes, Sir Alan, 127
Parthasarathy, M. D., 144
Patas monkeys, See Hussar monkeys
Patterson, R. S., 129
Peacock, Thomas Love, 284
Penant, Thomas, 203
Penthouse magazine, 130
Perachio, Adrian, 450, 452
Peroxidation, 457
Perrault, Claude, 203
Perry, John, 71–72, 74–75
Petiver, James, 60
Pheromones, 128–30
Philadelphia Academy of Natural Sciences, 392
Philadelphia Zoological Garden, 351, 392
Pi, Sabater, 199
Picasso, Pablo, 175
Pigtail macaques (Macaca nemestrina), 109, 130–34, 143, 145, 149
Pilbeam, David, 344, 354
Piltdown man, 355–58
Pinches (Oedipomidas), 69, 75
Pisani, Leo, 330
Pithecanthropus (Homo erectus), 253, 254, 255, 354, 355, 444, 445
Pitman, Charles, 400
Playboy magazine, 314
Play-fighting, 99
Play groups, 191
Pleistocene epoch, 157, 251
Pliny, 203, 388
Poliomyelitis vaccine, 125
Polo, Marco, 27, 280, 345
Pope, Alexander, 284
Pottos (Periodicticus), 49, 55–58
Prehensile tails, 86–87, 88, 94, 101, 105, 108
Premack, David and Ann, 324–25, 326, 328, 329, 333
Primate Centers, 447–48
Primate News, 47
Primates, 13–23, 24, 27, 70, 283
Proboscis monkeys, 108, 149, 232, 248
Proconsul, 251, 278, 435, 443
Proconsul africanus, 278
Proconsul major, 279
Proconsul nyanzae, 278–79
Propliopithecus, 250–51
Pro-Primates, 18–19
Prosimians, 24–25, 26
Psychopathology, 118
Purchas, Samuel, 388
Purchas His Pilgrimes (Purchas), 388
Purple-faced monkeys, 243
Pygmy chimpanzees (Pan paniscus), 279, 283, 289, 341–45

Quadrupeds (Gesner), 197
Quid Quid Voluaris (Flaubert), 346

Radiodating, 253, 357
Raffles, Sir Stamford, 21, 263
Rahaman, H., 144
Ramapithecus, 251, 256
Ray, John, 197–98
Reade, William Winwood, 396
Red-bellied lemur (Lemur rubiventer), 31
Red Book, 72
Red-handed tamarins, 71
Red hussar monkeys, 108
Red monkeys, See Military or red monkeys
Red stump-tail macaques (Macaca speciosa), 109, 146–52
Reproduction, control of, 459
Review of Zoology and Botany of Africa, 341
Reynolds, Vernon, 281–82, 283, 378, 429
Rhesus monkeys (Macaca mulatta), 108–09, 110–30, 140, 149, 150, 151, 176, 239, 275, 430
RH factor, 125–26
Rhine, J. B., 461
Rhythm, sense of, chimpanzees, 296–97
Rift Valley, 398
Ringling, John, 391
Ring-tailed lemur (Lemur catta), 29, 37, 41–45, 47
Ringtail monkeys, 87–93
Ripley, Suzanne, 150
Robinson, J. T., 432
Rogers, Charles, 118, 119, 333
Rose, Robert, 126, 127
Rousseau, Jean-Jacques, 284
Royal Geographical Society, British, 395
Royal Menagerie, Copenhagen, 197
Royal Menagerie, Paris, 203
Royal Society, British, 281
Ruffed lemur (Lemur variegatus), 31
Rumbaugh, Duane, 204, 274, 275, 330, 422

Sackett, Gene, 118–19, 145
Sacred baboons (Papio hamadryas), 173, 177, 178, 182
St.-Hilaire, Etiénne Geoffroy, 21
Sakis monkeys, 105; bearded, 75, 80
Sakiwinkis monkeys, 75, 77–80
San Antonio Zoo, 189
Sanderson, Ivan, 20, 70, 71, 80, 101–02, 162–64, 166, 177, 190, 193–95
San Diego Zoo, 422
Sarich, Vincent M., 253
Savage, Reverend, 389
Savage Africa (Reade), 396
Schaller, George B., 150, 360, 398, 399, 401–06, 428
Schiller, Paul, 302
Schouteden, H., 341
Schroeder, Dr., 226
Schultz, Adolf, 276, 399
Schwarz, Ernst, 341
Science magazine, 454
Scientific American, 325
Scriblerus Club, 283–84
Scruffy (Gallico), 160
Sepilok Forest Reserve, 378–82

Sex behavior, *See* Courtship and mating behavior
Sex hormones, 127, 128, 129
Sherrington, Sir Charles, 125
Shrews, ground, 17–18; tree, 16–18, 19–21, 23, 26
Siamangs, 257, 258, 262, 265
Sifaka (*Propithecus*), 35, 37–41
Sigmon, B. A., 430, 432
Sign language, chimpanzees, 317, 322–33
Silva, G. S. de, 359, 378, 379, 382
Simons, Elwyn, 344, 354
Slender lorises, 49, 53–54
Slow lorises, 49, 55
Smell, sense of, 23; green monkeys, 220–21; lemurs, 29, 39, 44; mouse lemurs, 31; rhesus monkeys, 127, 128–30; tarsiers, 65
Smith, Sir Grafton Elliot, 321, 432
Smith-Woodward, Sir Arthur, 355
Snub-nosed monkeys, 232, 244–48; Metawai leaf, 244–46; *Rhinopithecus*, 246–48
Social Life of Monkeys and Apes (Zuckerman), 223
Social relationship, lemurs, 45–47
Society for Experimental Biology, 453
Solomon, King, 174
Sooty mangabeys, 167
"Sounds of Silence, The" (Hall and Hall), 314
Southwick, Charles, 139–40
Space Monkey (Burt), 86
Spectral tarsiers, 59–65
Speech, chimpanzees, 317–21
Speke, John Hanning, 399
Spider monkeys (*Ateles*), 66, 87, 101–05, 108; woolly, 105
Spinal cord, 449
Squirrel monkeys, 66, 75, 81–86, 105–06
Stanley, Henry, 399
Stereoscopic vision, 26
Stone Age, 168, 354
Struhsaker, T. T., 150, 290
Stump-tail macaques, 429–30
Suomi, S., 118
Surrogate mothers, 115, 119, 145, 269
Swamp monkeys, 107–08, 166, 200, 209
Swift, Jonathan, 175, 284, 444
Swimming ability, 137, 148–49, 365–66, 433–34
Sykes' monkey, *See* White-throated guenons
Sylvian fissure, 440, 441
Systema naturae (Linnaeus), 283, 348

Tabraca (Siamese missionary), 262
Talapoin monkeys (*Myopithecus talapoin*), 149, 166, 200, 204
Tamarins, 66, 69, 71–75; bald-headed, 71; black-faced, 71; golden lion or maned, 71–75; red-handed, 71; titi, 66, 69, 75; white-faced or mustached, 71
Tappen, Neil, 230
Tarsiers (*Tarsius*), 26, 59–65, 250, 437, 438, 440, 441
Tati, Jacques, 316
Temporal lobes, 437

Teniers the Younger, 175
Testosterone level, 127–28
Theophilus, 174
Those About to Die (Mannix), 170
Thousand and One Nights, A, 170–71
Tiger at the Gate (Giraudoux), 176
Tilney, F., 437
Titi tamarins, 66, 69, 75
Titian, 175
Tools, use of, 428; capuchin monkeys, 88; chimpanzees, 290–91
Toque monkeys (*Macaca sinica*), 108, 145–46
Touch, sense of, 23
Toulouse-Lautrec, Henri, 175
Traill, H., 287
Trans-3-methyl hexanoic acid, 129
Tree shrews, 16–18, 19–21, 23, 26, 438
Trent, Hope, 224, 225
Truepenny, Ann, 216, 219
Truepenny, Charlotte, 215–20
Tulp, Nicolas, 281, 283, 345
Turgenev, Ivan, 444
Turpin, Richard, 262
Tutankhamen, 172
Two Years in the Jungle (Hornaday), 349
Tyson, Edward, 281–82, 345

Uakaris monkeys, 75, 80, 105

Van Gulick, R. H., 260, 261
Van Hooff, J., 314
Vervet monkeys, *See* Green monkeys
Vesalius, 282
Virus infections, 182–83
Vision, sense of, 23; gibbons, 273–74; stereoscopic, 26; woolly monkeys, 100–01
Vitamin C, 227, 369
Vitamin D, 29–39, 73, 178
Vitamin E, 458
Vocalization, *See* Communication

Walk of dominant primates, 149–51
Wallace, Alfred Russel, 349, 351, 366
Wallace, George, 449
Warner, Harold, 330, 332
Washburn, Sherwood, 191, 192, 276
Washing behavior, Japanese macaques, 137–38
Waterman, Walter, 390
Waters, Edward, 111
Wau-wau gibbons, 258, 264, 268–69
Weiner, Joe, 355, 356, 358
Welch, Raquel, 316
West, Clifford and Eleanor, 333
West, Mae, 316
West Foundation, 340
Whipsnade Zoo, 112
White-collored mangabeys, 167
White-faced tamarins, 71
White-handed marmosets, 75
White-throated guenon monkeys, 223–27
Wiener, Alexander, 125

Wilhelm of Sweden, Prince, 399
Williams, Leonard, 96–97, 98, 100, 101, 296
Wilson, Allen C., 253
Wilson, Mark, 340–41
Wilson, Reverend, 389
Wilson, Thomas, 111
Woolly lemurs, 35
Woolly monkeys, 87, 93–101, 108
Wong, James, 382
Worm Runner's Digest, 454
Wyman, Jeffries, 389

Year of the Gorilla, The (Schaller), 398, 401

Yellow baboons (*Papio cynocephalas*), 177
Yerkes, Robert M., 125, 264, 310, 344, 363, 368, 391, 400, 429
Yerkes Field Station, 93, 150, 222, 298
Yerkes Laboratories of Primate Biology, 303, 306, 400, 446
Yerkes Primate Research Center, 13, 30, 81, 86, 101, 114, 118, 133, 134, 149, 153, 156, 182, 189, 196, 248, 300–02, 303, 305–06, 333, 344, 368, 373, 382, 413, 447, 449, 456, 459
Young, Gordon, 150–51

Zinjanthropus, 255
Zuckerman, Solly, 196, 222–23, 305, 308

479